UNITEXT - La Matematica per il 3+2

Volume 101

Marco Baronti · Filippo De Mari
Robertus van der Putten · Irene Venturi

Calculus Problems

 Springer

Marco Baronti
Dipartimento di Matematica
Università di Genova
Genova
Italy

Robertus van der Putten
DIME
Università di Genova
Genova
Italy

Filippo De Mari
Dipartimento di Matematica
Università di Genova
Genova
Italy

Irene Venturi
Dipartimento di Matematica
Università di Genova
Genova
Italy

ISSN 2038-5722 ISSN 2038-5757 (electronic)
UNITEXT - La Matematica per il 3+2
ISBN 978-3-319-15427-5 ISBN 978-3-319-15428-2 (eBook)
DOI 10.1007/978-3-319-15428-2

Library of Congress Control Number: 2016943794

This Springer imprint is published by Springer Nature
The registered company is Springer International Publishing AG Switzerland

Die mathematische Analysis
gewissermassen eine einzige Symphonie
des Unendlichen ist

(D. Hilbert, Über das Unendliche, 1927)

In a certain sense, mathematical analysis
is a symphony of the infinite

(D. Hilbert, On the Infinite, 1927)

Preface

The aim of this book is to provide a practical working tool for students in Engineering, Mathematics, and Physics, or in any other field where rigorous Calculus is needed. The emphasis is thus on problems that enhance students' skill in solving standard exercises with a careful attitude, encouraging them to devote an attentive eye to what may or may not be done in manipulating formulae or deriving correct conclusions, while maintaining, whenever possible, a fresh approach, that is, seeking guiding ideas.

Every chapter starts with a summary of the main results that should be kept in mind and used for the exercises of that chapter; this is followed by a selection of guided exercises. The theoretical preamble is meant to recapitulate the main definitions and results and should also offer a bird's-eye view on the topic treated in the chapter. Hence, the student can quickly review the main theoretical facts and then, most importantly, "learn by examples," becoming acquainted with the specific techniques by seeing them applied directly to the problems. Each exercise ends with a short comment which underlines the main issues of that specific exercise, the leading ideas, and the main techniques. A selection of problems closes each chapter, the answers to which are all listed in Solutions. The reader is urged to try to solve some of these problems, which are similar, but not always trivially analogous, to those that have been presented in detail. Mathematics is never just an application of rules, but requires understanding, clear thought, and a bit of imagination. A different problem, a new question, a slightly skew formulation: this is where one really begins to master a technique and to consolidate it. So, rather than feeling discouraged, the student should develop curiosity and be aware that any significant progress does require some effort, and a little sweat is really part of the game.

Perhaps the most distinctive feature of this book is that our approach is very direct and refers to a concrete experience. The material is in fact mostly taken from actual written tests that have been delivered in the years 2000–2013 at the Engineering School of the University of Genova. Literally, thousands of students have worked on these problems, so our first and foremost acknowledgment goes to them, because they have helped us greatly over the years, tuning our views and letting us see where the main difficulties really are, those that need both clear statements and specifically designed exercises. Their fellow colleagues, the present and future students, are of course our public and our intended readers. Some complementary standard material has also been added, especially where the main thrust is in the direction of unraveling the details of basic techniques and achieving a reasonably complete panorama of possible scenarios.

The book ends with a chapter of problems that are not designed with a single issue in mind but rather require a variety of techniques, and should perhaps be addressed as a final check on the global preparation. Indeed, they have all been assigned in written tests and have all been worked on by large numbers of students. Intentionally, no solutions are given or even hinted at, and they are not ordered according to increasing difficulty. The student who wants to challenge him- or herself with questions that many other students have faced as a final exam should look with particular interest at Chap. 16.

The topics covered are those that are typically taught in a first-year engineering undergraduate Calculus course in Italy, with possible variants. The basic focus is on functions of one real variable. As for basic ordinary differential equations, separation of variables, linear first-order, and constant coefficients ODEs are discussed.

We believe that anyone who can solve the suggested problems with a reasonable degree of accuracy is in a safe position to achieve a positive result in most Italian universities. Our international experience also tells us that the same may be claimed for most universities around the world, for undergraduate Calculus or Advanced Calculus.

Genova, Italy   Marco Baronti
May 2016   Filippo De Mari
                                                            Robertus van der Putten
                                                                      Irene Venturi

Contents

About the Authors

Marco Baronti was born in Genova in 1956. Since 1990, he is associate professor in Mathematical Analysis at the University of Genova. His scientific interests are mainly in Functional Analysis and in particular in Geometry of Banach Spaces.

Filippo De Mari was born in Genova in 1959. In 1987 he received his Ph.D. from Washington University in St. Louis, USA. Since 1998, he is associate professor in Mathematical Analysis at the University of Genova. His scientific interests are mainly in Harmonic Analysis, Representation Theory and Lie Groups.

Robertus van der Putten was born in Sanremo in 1959. In 1989 he received his Ph.D. from the University of Milan. Since 1990, he is researcher in Mathematical Analysis at the University of Genova. His scientific interests are mainly in Calculus of Variations.

Irene Venturi was born in Viareggio in 1978. In 2009 she received her Ph.D. from the University of Genova and in 2011 a Master's in Security Safety and Sustainability in Transportation Systems. She is a teacher in Mathematics and has several editorial collaborations.

Chapter 1
Manipulation of Graphs

1.1 Operations on Graphs

A *coordinate system* determines a bijective correspondence between $\mathbb{R}^2$ and the plane. Assume that an orthogonal system of coordinates has been fixed and henceforth systematically identify $\mathbb{R}^2$ with the plane. Given a subset I of $\mathbb{R}$, for example an interval, and a function $f : I \rightarrow \mathbb{R}$, the *graph* of f is the subset

$$\Gamma(f) = \big\{(x, f(x)) : x \in I\big\}$$

of $\mathbb{R}^2$. If the graph of f is known, it is possible to understand the graphs of the functions that can be obtained from f using the sum and the product, that is, the functions $f(x + a), f(x) + a, f(ax)$ and $af(x)$, where a is a fixed real number. More precisely, if $f : I \rightarrow \mathbb{R}$ and $a \in \mathbb{R}$ are given, then the function $\tau_a f$ defined on the set

$$I + a = \{x + a : x \in I\}$$

by the relation
$$\tau_a f(x) = f(x - a)$$

is the *translation* of f by a. The graph of $\tau_a f$ is obtained from the graph of f by translating it to the right by a if $a > 0$ and to the left by $|a|$ if $a < 0$, see Fig. 1.1.

The graph of the function $f + a : I \rightarrow \mathbb{R}$, defined by

$$(f + a)(x) = f(x) + a,$$

is obtained from $\Gamma(f)$ by translating it upwards by a if $a > 0$ and downwards by $|a|$ if $a < 0$, see Fig. 1.2.

Similarly, if $a > 0$, the *dilate* of f by a is the function defined on the set

$$aI = \{ax : x \in I\}$$

© Springer International Publishing Switzerland 2016

M. Baronti et al., *Calculus Problems*, UNITEXT - La Matematica per il 3+2 101,

DOI 10.1007/978-3-319-15428-2_1

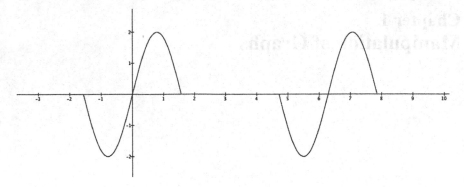

Fig. 1.1   The graph of f *(left)* and the graph of $\tau_{2\pi}f$ *(right)*

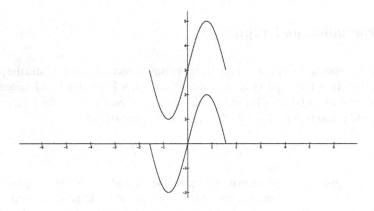

Fig. 1.2   The graph of f *(below)* and the graph of $f + 3$ *(above)*

by

$$\delta_a f(x) = f(a^{-1}x).$$

The graph of $\delta_a f$ is obtained from the graph of f by a horizontal dilation or contraction, according as $a > 1$ (dilation) or $a < 1$ (contraction), see Fig. 1.3.

The graph of the function $af : I \to \mathbb{R}$ defined by

$$(af)(x) = af(x)$$

is obtained from $\Gamma(f)$ by a vertical dilation or contraction, according as $a > 1$ (dilation) or $a < 1$ (contraction), see Fig. 1.4.

In many applications the notions of *even* function and that of *odd* function are important. In order to avoid formal complications, it is advisable to consider functions defined on *symmetric* domains.

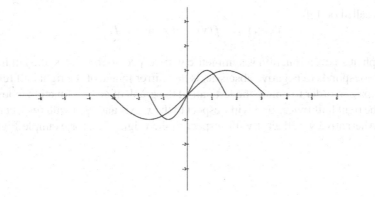

Fig. 1.3 The original domain is $[-\pi/2, \pi/2]$, that of $\delta_2 f$ is $[-\pi, \pi]$

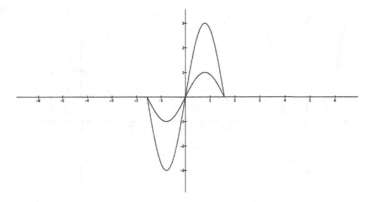

Fig. 1.4 The graph of $3f$ is three times wider than that of f

Definition 1.1 A subset S of $\mathbb{R}$ is symmetric with respect to the origin if the following property holds: if $x \in S$, then $-x \in S$. It is symmetric with respect to $x_0 \in \mathbb{R}$ if the shifted set $S - x_0 = \{s - x_0 : s \in S\}$ is symmetric with respect to the origin.

Intervals like $[-a, a]$ or $(-a, a)$ are obviously symmetric with respect to the origin. A symmetric set with respect to the origin does not necessarily contain the origin itself, for instance $(-b, -a) \cup (a, b)$. The interval $S = [1, 3]$ is symmetric with respect to $x_0 = 2$ because $S - 2 = [1, 3] - 2 = [-1, 1]$ is symmetric with respect to 0.

Definition 1.2 Let $I \subseteq \mathbb{R}$ be a symmetric set with respect to the origin and take $f : I \to \mathbb{R}$. The function f is called even if

$$f(-x) = f(x) \qquad \text{for all } x \in I,$$

and it is called odd if

$$f(-x) = -f(x) \qquad \text{for all } x \in I.$$

The graph of an even function is symmetric with respect to the y-axis: the left half of it, that corresponds to negative abscissæ, is the mirror image of the right half relative to the y-axis. In order to obtain the left half of the graph of an odd function, one must reflect the right half twice: first with respect to the x-axis and then with respect to the y-axis. Alternatively, reflect it with respect to the origin. See for example Figs. 1.5 and 1.6.

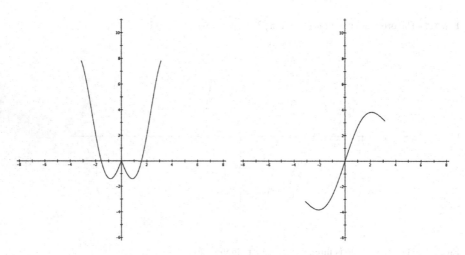

Fig. 1.5 The graph of an even function *(left)* and that of an odd function *(right)*

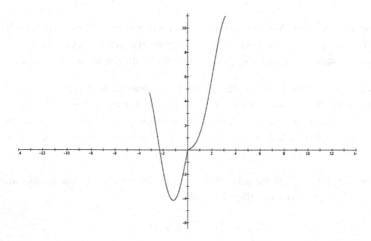

Fig. 1.6 Graph of the function whose even and odd parts are those of Fig. 1.5

Definition 1.3 Let $I \subseteq \mathbb{R}$ be a symmetric set with respect to the origin and take $f : I \to \mathbb{R}$. The functions f_e and f_o defined on I by

$$f_e(x) = \frac{1}{2}\left(f(x) + f(-x)\right), \qquad f_o(x) = \frac{1}{2}\left(f(x) - f(-x)\right) \qquad (1.1)$$

are called the *even* and *odd* parts of f, respectively.

The properties of f_e and f_o are described next.

Proposition 1.1 *Let $I \subseteq \mathbb{R}$ be a symmetric set with respect to the origin and take $f : I \to \mathbb{R}$. Then f_e is even, f_o is odd and $f = f_e + f_o$. This decomposition is unique, in the sense that if $f = G + H$ for some functions $G, H : I \to \mathbb{R}$ that are even and odd, respectively, then $G = f_e$ and $H = f_o$.*

1.2 Guided Exercises on Graphs

1.1 Consider the function $f : [-2, 2] \to \mathbb{R}$ whose graph is depicted in Fig. 1.7. Find the domains and draw the graphs of the following functions:

$$f_1(x) = |f(x)| \qquad\qquad f_2(x) = f(|x|)$$
$$f_3(x) = f(1-x) \qquad\qquad f_4(x) = 1 - f(x)$$
$$f_5(x) = 4 + f(x+6) \qquad\qquad f_6(x) = f(2x)$$
$$f_7(x) = 2f(x).$$

Fig. 1.7 Graph of the function in Exercise 1.1

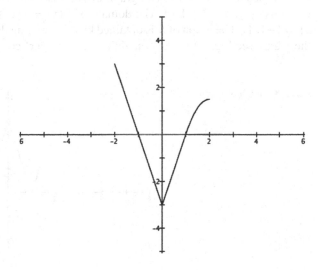

Answer. Since the absolute value is defined on $\mathbb{R}$, the domain of f_1 coincides with the domain of f, hence $\mathrm{Dom}(f_1) = [-2, 2]$. Furthermore, from the very definition of absolute value of a real number, it follows that

$$f_1(x) = \begin{cases} f(x) & \text{if } f(x) \geq 0 \\ -f(x) & \text{if } f(x) < 0 \end{cases}$$

and hence the graph of f_1 is the same as that of f if $f(x) \geq 0$ and is its mirror image with respect to the x-axis if $f(x) < 0$. See Fig. 1.8.

The domain of $f_2(x) = f(|x|)$ consists of those $x \in \mathbb{R}$ for which $|x| \in \mathrm{Dom}(f) = [-2, 2]$. Hence $\mathrm{Dom}(f_2) = [-2, 2]$. By definition of absolute value, $f_2(x) = f(x)$ if $x \in [0, 2]$. Therefore, the graph of f_2 in the right half plane, i.e. the portion that is contained in $\{(x, y) \in \mathbb{R}^2 : x \geq 0\}$, coincides with the graph of f that lies in the same half plane. Since f_2 is even, the portion of its graph that lies in the left half plane is the mirror image relative to the y-axis of the portion of the graph that lies on the right half plane, see Fig. 1.9.

The domain of $f_3(x) = f(1-x)$ consists of those $x \in \mathbb{R}$ for which $1-x \in \mathrm{Dom}(f)$. This is equivalent to $-2 \leq 1-x \leq 2$, whence $x \in [-1, 3]$, so that $\mathrm{Dom}(f_3) = [-1, 3]$. To draw the graph of f_3, look first at the graph of $f(-x)$, depicted in Fig. 1.10. Translating this to the right by 1, one gets the graph of $f(1-x)$, depicted in Fig. 1.11.

The domain of $f_4(x) = 1 - f(x)$ is the same as the domain of f, namely $[-2, 2]$. Consider first $\Gamma(-f)$, which is the mirror image relative to the x-axis of $\Gamma(f)$. The graph of f_4 is obtained from $\Gamma(-f)$ translating it upwards by 1, see Fig. 1.12.

Analogously to the case of f_3, the domain of $f_5(x) = 4 + f(x+6)$ consists of those $x \in \mathbb{R}$ for which $-2 \leq x+6 \leq 2$. Hence $\mathrm{Dom}(f_5) = [-8, -4]$. Consider first the graph of $f(x+6)$, which is obtained by translation to the left by 6 from the graph of f, depicted in Fig. 1.13. The graph of $f_5(x)$ is obtained by translating the latter upwards by 4, see Fig. 1.14. The domain of $f_6(x) = f(2x)$ is $\{x \in \mathbb{R} : -2 \leq 2x \leq 2\} = [-1, 1]$. The graph of f_6 is obtained by compressing that of f by a factor 2 along the x-axis, see Fig. 1.15. The domain of $f_7(x) = 2f(x)$ coincides with the domain of

Fig. 1.8 Graph of $|f(x)|$

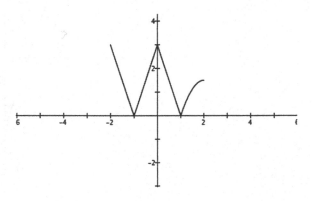

Fig. 1.9 Graph of $f(|x|)$

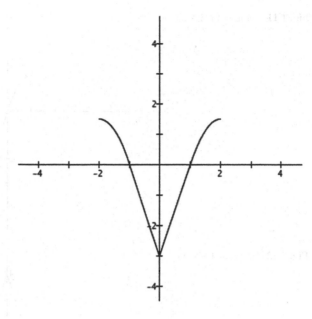

Fig. 1.10 Graph of $f(-x)$

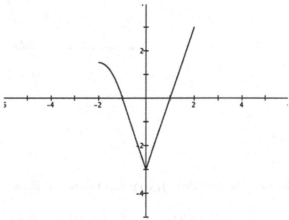

f. The graph of f_7 is by dilating the graph of f by a factor 2 in the y direction, see Fig. 1.16.

1.2 Consider the function $f : [-1, 2] \to \mathbb{R}$ whose graph is depicted in Fig. 1.17. Find the domains and draw the graphs of the following functions:

$$f_1(x) = f(|x| - 2), \qquad f_2(x) = f(2|x| - 3), \qquad f_3(x) = 1 - f(|x + 3|).$$

Fig. 1.11 Graph of $f(1-x)$

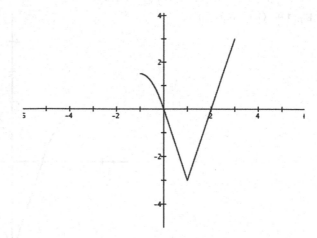

Fig. 1.12 Graph of $1-f(x)$

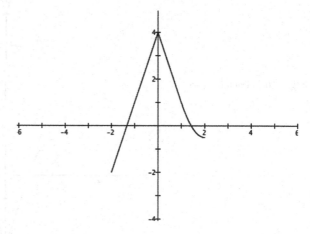

Answer. The domain of f_1 consists of those $x \in \mathbb{R}$ for which $|x| - 2 \in [-1, 2]$, i.e.

$$\text{Dom}(f_1) = \{x \in \mathbb{R} : 1 \le |x| \le 4\} = [-4, -1] \cup [1, 4].$$

The graph of $f_1(x) = f(|x| - 2)$ will be determined in two steps. Consider first the function $f(x - 2)$. Its graph is obtained from that of f by translation to the right by 2, see Fig. 1.18. Since $f_1(x) = f(|x| - 2)$ is an even function, the part of its graph that lies in the left hand side is obtained from the graph of $f(x - 2)$ by symmetry with respect to the y-axis, see Fig. 1.19.

Similarly to the previous case, the domain of $f_2(x) = f(2|x| - 3)$ is given by those $x \in \mathbb{R}$ for which $2|x| - 3 \in [-1, 2]$, namely

$$\text{Dom}(f_2) = \{x \in \mathbb{R} : 1 \le |x| \le \frac{5}{2}\} = [-\frac{5}{2}, -1] \cup [1, \frac{5}{2}].$$

Fig. 1.13 Graph of $f(x+6)$

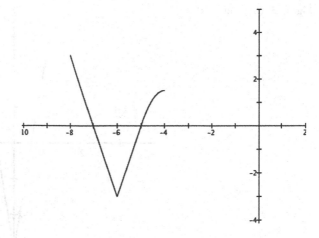

Fig. 1.14 Graph of
$4 + f(x+6)$.

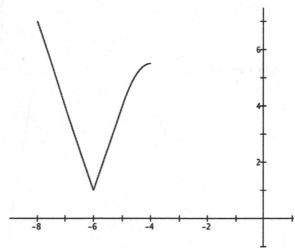

One can proceed in three steps. Consider first $x \mapsto f(x-3)$, the graph of which is a translation to the right by three units of the graph of f, see Fig. 1.20.

A contraction by a factor $1/2$ relative to the x axis yields $f(2x-3)$, and a final symmetrization with respect to the y-axis produces the graph of $f_2(x) = f(2|x|-3)$, see Fig. 1.21.

Finally, consider $f_3(x) = 1 - f(|x+3|)$. The domain of f_3 is

$$\text{Dom}(f_3) = \{x \in \mathbb{R} : -1 \le |x+3| \le 2\} = \{x \in \mathbb{R} : |x+3| \le 2\} = [-5, -1]$$

One starts with $f(|x|)$, see Fig. 1.22. Then, a translation to the left by 3 gives the graph of $f(|x+3|)$, see Fig. 1.23, and a reflection with respect to the x-axis gives

Fig. 1.15 Graph of $f(2x)$

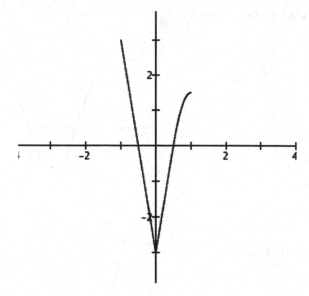

Fig. 1.16 Graph of $2f(x)$

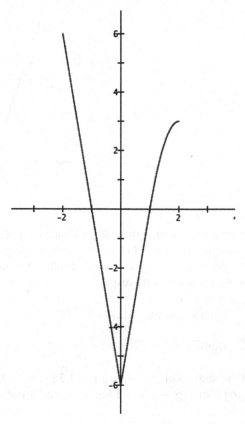

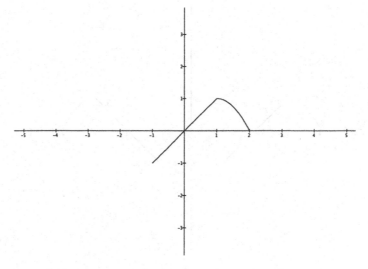

Fig. 1.17 Graph of the function f in Exercise 1.2

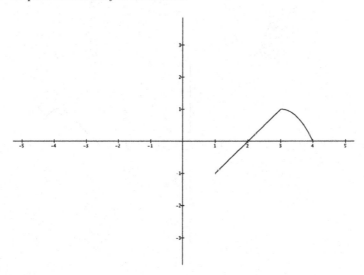

Fig. 1.18 Graph of $f(x - 2)$

$-f(|x + 3|)$, see Fig. 1.24. A final upwards translation by 1 produces the graph of $f_3(x) = 1 - f(|x + 3|)$, see Fig. 1.25.

1.3 Consider the function $f(x) = x^2 + x$. Draw the graphs of

$$f_1(x) = -f(x) \quad f_2(x) = f(-x) \quad f_3(x) = -f(-x).$$

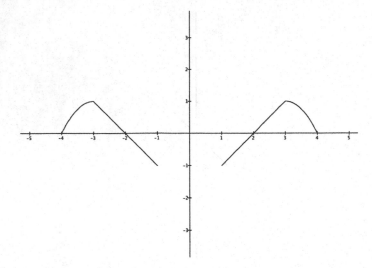

Fig. 1.19 Graph of $f(|x| - 2)$

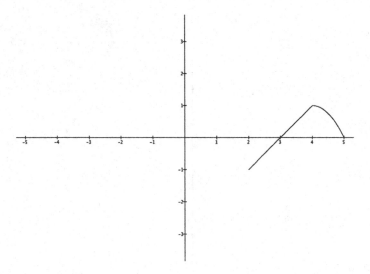

Fig. 1.20 Graph of $f(x - 3)$

Answer. The graph of f is a parabola. Its vertex is $V = (-1/2, -1/4)$ and it intersects the x-axis in the origin and in the point $(-1, 0)$. If $x \in \mathbb{R}$, then $(x, f_1(x)) = (x, -f(x)) \in \Gamma(f_1)$. Therefore, the graph of f_1 is obtained from that of f by symmetry with respect to the x-axis, see Fig. 1.26. If $x \in \mathbb{R}$, then $(-x, f(-x)) \in \Gamma(f)$ whereas $(x, f_2(x)) = (x, f(-x)) \in \Gamma(f_2)$. Hence, the graph of

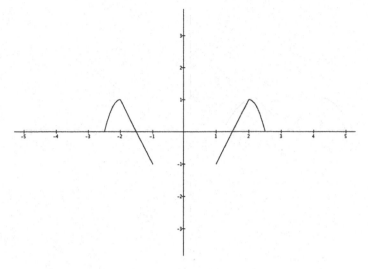

Fig. 1.21 Graph of $f(2|x| - 3)$

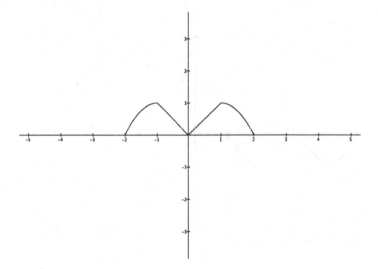

Fig. 1.22 Graph of $f(|x|)$

f_2 is obtained from that of f by symmetry with respect to the y-axis, see Fig. 1.27. If $x \in \mathbb{R}$, then $(x, f_3(x)) = (x, -f(-x)) \in \Gamma(f_3)$ whereas $(x, f(-x)) \in \Gamma(f_2)$. Hence, the graph of f_3 is obtained from that of f_2 by symmetry with respect to the x-axis, see Fig. 1.28.

1.4 Consider the function $f : \mathbb{R} \to \mathbb{R}$ the graph of which is depicted in Fig. 1.29. Draw the graphs of the functions:

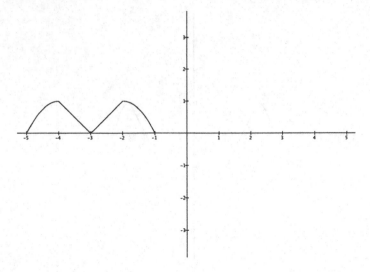

Fig. 1.23   Graph of $f(|x + 3|)$

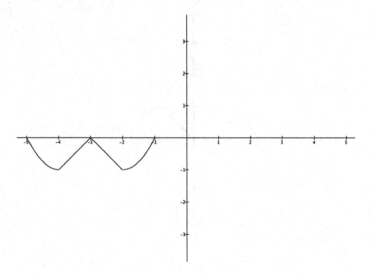

Fig. 1.24   Graph of $-f(|x + 3|)$

$$f_1(x) = f(-x) \qquad\qquad f_2(x) = |f(-x)|$$
$$f_3(x) = f(x + 1) \qquad\qquad f_4(x) = f(x - 1)$$
$$f_5(x) = \frac{1}{2}f(x) \qquad\qquad f_6(x) = f(2x).$$

Answer. The graph of f_1 is obtained from that of f by symmetry with respect to the x-axis, see Fig. 1.30. The graph of $f_2(x) = |f_1(x)|$ is in Fig. 1.31. The graph of f_3 is

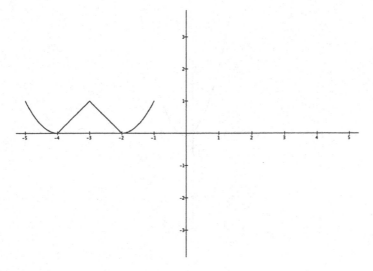

Fig. 1.25 Graph of $1 - f(|x + 3|)$

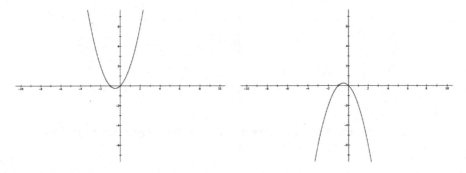

Fig. 1.26 Graphs of f and of $-f$

obtained from that of f by horizontal translation, but in the negative direction, see Fig. 1.32. Analogously, one obtains the graph of f_4, see Fig. 1.33.

The graph of f_5, Fig. 1.34, is obtained from that of f by vertically compression by $1/2$, and that of f_6, Fig. 1.35, by horizontal contraction by 2.

1.5 Draw the graphs of the following functions:

$$f_1(x) = |x + 2| \qquad f_2(x) = |x - 3| \qquad f_3(x) = |x| - 2$$
$$f_4(x) = 1 - |x| \qquad f_5(x) = |3x| \qquad f_6(x) = |2 - x^2|.$$

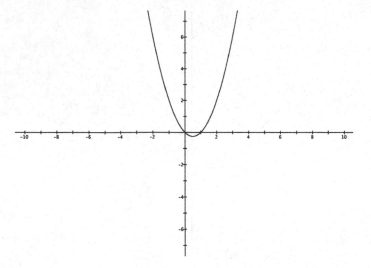

Fig. 1.27   Graph of $f(-x)$

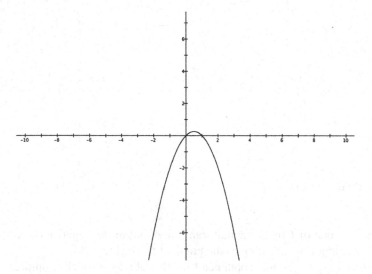

Fig. 1.28   Graph of $-f(-x)$

Answer. From the definition of absolute value,

$$f_1(x) = \begin{cases} x+2 & x \geq -2 \\ -x-2 & x < -2. \end{cases}$$

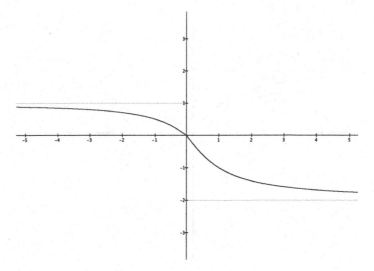

Fig. 1.29 Graph of the function f of Exercise 1.4

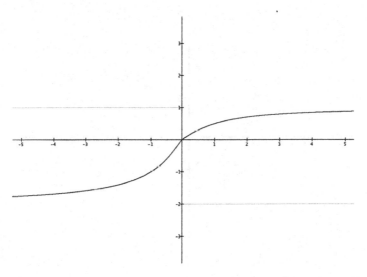

Fig. 1.30 Graph of $f(-x)$

It follows that $\Gamma(f_1) = \{(x, x + 2) : x \geq -2\} \cup \{(x, -x - 2) : x < -2\}$ and hence one obtains the graph in Fig. 1.36. Alternatively, it is possible to start from the graph of $|x|$ and then to translate it to the left by 2.

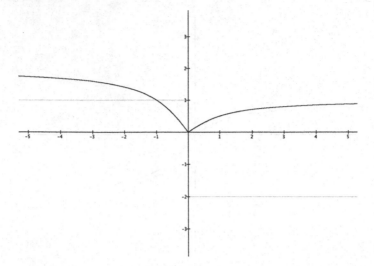

Fig. 1.31   Graph of $|f(-x)|$

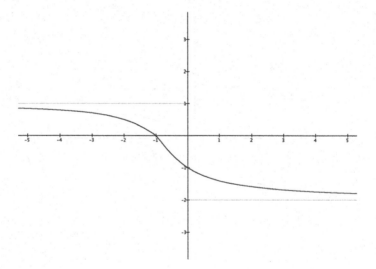

Fig. 1.32   Graph of $f(x+1)$

The graphs of the functions $f_2, \ldots, f_5$, depicted below in Figs. 1.37, 1.38, 1.39, 1.40 and 1.41, respectively, are obtained similarly, that is, starting from the graph of $|x|$ and then operating with the appropriate translations and symmetries. The graph of f_6 is obtained by first drawing the graph of x^2, then that of $-x^2$, then that of $2 - x^2$ and finally by taking the absolute value, which amounts to reflecting the portion of the graph that lies below the x-axis with respect to the x-axis.

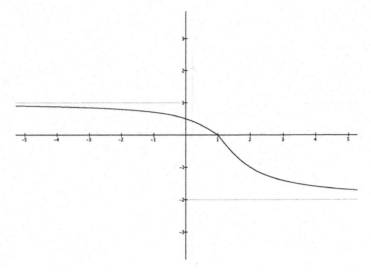

Fig. 1.33 Graph of $f(x-1)$

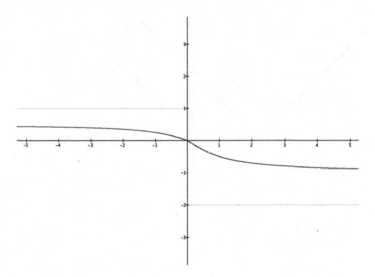

Fig. 1.34 Graph of $f(x)/2$

1.6 Consider the function

$$f(x) = \begin{cases} 2 & x \le -1 \\ -2x & -1 < x < 0 \\ 2(x^2 - 1) & 0 \le x \le 1 \\ x - 1 & 1 < x \le 2 \\ 1 & x > 2. \end{cases}$$

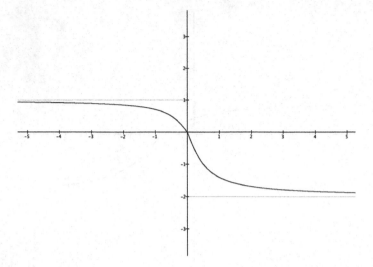

Fig. 1.35 Graph of $f(2x)$

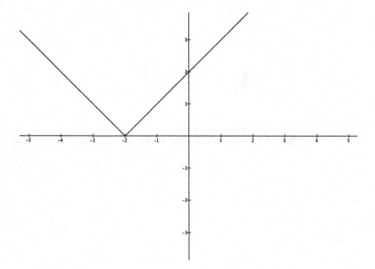

Fig. 1.36 Graph of $|x + 2|$

Find the decomposition of f into its even and odd parts, and draw their graphs.

Answer. First of all, the graph of f is as depicted below in Fig. 1.42.

The formulae that define f_e and f_o are best analysed in each of the following subsets of $\mathbb{R}$ separately:

$$(-\infty, -2), \quad \{-2\}, \quad (-2, -1), \quad \{-1\}, \quad (-1, 0), \quad \{0\}.$$

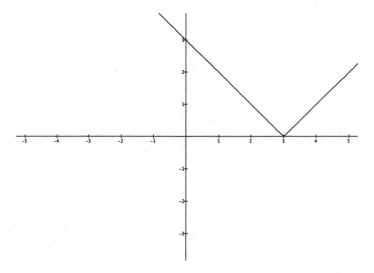

Fig. 1.37 Graph of $|x - 3|$

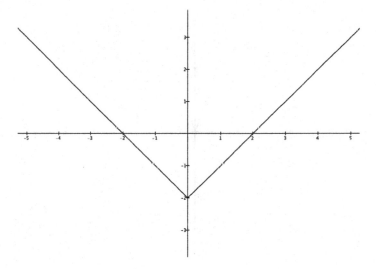

Fig. 1.38 Graph of $|x| - 2$

The values of $f_e(x)$ will first be obtained for $x \in (-\infty, 0)$. Symmetry considerations will then guide the analysis in $(0, +\infty)$. If $x \in (-\infty, -2)$, then $-x \in (2, +\infty)$. Hence $f(x) = 2$ and $f(-x) = 1$. Formula (1.1) gives:

$$f_e(x) = \frac{1}{2}\left(f(x) + f(-x)\right) = \frac{1}{2}(2 + 1) = \frac{3}{2}.$$

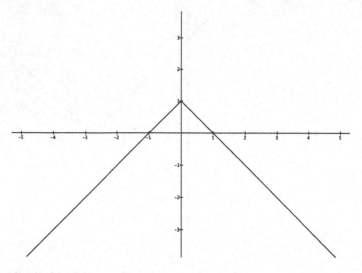

Fig. 1.39   Graph of $1 - |x|$

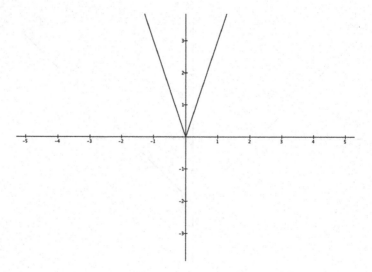

Fig. 1.40   Graph of $|3x|$

As for the point -2,

$$f_e(-2) = \frac{1}{2}\left(f(-2) + f(2)\right) = \frac{1}{2}(2 + 1) = \frac{3}{2}.$$

Similar arguments in the remaining sets yield:

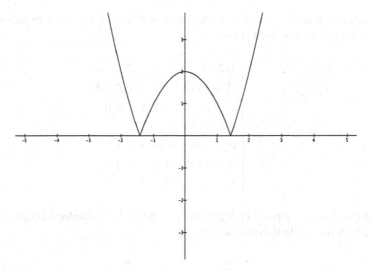

Fig. 1.41 Graph of $|2 - x^2|$

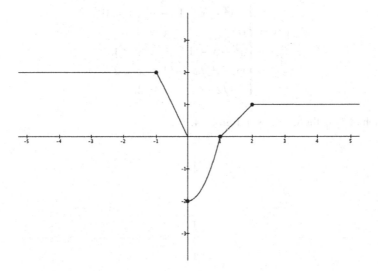

Fig. 1.42 Graph of the function f of Exercise 1.6

$$f_e(x) = \frac{1}{2}\left(2 + ((-x) - 1)\right) = \frac{1}{2}(1 - x) \qquad\qquad x \in (-2, -1)$$

$$f_e(-1) = \frac{1}{2}(2 + 0) = 1$$

$$f_e(x) = \frac{1}{2}\left(-2x + 2\left((-x)^2 - 1\right)\right) = x^2 - x - 1 \qquad\qquad x \in (-1, 0)$$

$$f_e(0) = \frac{1}{2}\left(f(0) + f(-0)\right) = f(0) = -2.$$

By symmetry, that is, by appealing to the fact that $f_e(x) = f_e(-x)$, the expression of f_e in $(0, +\infty)$ can be obtained at once:

$$
f_e(x) = \begin{cases}
3/2 & x \leq -2 \\
(1-x)/2 & -2 < x \leq -1 \\
x^2 - x - 1 & -1 < x < 0 \\
-2 & x = 0 \\
x^2 + x - 1 & 0 < x < 1 \\
(1+x)/2 & 1 \leq x < 2 \\
3/2 & x \geq 2.
\end{cases}
$$

The graph of the even part of f is therefore as in Fig. 1.43. Analogous calculations give the following explicit expression of f_o:

$$
f_o(x) = \begin{cases}
1/2 & x \leq -2 \\
(3+x)/2 & -2 < x \leq -1 \\
-x^2 - x + 1 & -1 < x < 0 \\
0 & x = 0 \\
x^2 - x - 1 & 0 < x < 1 \\
(x-3)/2 & 1 \leq x < 2 \\
-1/2 & x \geq 2.
\end{cases}
$$

The graph of f_o is therefore as in Fig. 1.44.

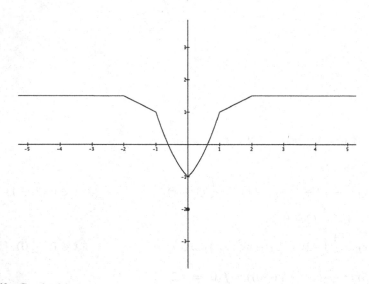

Fig. 1.43   Graph of f_e

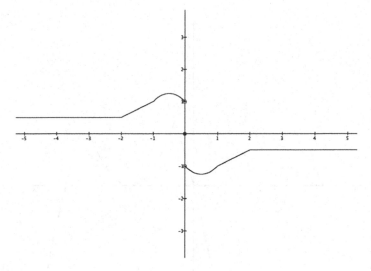

Fig. 1.44 Graph of f_o

1.3 Problems on Graphs

1.7 Draw the graphs of the following functions:

$$f_1(x) = |x + 1| - |2(x + 1)|$$

$$f_2(x) = \frac{|x|}{x} + 2|x| - \left|1 - |x|\right|$$

$$f_3(x) = \left|1 - \left|1 - |x|\right|\right|$$

$$f_4(x) = |x + 2| - (x + 2).$$

1.8 Draw the graph of $f(x) = x/(x - 2)$.

1.9 Draw the graphs of the functions

$f_1(x) = -f(x)$	$f_2(x) = f(-x)$	$f_3(x) = |f(x)|$
$f_4(x) = f(|x|)$	$f_5(x) = f(x - 1)$	$f_6(x) = f(x + 2)$
$f_7(x) = f(x) - 2$	$f_8(x) = 2f(x)$	$f_9(x) = f(x)/2$
$f_{10}(x) = f(3x)$	$f_{11}(x) = f(x/2)$	

where $y = f(x)$ is the graph in Fig. 1.45.

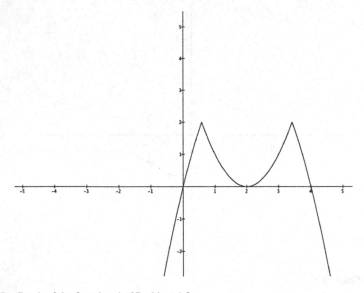

Fig. 1.45   Graph of the function f of Problem 1.9

1.10   Consider the function $f : \mathbb{R} \to \mathbb{R}$ defined by:

$$f(x) = \begin{cases} x & x > 0 \\ 1 & x = 0 \\ -1 & x < 0. \end{cases}$$

Find the explicit expression and then draw the graphs of the following functions:

$$f_1(x) = f(-x) \qquad f_2(x) = f(x+1) \qquad f_3(x) = f(|x|).$$

1.11   Consider the function $f : \mathbb{R} \to \mathbb{R}$ defined by:

$$f(x) = \begin{cases} \dfrac{x^2}{2} & x \geq 0 \\[2mm] \dfrac{x^2 - 6x}{2} & x < 0. \end{cases}$$

Draw the graphs of the functions f, f_e and f_o.

1.12   Discuss, with graphic arguments, the inequality $\sqrt{|x|} \geq k$ as k ranges in $\mathbb{R}$.

1.13 Consider the function $f : \mathbb{R} \to \mathbb{R}$ defined by:

$$f(x) = \begin{cases} x+2 & x < 1 \\ 0 & x = 1 \\ \dfrac{3}{4}(x-3)^2 & x > 1,\ x \neq 5 \\ 2 & x = 5. \end{cases}$$

Draw the graphs of the functions f_e and f_o.

1.14 Consider the function $f : \mathbb{R} \to \mathbb{R}$ defined by:

$$f(x) = \begin{cases} \dfrac{\log x - x^3}{2} & x > 0 \\ 0 & x = 0 \\ \dfrac{\log(-x) - x^3}{2} & x < 0. \end{cases}$$

Draw the graphs of the functions f_e and f_o.

1.15 Consider the function $f : \mathbb{R} \to \mathbb{R}$ defined by:

$$f(x) = \alpha x^3 + \beta x^2 + \gamma x + \delta \qquad \alpha, \beta, \gamma, \delta \in \mathbb{R}.$$

Find $\alpha, \beta, \gamma, \delta \in \mathbb{R}$ in such a way that:

(a) f is even;
(b) f is odd;
(c) $f(1) = f(-1) = 0$ and $f(2) = f(-2) = 1$.

1.16 Let $g : [-1, 3] \to \mathbb{R}$ be the function whose graph is in depicted in Fig. 1.46.
Determine the graph of each of the following functions:

$$g_1(x) = f(-x) \qquad\qquad g_2(x) = f(|x|) \qquad\qquad g_3(x) = f(1-x)$$
$$g_4(x) = 1 - f(x) \qquad\qquad g_5(x) = 3 + f(x+3) \qquad g_6(x) = f(3x)$$
$$g_7(x) = 2f(x) \qquad\qquad g_8(x) = -|f(x)|.$$

1.17 Let $p : [-1, 4] \to \mathbb{R}$ be the function whose graph is in Fig. 1.47. Draw the graph of $p(4 - |x|)$ and of $4 - |p(2 - x)|$.

1.18 Let $q : (-3, 1) \to \mathbb{R}$ be the function whose graph is in Fig. 1.48. Draw the graph of $|q(-x)|$ and of $|3 - q(x+2)|$.

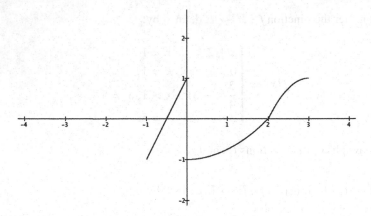

Fig. 1.46   Graph of the function *g* in Problem 1.16

Fig. 1.47   Graph of the
function *p* in Problem 1.17

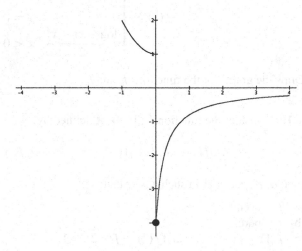

Fig. 1.48   Graph of the
function *q* in Problem 1.18

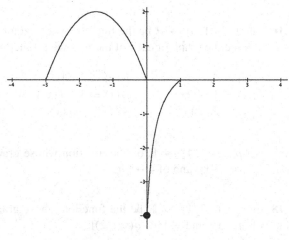

Chapter 2
Invertible Mappings

2.1 Injective, Surjective and Bijective Mappings

Given the map $f : A \to B$, and $I \subset A$, the set

$$f(I) = \{f(x) : x \in I\}$$

is called the image of I under f. If $I = A$, then $f(A)$ is called the *image* of f, or the *range* of f, and denoted $\operatorname{Im}(f)$. Observe that $f(A) \subset B$ but that, in general, $f(A) \neq B$.

Definition 2.1 The map $f : A \to B$ is called *surjective* if $f(A) = B$, that is, if for every $b \in B$ there exists $a \subset A$ such that $f(a) = b$, and it is called *injective* if it never sends distinct points into the same point, that is, if $f(a_1) \neq f(a_2)$ for any $a_1, a_2 \in A$ with $a_1 \neq a_2$. Finally, f is called *bijective* if it is both injective and surjective.

From the definition of injective map it follows at once that f is not injective whenever there exist $a_1, a_2 \in A$ with $a_1 \neq a_2$ such that $f(a_1) = f(a_2)$. Hence, even functions, defined on symmetric subsets of $\mathbb{R}$, are never injective. Similarly, f is not surjective if there exists $b \in B$ that is not in the image of f.

Definition 2.2 Let $f : A \to B$ be a map and take a subset $J \subseteq B$. The *inverse image*, or *preimage*, of J under f is the set of points of A that are sent by f into J, that is

$$f^{-1}(J) = \{a \in A : f(a) \in J\} \subseteq A.$$

Evidently, f is injective exactly when the inverse image of any singleton (that is, a set of the form $J = \{b\}$, for some $b \in B$) is either a singleton (a set of the form $\{a\}$, for some $a \in A$) or empty, and is surjective when the inverse image of any singleton is not empty.

It is possible and useful to interpret the notions of injectivity and surjectivity for maps that are defined on subsets of $\mathbb{R}$ into $\mathbb{R}$ in terms of their graphs. Take such a

© Springer International Publishing Switzerland 2016   29
M. Baronti et al., *Calculus Problems*, UNITEXT - La Matematica per il 3+2 101,
DOI 10.1007/978-3-319-15428-2_2

map $f : I \subseteq \mathbb{R} \to \mathbb{R}$ and denote as usual by $\Gamma(f)$ its graph. For any $y_0 \in \mathbb{R}$,

$$f^{-1}(\{y_0\}) = \{x \in I : (x, y_0) \in \Gamma(f)\}.$$

Therefore, the preimage of a point is found by considering the horizontal line $y = y_0$ and then by collecting all the abscissae of the points that lie on the intersection between the horizontal line and $\Gamma(f)$. It follows in particular that f is injective if and only if every horizontal line intersects $\Gamma(f)$ in at most one point and it is surjective if and only if every horizontal line intersects $\Gamma(f)$ in at least one point. Therefore, f is bijective if and only if every horizontal line intersects $\Gamma(f)$ in exactly one point.

If one considers a function $f : I \subset \mathbb{R} \to J$, where J is a prescribed subset of $\mathbb{R}$, then the previous graphical interpretations must be modified by taking horizontal lines of the form $y = y_0$ only for the values y_0 that belong to J. For example, the function $f : [0, 1] \to [0, 1]$ defined by $x \mapsto x$ is bijective, as well as $x \mapsto x^n$ for any positive integer n. Indeed, for any positive integer n, and every horizontal line with equation $y = y_0$ with $y_0 \in [0, 1]$ intersects the graph of the function $x \mapsto x^n$ in the single point $x_0 = \sqrt[n]{y_0} \in [0, 1]$. Further, the mapping $\varphi : [-\pi/2, \pi/2] \to [-1, 1]$ defined by $\varphi(x) = \sin x$ is also bijective, as the reader is urged to check with a simple drawing and then appealing to elementary trigonometry, whereas the map $\psi : \mathbb{R} \to \mathbb{R}$ defined by $\psi(x) = \sin x$ is neither injective (because $\sin x = \sin(x + 2k\pi)$ for any $k \in \mathbb{Z}$) nor surjective (because $2 \in \mathbb{R}$ is not in the image of ψ).

The property of being injective is somehow intrinsic to a map, whereas the property of being surjective can always be achieved by suitably changing the codomain. Indeed, given $f : A \to B$, the new map

$$\tilde{f} : A \to f(A), \qquad \tilde{f}(a) = f(a) \tag{2.1}$$

is defined by the same law and on the same set as f, and is automatically surjective. Occasionally, $\tilde{f}$ will be referred to as the *surjective map naturally associated* with f.

2.2 Inversion of a Map

Definition 2.3 Take any set A. The map $\mathrm{id}_A : A \to A$ defined by $\mathrm{id}_A(a) = a$ for every $a \in A$ is called the *identity mapping* of A.

Definition 2.4 The map $f : A \to B$ is called *invertible* if there is a map $g : B \to A$ such that:

(i) $g \circ f = \mathrm{id}_A$;
(ii) $f \circ g = \mathrm{id}_B$.

In this case, the map g, necessarily unique, is called the *inverse map* of f and is denoted $g = f^{-1}$.

Proposition 2.1 *The map $f : A \to B$ is invertible if and only if it is bijective.*

Although Proposition 2.1 clarifies that only the bijective maps are invertible, it is customary to relax the notion of invertibility in view of the fact that surjectivity can always be achieved, as discussed at the end of the previous section. In what follows, the notion of invertible map is used to mean that f is injective. If this is the case, the surjective map $\tilde{f}$ naturally associated with f by (2.1) is actually bijective and hence invertible in the strict sense. This slight ambiguity is best circumvented by requiring to explicitly determine the image of f, which coincides with the domain of the inverse (whenever f is injective), and then, with slight abuse of notation, to identify f with $\tilde{f}$.

Another issue that often occurs naturally is local invertibility. By this it is meant that a function f might fail to be injective on its domain, for example $f(x) = x^2$ is not injective on $\mathbb{R}$, but perhaps its restriction to a proper subset of its domain is injective and thus, in the broader sense just discussed, invertible. For example the restriction of $f(x) = x^2$ to $[0, +\infty)$ is injective. This justifies the following definition.

Definition 2.5 If $f : A \to B$ is a map and $I \subset A$ is a subset of A such that the restriction $f|_I$ is injective, then f is said to be invertible on I (onto its image).

A map $f : A \to B$ is invertible on I onto its image if and only if for any b in $f(I)$ the equation $f(a) = b$ has a unique solution $a \in I$.

2.3 Monotone Functions

Definition 2.6 The function $f : I \subset \mathbb{R} \to \mathbb{R}$ is called:

  (i) *increasing* if, whenever $x_1, x_2 \in I$ are such that $x_1 < x_2$, then $f(x_1) < f(x_2)$;
 (ii) *nondecreasing* if, whenever $x_1, x_2 \in I$ are such that $x_1 < x_2$, then $f(x_1) \leq f(x_2)$;
(iii) *decreasing* if, whenever $x_1, x_2 \in I$ are such that $x_1 < x_2$, then $f(x_1) > f(x_2)$;
 (iv) *nonincreasing* if, whenever $x_1, x_2 \in I$ are such that $x_1 < x_2$, then $f(x_1) \geq f(x_2)$.

If f satisfies either of the above conditions, then it is called *monotone* or *monotonic*. If it satisfies either (a) or (c), then it is called *strictly monotone*, or *strictly monotone*. The strictly monotone functions are those that either preserve or invert the order relations. Sometimes the increasing functions are called *strictly increasing* and the decreasing functions are called *strictly decreasing*.

Observe that the composition of monotone maps is always monotone. The point is that, loosely speaking, a monotone map either preserves or inverts the order, either in the strong sense (strictly monotone maps) or in the weak sense (kinds (ii) and (iv) in the definition), so that in the end the order is either preserved or reversed (strongly or weakly), according to which kind of monotoneities were involved. The reader is urged to check which compositions lead to which monotone maps.

Table 2.1 monotoneity of $f + g$ with $f, g : I \to \mathbb{R}$

f	g	$f + g$
Increasing	Increasing	Increasing
Decreasing	Decreasing	Decreasing

Table 2.2 monotoneity of fg with $f, g : I \to \mathbb{R}$

f	Sign of f	g	Sign of g	fg
Increasing	Positive	Increasing	Positive	Increasing
Increasing	Positive	Decreasing	Negative	Decreasing
Increasing	Negative	Decreasing	Positive	Increasing
Decreasing	Positive	Increasing	Negative	Increasing
Decreasing	Negative	Increasing	Positive	Decreasing
Decreasing	Negative	Decreasing	Negative	Increasing

Table 2.3 monotoneity of $f \circ g$, with $\mathrm{Im}(g) \subset \mathrm{Dom}(f)$

f	g	$f \circ g$
Increasing	Increasing	Increasing
Increasing	Decreasing	Decreasing
Decreasing	Decreasing	Increasing
Decreasing	Increasing	Decreasing

Proposition 2.2 *If $f : I \subset \mathbb{R} \to \mathbb{R}$ is strictly monotone, then f is injective.*

Notice that the reverse implication is false, for example the function $f(x) = 1/x$ is injective but not monotone on its natural domain $\mathbb{R} \setminus \{0\}$. Observe also that the inverse of a strictly monotone map is again a strictly monotone map, with the same type of monotoneity.

Monotoneity of sums, products and compositions of functions can be inferred, but not always. The results are summarized in Tables 2.1, 2.2 and 2.3 in the cases in which a conclusion can be drawn. In each of the remaining cases it is possible to produce examples with different behaviours.

2.4   Guided Exercises on Invertible Mappings

2.1 Consider the function $f : [0, +\infty) \to [0, \pi/4)$ defined by $f(x) = \arctan \dfrac{x}{x+1}$. Prove that f is invertible and write the explicit expression of its inverse.

Answer. Consider the auxiliary maps $h : [0, +\infty) \to [0, 1)$ and $g : [0, 1) \to [0, \pi/4)$ defined by

$$h(x) = \frac{x}{x+1}, \qquad g(x) = \arctan x.$$

Then f is the composition $f(x) = g(h(x)) = g \circ h(x)$. Now, both h and g are strictly increasing maps, so that f is also strictly increasing, hence invertible. Indeed, the arctangent map is monotone, being the inverse map of a strictly increasing map (the restriction of the tangent to $(-\pi/2, \pi/2)$), whereas to see that h is monotone just observe that

$$h(x) = \frac{x+1-1}{x+1} = 1 - \frac{1}{x+1}.$$

In order to find the expression of the inverse, for any given $y \in [0, \pi/4)$ one must find $x \in [0, +\infty)$ such that $f(x) = y$. Now

$$\arctan \frac{x}{x+1} = y \implies \tan y = \frac{x}{x+1}$$
$$\implies x(1 - \tan y) = \tan y$$
$$\implies x = \frac{\tan y}{1 - \tan y}$$

Therefore $f^{-1} : [0, \pi/4) \to [0, +\infty)$ is defined by $f^{-1}(y) = \tan y/(1 - \tan y)$.

This exercise is a direct application of two properties: the composition of strictly monotone maps is strictly monotone and a strictly monotone map is invertible. In the case at hand, the search of the inverse map leads to an answer by "undoing" each operation, in the correct order, or, more technically, observing that

$$f = g \circ h \implies f^{-1} = h^{-1} \circ g^{-1}.$$

Evidently, here $g^{-1}(x) = \tan x$ and $h^{-1}(x) = x/(1-x)$.

2.2 Determine if the function

$$f(x) = \begin{cases} -x^2 - 1 & x < 0 \\ 0 & x = 0 \\ x + 1 & x > 0 \end{cases}$$

is invertible and, if yes, find an explicit expression of its inverse.

Answer. The restriction $f_1 = f\big|_{(-\infty,0]} \to (-\infty, -1) \cup \{0\}$ is a bijection. Indeed, if $x_1, x_2 \in (-\infty, 0)$ are such that $x_1 < x_2$, then

$$f(x_1) - f(x_2) = -x_1^2 + x_2^2 = (x_2 - x_1)(x_1 + x_2) \neq 0,$$

and, more precisely, $f_1(x_1) - f_1(x_2) < 0$, so that f_1 is strictly increasing. Take now $y \in (-\infty, -1)$. Then $y \in \text{Im}(f_1)$ because the negative number $x = -\sqrt{-1-y}$ satisfies

$$-x^2 - 1 = -\left(-\sqrt{-1-y}\right)^2 - 1 = y.$$

Observe further that $f_1(0) = 0$, so that f_1 is a bijection.

Next, $f_2 = f\big|_{(0,+\infty)} \to (1, +\infty)$, which is defined by $f_2(x) = x + 1$, is clearly also a bijection and hence $f : \mathbb{R} \to (-\infty, 1) \cup \{0\} \cup (1, +\infty)$ is a bijection, with inverse

$$f^{-1}(x) = \begin{cases} -\sqrt{-1-x} & x < -1 \\ 0 & x = 0 \\ x - 1 & x > 1. \end{cases}$$

In this exercise the given function has different expressions in different intervals and therefore needs to be analyzed in each of them separately. Now, 0 goes to 0, and it is rather clear that in fact the negative real numbers are sent to negative real numbers and likewise for the positive real numbers, so that in the end the map is a bijection. All remains to be done is to write the explicit inverse mappings.

2.3 Prove that the map $f : [1, +\infty) \to [1, +\infty)$ defined by $f(x) = e^{\log^2 x}$ is invertible, and write the explicit expression of f^{-1}.

Answer. Take x_1 and x_2 with $1 \le x_1 < x_2$. Then

$$0 \le \log x_1 < \log x_2 \implies e^{\log^2 x_1} < e^{\log^2 x_2} \implies f(x_1) < f(x_2).$$

Therefore f is strictly increasing, hence invertible. Take now $y \ge 1$. The point $x \ge 1$ satisfies $f(x) = y$ provided that

$$e^{\log^2 x} = y = e^{\log y} \implies \log^2 x = \log y \implies \log x = \sqrt{\log y} \implies x = e^{\sqrt{\log y}}.$$

Evidently $\sqrt{\log y} \ge 0$ and hence its exponential is in $[1, +\infty)$. Therefore the image of f is $[1, +\infty)$ and $f^{-1} : [1, +\infty) \to [1, +\infty)$ is given by $f^{-1}(y) = e^{\sqrt{\log y}}$.

This exercise is standard and simply requires to see that the given function actually maps the set $[1, +\infty)$ bijectively onto itself. For injectivity, it is immediately seen that f is increasing. For surjectivity, it is easy to find the solution of $f(x) = y$, that is, to find the inverse map.

2.4 Consider the function $g(x) = \dfrac{1}{4 \arcsin x - \pi}$.

(a) Find the domain of g.
(b) Determine the image of g.
(c) Prove that the restriction of g to $(-1, 1/\sqrt{2})$ is invertible and write its inverse.

Answer. (a) Put $f(x) = \arcsin x$ and $h(y) = 1/(4y - \pi)$, so that $g = h \circ f$. The domain of g is therefore:

$$\mathrm{Dom}(g) = \{x \in \mathbb{R} : x \in \mathrm{Dom}(f),\ f(x) \in \mathrm{Dom}(h)\}$$
$$= \{x \in \mathbb{R} : x \in [-1, 1],\ \arcsin x \neq \frac{\pi}{4}\}$$
$$= \{x \in \mathbb{R} : x \in [-1, 1],\ x \neq \frac{\sqrt{2}}{2}\}$$
$$= [-1, \frac{\sqrt{2}}{2}) \cup (\frac{\sqrt{2}}{2}, 1].$$

(b) The image of g consists of those $y \in \mathbb{R}$ for which there exists $x \in \mathrm{Dom}(g)$ such that $y = g(x)$. Thus

$$y \in \mathrm{Im}(g) \iff \text{there exists } x \in \mathrm{Dom}(g) \text{ such that: } y = (4 \arcsin x - \pi)^{-1}$$
$$\iff \text{there exists } x \in \mathrm{Dom}(g) \text{ such that: } 4y \arcsin x - \pi y = 1.$$

From the latter it follows that

$$y \neq 0 \text{ and } y \in \mathrm{Im}(g) \iff \text{there exists } x \in \mathrm{Dom}(g) \text{ such that: } \arcsin x = \frac{\pi y + 1}{4y}$$
$$\iff \frac{\pi y + 1}{4y} \in \mathrm{Im}(f) \setminus \{f(\frac{\sqrt{2}}{2})\} = [-\frac{\pi}{2}, \frac{\pi}{2}] \setminus \{\frac{\pi}{4}\},$$

where again $f(x) = \arcsin x$. Hence, if $y \neq 0$ and $y \in \mathrm{Im}(g)$, then

$$-\frac{\pi}{2} \leq \frac{\pi y + 1}{4y} \leq \frac{\pi}{2} \quad \text{and} \quad \frac{\pi y + 1}{4y} \neq \frac{\pi}{4}.$$

It follows that $y \in (-\infty, -1/(3\pi)] \cup [1/\pi, +\infty)$ and hence

$$\mathrm{Im}(g) = (-\infty, -\frac{1}{3\pi}] \cup [\frac{1}{\pi}, +\infty).$$

(c) The function $f(x) = \arcsin x$ is increasing. Hence

$$-\frac{\pi}{2} = f(-1) < f(x) < f\left(\frac{1}{\sqrt{2}}\right) = \frac{\pi}{4},$$

namely $f(x) \in (-\pi/2, \pi/4)$, for every $x \in (-1, 1/\sqrt{2})$. Further, $h(y)$ is decreasing in $(-\pi/2, \pi/4)$. Since g is the composition of f and h, which are both strictly monotone but with opposite monotoneity, g is decreasing on $(-1, 1/\sqrt{2})$ and hence invertible on this interval. Finally, from (b) it follows that

$$y = g(x) \iff \arcsin x = \frac{\pi y + 1}{4y}, \qquad \frac{\pi y + 1}{4y} \in [-\frac{\pi}{2}, \frac{\pi}{2}].$$

From this it follows that $x = \sin((\pi y + 1)/(4y))$, and finally

$$g^{-1}(y) = \sin\left(\frac{\pi y + 1}{4y}\right).$$

In this exercise, the basic idea is again to view g as a composite function. Once this is done, then finding the domain and the image is achieved by carefully following what each map (f and h) does. Most of the effort actually goes into finding the image of g. Inversion is done by inverting each of f and h.

2.5 Consider the function $f(x) = |x| + x^2 - 1$.

(a) Establish if $-5/4 \in \mathrm{Im}(f)$.
(b) Find the largest neighborhood of $x_0 = 1$ on which the function is invertible, and write the explicit analytic expression of the inverse.

Answer. (a) Observe that $-5/4 \in \mathrm{Im}(f)$ if and only if the equation

$$-\frac{5}{4} = |x| + x^2 - 1$$

has a solution in $\mathbb{R}$, the domain of f. However, for every $x \in \mathbb{R}$

$$f(x) = |x| + x^2 - 1 \geq -1 > -\frac{5}{4},$$

so that $-5/4 \notin \mathrm{Im}(f)$.

(b) Since f is an even function, it is not injective, hence not invertible. Consider the restriction of f to the interval $[0, +\infty)$ and denote it by g, explicitly $g(x) = x + x^2 - 1$. It is easy to see that g is increasing in its domain, for if $0 \leq x_1 \leq x_2$, then $x_1^2 < x_2^2$ and hence

$$g(x_1) = x_1^2 + x_1 - 1 < x_2^2 + x_2 - 1 = g(x_2).$$

Therefore g is invertible. In order to find the explicit analytic expression of the inverse of (the surjective map naturally associated with) g, the second order equation $y = x + x^2 - 1$ must be solved for x as a function of y. The roots of the polynomial $x^2 + x - 1 - y$ are

$$x_1(y) = \frac{-1 - \sqrt{5 + 4y}}{2}, \qquad x_2(y) = \frac{-1 + \sqrt{5 + 4y}}{2}.$$

Clearly, $x_1(y) < 0$, whereas $x_2(y) > 0$ because $y > -1$. Therefore

$$g^{-1}(y) = x_2(y) = \frac{-1 + \sqrt{5 + 4y}}{2}$$

and it follows that $\mathrm{Dom}(g^{-1}) = \mathrm{Im}(g) = [-1, +\infty)$.

This is a basic exercise on invertible mappings, where it is required to show that globally the function is not invertible but it is so if properly restricted. The presence of both an absolute value and a quadratic term imply that f is actually even, hence non invertible. But if one looks at one of the "branches" of f, namely $[0, +\infty)$, then f is monotone hence invertible. The explicit expression comes from taking the appropriate square root.

2.6 Consider the function $f(x) = \dfrac{1}{\log_{1/3}(x - 2)} - 1$.

(a) Find the domain of f.
(b) Establish if f is invertible for $x > 3$ and, if yes, find the explicit analytic expression of the inverse g^{-1}, where $g = f|_{(3,+\infty)}$, specifying its domain.

Answer. (a) In order for $\log_{1/3}(x - 2)$ to be well defined, it must be $x - 2 > 0$, that is $x > 2$. Further, $\log_{1/3}(x - 2) \neq 0$ if $x - 2 \neq 1$. Therefore $\mathrm{Dom}(f) = (2, 3) \cup (3, +\infty)$.

(b) For $x > 3$, f is increasing. Indeed, $x \mapsto \log_{1/3}(x - 2)$ is decreasing and positive. Hence $x \mapsto 1/\log_{1/3}(x - 2)$ is increasing and such is also the function $x \mapsto (1/\log_{1/3}(x - 2)) - 1$. As f is increasing on $(3, +\infty)$, the restriction g of f to this interval is invertible. Since $x > 3$, one has $\log_{1/3}(x - 2) < 0$ and $f(x) < -1$. Therefore, $f((3, +\infty)) \subset (-\infty, -1)$. Furthermore, if $y < -1$, then the equation $y = f(x)$ has a solution if and only if

$$\frac{1}{\log_{1/3}(x - 2)} - 1 = y \iff \frac{1}{\log_{1/3}(x - 2)} = y + 1$$

$$\iff \log_{1/3}(x - 2) = \frac{1}{y + 1}$$

$$\iff x - 2 = \left(\frac{1}{3}\right)^{\frac{1}{y+1}}$$

$$\iff x = 2 + \left(\frac{1}{3}\right)^{\frac{1}{y+1}}.$$

Since $y < -1$, it follows that $(1/3)^{\frac{1}{x+1}} > 1$ and $x > 3$, that is $(-\infty, -1) \subset f((3, +\infty))$. Hence $g^{-1}(x) = 2 + (1/3)^{\frac{1}{x+1}}$.

Here the proof that the appropriate restriction of f is increasing can be carried out by viewing f as a composition of functions. The final formula is obtained undoing each of the several functions in the correct order.

2.5   Problems on Invertible Mapings

2.7 Consider the function $f(x) = 1/(1 - 3^x)$.

(a) Find the domain and the image of f.
(b) Study the monotoneity of f and establish if f is invertible.
(c) Write f^{-1}, if it exists, and specify both its domain and its image.

2.8 Consider the function $f(x) = -1/(\log_2 |x| + 1)$.

(a) Find the domain and the image of f.
(b) Study the monotoneity of f and establish if f is invertible.
(c) Denote by g the restriction of f to $(1/2, +\infty)$. Determine whether g is invertible and, if yes, write g^{-1} explicitly, specifying its domain.

2.9 Consider the function $f(x) = \log\left(x^2 - 3x + 1\right)$.

(a) Find the domain of f.
(b) Find the largest interval I containing $x_0 = 3$ on which f is injective.
(c) Determine $J = f(I)$ and compute the inverse $(f|_I)^{-1} : J \to I$.

2.10 Consider the function

$$f(x) = \frac{1}{x^2 - x + 3} - \frac{3}{11}$$

and denote by g the restriction of f to $[1/2, +\infty)$. Determine g^{-1}, if it exists, and specify its domain.

2.11 Consider the function $f : \mathbb{R} \to [-1, 1]$ defined by

$$f(x) = \begin{cases} \dfrac{1}{x} & x \in (-\infty, -1] \\ x & x \in (-1, 0) \\ 1 - x & x \in [0, 1) \\ -\dfrac{1}{x} & x \in [1, +\infty). \end{cases}$$

(a) Draw the graph of f.
(b) Establish whether f is injective and/or surjective.
(c) Establish in which intervals f is increasing.
(d) Draw the graph of $f(|x|)$.

2.12 Consider the function $f(x) = \dfrac{1}{1 + \sqrt{|x - 2|}}$.

(a) Find the domain of f.
(b) Check if f is invertible for $x > 2$ and, if yes, find the inverse function g^{-1} of the restriction $g = f|_{(2, +\infty)}$.
(c) Find the even and odd parts of $F(x) = f(x + 2)$.

2.13 On which maximal intervals, if any, is $f(x) = \dfrac{1}{\sqrt{1 - \log x}}$ injective?

2.14 Show that the restriction of $f(x) = x - \dfrac{1}{x}$ to $(0, +\infty)$ is invertible and find an explicit expression of the inverse.

2.15 Consider the function $f(x) = \dfrac{e^x}{\sqrt{e^x - 1}}$. Find an interval I on which f is invertible and write the explicit expression of $(f|_I)^{-1}$.

2.16 Consider the function $f : (-\infty, +\infty) \to (-\pi/2, \pi/2]$ defined by

$$f(x) = \begin{cases} \arctan x & x \in (-\infty, -\frac{\pi}{4}] \cup [\frac{\pi}{4}, +\infty) \\ 2x + \dfrac{\pi}{2} & x \in (-\frac{\pi}{4}, 0] \\ -2x & x \in (0, \frac{\pi}{4}). \end{cases}$$

(a) Draw the graph of f.
(b) Establish whether f is injective and/or surjective.
(c) Establish in which intervals f is increasing.

2.17 On which intervals, if any, is the function $f(x) = \dfrac{e^{-x}}{\sqrt{\log x - 5}}$ injective?

2.18 Consider the function $f(x) = \dfrac{e^x}{e^{2x} + e^x + k}$.

(a) For which values of the real parameter k the function f is defined on $\mathbb{R}$?
(b) Put $k = 1$. Find a neighborhood of $x_0 = -1$ in which f is invertible and write an explicit analytic expression of the inverse.

2.19 Is the function

$$f(x) = \begin{cases} \dfrac{1}{2}\left[\left(\dfrac{2}{3}\right)^x + \dfrac{1}{x+1}\right] & x \geq 0 \\ \dfrac{2x+3}{x+1} & x < -1 \end{cases}$$

invertible in its domain?

2.20 Consider the function $g(x) = \sqrt{\log_2 x - \log_4(x-1)^2}$.

(a) Find the domain of g.
(b) Establish if g is invertible in its domain.

2.21 Consider the function

$$f(x) = \begin{cases} 2^{\frac{x^2+1}{x}} & x < 0 \\ -\frac{1}{5}\cos(2\arctan x) & x \geq 0. \end{cases}$$

(a) Find the image of f.
(b) Put $I = (-1, 1)$. Establish if $g = f|_I$ is invertible and, if yes, find g^{-1}.
(c) Determine a neighborhood of $x_0 = -2$ in which f is invertible.

Chapter 3
Maximum, Minimum, Supremum, Infimum

3.1 Upper and Lower Bounds, Maximum and Minimum

Definition 3.1 Let A be a non-empty subset of $\mathbb{R}$.

(i) The real number M is called an *upper bound* of A if $a \leq M$ for every $a \in A$. In this case, A is said to be *bounded above* by M.
(ii) The real number m is called a *lower bound* of A if $a \geq m$ for every $a \in A$. In this case, A is said to be *bounded below* by m.

If A is bounded above and below, then A is said to be *bounded*. The set A is bounded if and only if there exists $K > 0$ such that for every $a \in A$

$$|a| \leq K.$$

Many properties are more easily written by introducing the sets of upper and lower bounds of the given set A, namely[1]

$$A^* = \{M \in \mathbb{R} : M \geq a \text{ for every } a \in A\},$$
$$A_* = \{m \in \mathbb{R} : m \leq a \text{ for every } a \in A\}.$$

Thus, A is bounded above if and only if $A^* \neq \emptyset$ and it is bounded below if and only if $A_* \neq \emptyset$. Clearly, if $M \in A^*$, then every number greater than M is also in A^*, and similar considerations hold for A_*. Obviously it is not true that any number smaller than M is an upper bound, and analogous remarks hold for lower bounds. For example, if $A = (0, 1)$, then $1 \in A^*$ as any other number greater than 1, but no number less than 1 is an upper bound. In particular, no upper bound is in A. But if $B = (0, 1]$, then 1 is an upper bound of B and it belongs to B.

Definition 3.2 Let A be a non-empty subset of $\mathbb{R}$.

---

[1] The notation A^* and A_*, though common, is not universally adopted.

© Springer International Publishing Switzerland 2016
M. Baronti et al., *Calculus Problems*, UNITEXT - La Matematica per il 3+2 101,
DOI 10.1007/978-3-319-15428-2_3

(i) The real number M is called the *maximum* of A if $M \in A$ and if $M \geq a$ for every $a \in A$. In this case, one writes $M = \max A$.

(ii) The real number m is called the *minimum* of A if $m \in A$ and if $m \leq a$ for every $a \in A$. In this case, one writes $m = \min A$.

As observed above, a set may be bounded above and yet have no maximum, and it may be bounded below and yet have no minimum. However, whenever existing, the maximum is unique, and the same holds for the minimum.

3.2 Supremum, Infimum

It is an important and deep result that if $A \subset \mathbb{R}$ is bounded above, then the set A^* of upper bounds of A has a minimum. The fact that for any non-empty set $A \subset \mathbb{R}$

$$A^* \neq \emptyset \quad \Longrightarrow \quad \min A^* \text{ exists}$$

is actually equivalent to

$$A_* \neq \emptyset \quad \Longrightarrow \quad \max A_* \text{ exists},$$

and both are equivalent to the completeness of $\mathbb{R}$. Since $\mathbb{R}$ is indeed a complete ordered field, the following definition makes full sense.

Definition 3.3 Let A be a non-empty subset of $\mathbb{R}$.

(i) If A is bounded above, the minimum upper bound of A is called the *supremum* and is denoted sup A. If A is not bounded above, then one writes $\sup A = +\infty$.

(ii) If A is bounded below, the maximum lower bound of A is called the *infimum* and is denoted inf A. If A is not bounded below, then one writes $\inf A = -\infty$.

One can rephrase the above definitions by writing

$$\sup A = \begin{cases} \min A^* & A^* \neq \emptyset \\ +\infty & A^* = \emptyset \end{cases}, \quad \inf A = \begin{cases} \max A_* & A_* \neq \emptyset \\ -\infty & A_* = \emptyset. \end{cases}$$

In the literature, the supremum is also called *lowest upper bound*, abbreviated l.u.b., and the infimum is also called *greatest lower bound*, abbreviated g.l.b..

Proposition 3.1 *Let A be a non-empty subset of $\mathbb{R}$.*

(i) *If A is bounded above, then $\alpha \in \mathbb{R}$ is the supremum of A if and only if*

- α *is an upper bound of A;*
- *for every $\varepsilon > 0$ there exists $a \in A$ such that $a > \alpha - \varepsilon$.*

(ii) *If A is bounded below, then $\beta \in \mathbb{R}$ is the infimum of A if and only if*

- β is an lower bound of A;
- for every $\varepsilon > 0$ there exists $a \in A$ such that $a < \beta + \varepsilon$.

In general, given a non-empty $A \subset \mathbb{R}$ it may happen that even if A is bounded above, then it has no maximum, as when $A = (0, 1)$. However, if the maximum exists, then it coincides with the supremum. Hence the following are the possible cases:

$$A^* = \emptyset \quad \Longrightarrow \quad \sup A = +\infty \text{ and } \max A \text{ does not exist}$$

$$A^* \neq \emptyset \quad \Longrightarrow \quad \sup A \in \mathbb{R} \text{ and } \begin{cases} \text{either } \max A = \sup A \\ \text{or } \max A \text{ does not exist.} \end{cases}$$

Similarly, for any non-empty $A \subset \mathbb{R}$

$$A_* = \emptyset \quad \Longrightarrow \quad \inf A = -\infty \text{ and } \min A \text{ does not exist}$$

$$A_* \neq \emptyset \quad \Longrightarrow \quad \inf A \in \mathbb{R} \text{ and } \begin{cases} \text{either } \min A = \inf A \\ \text{or } \min A \text{ does not exist.} \end{cases}$$

The reader is urged to find an example for each of the above situations.

3.3 Guided Exercises on Maximum, Minimum, Supremum, Infimum

3.1 Determine, if existing, max A, min A, sup A and inf A, where

$$A = \left\{ \frac{x^2 - |x|}{x + 1} : x \in \mathbb{R}, \ x \neq -1 \right\}.$$

Answer. The set A is not bounded above because for any $M > 0$ there exists a positive $x > 2M + 1$, and for any such x it is $x^2 > x(2M + 1)$ and also $x + 1 < 2x$, so that

$$\frac{x^2 - |x|}{x + 1} = \frac{x^2 - x}{x + 1} > \frac{2Mx}{x + 1} > \frac{2Mx}{2x} = M.$$

Hence no $M > 0$ is an upper bound for A. Furthermore, for $x < 0$ it is

$$\frac{x^2 - |x|}{x + 1} = \frac{x^2 + x}{x + 1} = x$$

and hence any negative $x \neq -1$ belongs to A. Thus A does not have lower bounds either. In conclusion, sup $A = +\infty$, inf $A = -\infty$ and neither max A nor min A exists.

This simple exercise asks to realize that the numbers in A get arbitrarily large because of the term x^2 in the numerator, and arbitrarily small because in fact all negative real numbers except -1 belong to A.

3.2 Consider the sets $A = \left\{ \frac{x+5}{2x} : x < 0 \right\}$ and $B = \left\{ x : \frac{x+5}{2x} < 0 \right\}$. Is it true that $\sup A = \sup B$?

Answer. First of all, B is the set of solutions of the two systems

$$\begin{cases} x + 5 > 0 \\ x < 0 \end{cases} \qquad \begin{cases} x + 5 < 0 \\ x > 0 \end{cases}$$

and hence $B = (-5, 0)$. Therefore $\sup B = 0$. Next, the upper bounds of A are to be found. Since $x < 0$, it follows that

$$\frac{x+5}{2x} \le M \iff x + 5 \ge 2Mx \iff x(1 - 2M) \ge -5.$$

Now, if $M = 1/2$, then the above inequality is equivalent to $0 \ge -5$, which is obviously true. Hence $M = 1/2$ is an upper bound and, *a fortiori* such is any $M > 1/2$. If instead $M < 1/2$, then it is equivalent to

$$x \ge \frac{5}{2M - 1}.$$

This cannot hold true for every $x < 0$. Therefore, the set of upper bounds of A is $[1/2, +\infty)$ and $\sup A = 1/2 \ne 0 = \sup B$.

The issue in this exercise is to compare the image of the negative reals, that is A, with the inverse image of the negative reals, that is B, under a simple rational function. The latter set is easy to describe as the union of two sets of solutions of very simple systems, one of which (the second) has actually no solutions at all. For the former, it is enough to write down the basic condition for an upper bound M. A little ingenuity permits to prove that this condition is met if and only if $M \ge 1/2$, so that the required supremum is easily found.

3.3 Compute, if existing, $\max A$, $\min A$, $\sup A$ and $\inf A$, where

$$A = \left\{ x \in \mathbb{R} : \text{ either } 0 \le x < 1 \text{ or } x = \frac{2n - 3}{n - 1}, \ n \in \mathbb{N}, \ n \ge 2 \right\}.$$

Answer. First of all observe that

$$\frac{2n - 3}{n - 1} \ge 1 \iff 2n - 3 \ge n - 1 \iff n \ge 2$$

so that all elements of the form $(2n - 3)/(n - 1)$ with $n \geq 2$ are greater than 1. It follows that $0 = \min A = \inf A$. Further, 2 is an upper bound of A because

$$2 \geq \frac{2n - 3}{n - 1} \iff 2n - 2 \geq 2n - 3 \iff -2 \geq -3.$$

Take now $\varepsilon > 0$. Then there exists $n \geq 2$ such that

$$\frac{2n - 3}{n - 1} > 2-\varepsilon \iff 2n-3 > 2n-2-\varepsilon n+\varepsilon \iff \varepsilon n > 1+\varepsilon \iff n > \frac{1+\varepsilon}{\varepsilon},$$

which is true because $\mathbb{R}$ satisfies the Archimedean property. Therefore, by Proposition 3.1, $2 = \sup A$. If it were $2 = \max A$ then there would exist $n \geq 2$ for which

$$\frac{2n - 3}{n - 1} = 2 \iff 2n - 3 = 2n - 2 \iff -3 = -2,$$

which is false. Hence the maximum of A does not exist.

The set A is the union of $[0, 1)$ with a subset of $\mathbb{R}$ which is easily checked to lie inside $(1, 2)$, as the particular form of the ratio $(2n - 3)/(n - 1)$ suggests. Thus, the minimum, hence the infimum, of A is 0, whereas a little attention on the remaining part of A needs to be paid. It is intuitively clear that as n gets very large the ratio $(2n - 3)/(n - 1)$ is nearly 2. Thus, one applies the definition of supremum an indeed shows that there are numbers $(2n - 3)/(n - 1)$ greater than $2 - \varepsilon$ for any given ε.

3.4 Is the set $A = \{n/\sqrt{n + 1} : n \in \mathbb{N}, \ n \geq 1\}$ bounded? Does it have a maximum and a minimum? If yes, determine them.

Answer. The elements of A are all positive, and hence A is bounded below by 0. As for the minimum, observe that it is natural to expect that the elements

$$a_n = \frac{n}{\sqrt{n + 1}} = \sqrt{\frac{n^2}{n + 1}} \qquad n = 1, 2, 3, \ldots$$

become larger as n increases, so that a_1 should be the minimum. This is true if and only if for every positive integer

$$\sqrt{\frac{n^2}{n + 1}} \geq \sqrt{\frac{1}{2}} \iff \frac{n^2}{n + 1} \geq \frac{1}{2}$$
$$\iff 2n^2 - n - 1 \geq 0$$
$$\iff (2n + 1)(n - 1) \geq 0,$$

which is true for $n \geq 1$. Therefore $\sqrt{2}/2$ is the minimum of A.

Suppose next that there exists $L \geq 0$ such that $n/\sqrt{n+1} \leq L$ for all positive integers n. Taking squares,

$$\frac{n}{\sqrt{n+1}} \leq L \implies \frac{n^2}{n+1} \leq L^2$$
$$\implies n^2 - L^2 n - L^2 \leq 0$$
$$\implies n \leq \frac{1}{2}(L^2 + L\sqrt{L^2 + 4}),$$

which would imply that $\mathbb{N}$ is bounded, a contradiction. Hence A is not bounded above and, in particular, it has no maximum.

This exercise asks to understand what happens qualitatively to the elements of A, each of which is indexed by a positive integer n and can thus be denoted by a_n, as n gets larger. A minute thought reveals that a_n also gets larger, hence the first element, a_1, should be the smallest. The question is whether a_n remains bounded, but the special form of a_n suggests that the ratio $n/\sqrt{n+1}$ becomes arbitrarily large, so that A should be unbounded above. This is proved by contradiction, showing that if this were not so, then $\mathbb{N}$ itself would be bounded above. This exercise anticipates one of the themes of Chap. 4, where this kind of issue can be treated in a systematic way (see in particular Theorem 4.5 of Chap. 4).

3.5 Consider the function

$$f(x) = \begin{cases} x & x < 1 \\ 0 & x = 1 \\ (x-1)^2 - 1 & x > 1. \end{cases}$$

Compute, if existing, max A_k and min A_k, where $A_k = \{x \in \mathbb{R} : |f(x) - 1| \leq k\}$ as k ranges in $[0, +\infty)$.

Answer. It is laborious but not hard to show that

$$A_k = \begin{cases} \{1 + \sqrt{2}\} & k = 0 \\ [1-k, 1) \cup [1 + \sqrt{2-k}, 1 + \sqrt{2+k}] & 0 < k < 1 \\ [1-k, 1] \cup [1 + \sqrt{2-k}, 1 + \sqrt{2+k}] & 1 \leq k < 2 \\ [1-k, 1 + \sqrt{2+k}] & k \geq 2. \end{cases}$$

The computation can be eased by drawing the graph of $x \mapsto |f(x) - 1|$ as in Fig. 3.1. It follows that for $k > 0$ it is min $A_k = 1 - k$ and max $A_k = 1 + \sqrt{2+k}$, while for $k = 0$ it is min $A_0 = $ max $A_0 = 1 + \sqrt{2}$.

This exercise requires patience. The first step is to gain a sound idea of what is asked, which is to determine the inverse image under $x \mapsto |f(x) - 1|$ of the set $\{y : y \leq k\}$, as k ranges in the non negative numbers. A drawing helps a lot, but is not the answer to the problem. It helps, for example, to understand that the

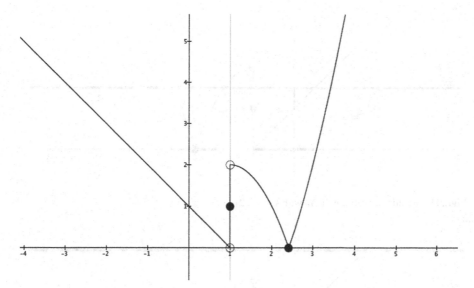

Fig. 3.1 Graph of $x \mapsto |f(x) - 1|$

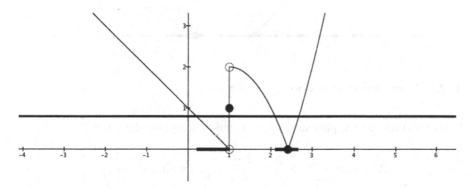

Fig. 3.2 Graphic description of A_k with $k \in (0, 1)$

problem is best analyzed by letting k varying in the chosen different intervals. Then, for each of them one sees what the set A_k looks like and ought to be, and finally one computes, checking that the inequalities are the correct ones. In Figs. 3.2, 3.3 and 3.4 the procedure is illustrated.

3.6 Compute, if existing, max A, min A, sup A and inf A, where

$$A = \left\{ \frac{2 - x^2}{x^2 + x + 1} : x < -\frac{1}{2} \right\}.$$

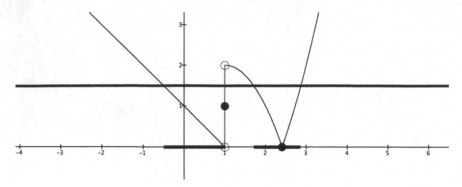

Fig. 3.3 Graphic description of A_k with $k \in (1, 2)$

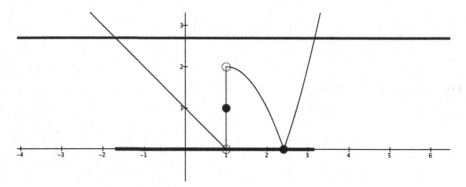

Fig. 3.4 Graphic description of A_k with $k \in (2, +\infty)$

Answer. In order to compute sup A, it is useful to determine the set

$$A^* = \left\{ M \in \mathbb{R} : M \geq \frac{2 - x^2}{x^2 + x + 1}, \ \text{for all } x < -\frac{1}{2} \right)\right\}$$

of upper bounds of A, of which sup A is the minimum. Observe first that the denominator is positive, so that it is enough to analyze the numerator. Now, $2 - x^2 > 0$ if $-\sqrt{2} < x < -1/2$, so that every upper bound $M \in A^*$ is necessarily positive. Furthermore, $M \in A^*$ if and only if for every $x < -1/2$ the inequality

$$x^2(M + 1) + Mx + M - 2 \geq 0$$

holds. Since $M + 1 > 0$, this inequality is true for every real number x provided that the discriminant Δ associated to it is less than or equal to 0. It follows that $\{M > 0 : \Delta \leq 0\} \subset A^*$. If instead $\Delta > 0$, then the set of solutions is

$$\left\{ x \in \mathbb{R} : x \leq \frac{-M - \sqrt{\Delta}}{2(M + 1)} \right\} \cup \left\{ x \in \mathbb{R} : x \geq \frac{-M + \sqrt{\Delta}}{2(M + 1)} \right\}.$$

In this case, the inequality is true for every $x < -1/2$ provided that

$$-\frac{1}{2} \leq \frac{-M - \sqrt{\Delta}}{2(M+1)},$$

namely, whenever $\sqrt{\Delta} \leq 1$. It follows that $A^* = \{M > 0 : \Delta \leq 1\}$. Solving $\Delta \leq 1$, yields $M \geq 7/3$ and the conclusion is sup $A = \min A^* = 7/3$. Finally, A has no maximum because $7/3 \notin A$, for if $7/3 \in A$ then there would exist $x < -1/2$ such that $(2 - x^2)/(x^2 + x + 1) = 7/3$ the solutions of which are $-1/2, -1/5$, which are not in A.

In order to compute inf A, it is useful to determine the set

$$A_* = \left\{ m \in \mathbb{R} : m \leq \frac{2 - x^2}{x^2 + x + 1}, \text{ for all } x < -\frac{1}{2} \right\}$$

of lower bounds of A, of which inf A is the maximum. Observe that if $m \in A_*$, then

$$(m + 1)x^2 + mx + m - 2 \leq 0$$

for every $x < -1/2$. Now, $-1 \in A$ and consequently every lower bound is less than or equal to -1. In fact, -1 is not a lower bound. Indeed, if -1 were a lower bound, then the second degree inequality would reduce to $-x - 3 \leq 0$, which is not satisfied by any $x < -1/2$. It follows that if $m \in A_*$, then $m < -1$. The second degree inequality is satisfied by any real number if the associated discriminant Δ is less than or equal to 0, so that $\{m < -1 : \Delta \leq 0\} \subset A_*$. If instead $\Delta > 0$, then the set of solutions is

$$\left\{ x \in \mathbb{R} : x \leq \frac{-m + \sqrt{\Delta}}{2(m+1)} \right\} \cup \left\{ x \in \mathbb{R} : x \geq \frac{-m - \sqrt{\Delta}}{2(m+1)} \right\}.$$

In this case, the inequality is true for every $x < -1/2$ provided that

$$-\frac{1}{2} \leq \frac{-m + \sqrt{\Delta}}{2(m+1)}$$

namely, whenever $-1 \geq \sqrt{\Delta}$. The latter, however, is never true. It follows therefore that $A_* = \{m < -1, \Delta \leq 0\}$. Solving $\Delta \leq 0$ yields $m < (2 - 2\sqrt{7})/3$ and the conclusion is inf $A = \max A_* = (2 - 2\sqrt{7})/3$. Finally, inf $A \in A$, because $(2 - 2\sqrt{7})/3 \in A$ if and only if there exists $x < -1/2$ such that

$$\frac{2 - x^2}{x^2 + x + 1} = \frac{2 - 2\sqrt{7}}{3},$$

the unique solution of which is $x = -3 - \sqrt{7} < -1/2$.

This is a slightly tricky exercise that asks for a separate analysis of the set of upper bounds A^* and the set of lower bounds A_* of the given set A. Both of them appear as sets of numbers out of which the coefficients of a second degree polynomial P are built. In the case A^*, the requirement is that P satisfies the inequality $P(x) \geq 0$ for every $x < -1/2$, where in the case A_* the requirement is $P(x) \leq 0$ for every $x < -1/2$. Thus, ultimately, one has to analyse under which conditions the interval $(-\infty, -1/2)$ lies to the left of the leftmost root of P, or else when P has no real roots and the inequality is identically satisfied.

3.4   Problems on Maximum, Minimum, Supremum, Infimum

3.7 Consider the inequality $\sqrt{x^2 - 4x + 3} \leq 3 - x$.

(a) Determine the set S of solutions.
(b) Compute, if existing, max S, min S, sup S and inf S.

3.8 Consider the set $A = \{x \in \mathbb{R} : x - \sqrt{|x - 1|} \leq 1\}$. Compute, if existing, maximum and minimum of A.

3.9 Determine the supremum of $A = \left\{\dfrac{2x - 1}{x + 1} : x \geq 0\right\}$.

3.10 Compute, if existing, max A, min A, sup A and inf A, where

$$A = \left\{\frac{n}{n^2 + 1} : n = 1, 2, \ldots\right\}.$$

3.11 Compute, if existing, max A, min A, sup A and inf A, where

$$A = \left\{(-1)^n \frac{n}{n^2 + 1} : n \in \mathbb{N}, n \geq 2\right\}.$$

3.12 Compute, if existing, max, min, sup and inf, of the following sets:

$$A = \left\{\frac{x + 1}{x^2 + 1} : x \in \mathbb{R}\right\},$$

$$B = \left\{\frac{n^2 - 4n}{2n - 9} : n \in \mathbb{N}\right\},$$

$$C = \left\{(-1)^n \frac{n + 5}{2n^2 + 1} : n \in \mathbb{N}\right\}.$$

3.13 Compute, if existing, max A, min A, sup A and inf A, where

$$A = \left\{(-1)^n \left(\frac{2}{|2n - 11|} + \frac{1}{n}\right) : n \in \mathbb{N}, n \neq 0\right\}.$$

3.14 Given two non-empty subsets A and B of $\mathbb{R}$, write $A + B = \{a + b : a \in A,$ $b \in B\}$. Prove that $\sup(A + B) = \sup A + \sup B$, with the understanding that "$+\infty + \infty = +\infty$" and "$S + \infty = +\infty$" for every $S \in \mathbb{R}$.

Chapter 4
Sequences

4.1 Lists of Real Numbers

Among the many possible functions, *sequences* play a prominent role in Analysis. They are meant to describe *ordered lists* of real numbers, with the understanding that any such list is allowed to, and in general will, consist of infinitely many numbers. Technically, they are defined as follows.

Definition 4.1 A sequence is a map $a : \mathbb{N} \to \mathbb{R}$, usually denoted $(a_n)_{n \geq 0}$. It is customary to write a_n in place of $a(n)$. Each individual a_n is referred to as the nth *term* of the sequence.

It is important to notice that a sequence is an ordered list, the first term of which is a_0, the second a_1, and so on. In many books, the notation $\{a_n\}_{n \geq 0}$ is used in place of $(a_n)_{n > 0}$. This may lead to confusion because in mathematics round brackets are used to emphasize order, so that for example $(1, 2) \neq (2, 1)$, whereas curly brackets are used to mean sets, that is, unordered collections, so that $\{1, 2\} = \{2, 1\}$.

Since the main idea that is conveyed in the notion of sequence is that of an ordered list, real valued maps defined on subsets of the type $\{n \in \mathbb{N} : n \geq n_0\}$ for a fixed positive integer n_0 are also considered sequences.

Many sequences are defined by declaring an explicit formula for a_n, but it is often important to consider sequences that are defined by a *recurrence* relation, namely

$$a_0 = x_0, \qquad a_{n+1} = f(a_n), \tag{4.1}$$

where $x_0 \in \mathbb{R}$ and $f : \mathbb{R} \to \mathbb{R}$ is a given function. Informally speaking, the meaning of (4.1) is that each term of the sequence is constructed from the previous one by means of f, starting from x_0. For this reason one says that the thus constructed sequence is *initialized* at x_0. Practically,

$$a_0 = x_0, \; a_1 = f(x_0), \; a_2 = f(a_1) = f(f(x_0)), \; \ldots$$

© Springer International Publishing Switzerland 2016
M. Baronti et al., *Calculus Problems*, UNITEXT - La Matematica per il 3+2 101,
DOI 10.1007/978-3-319-15428-2_4

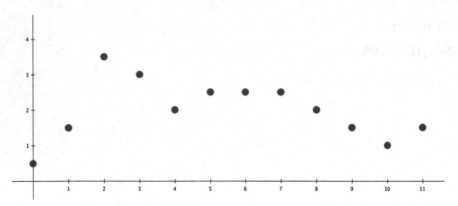

Fig. 4.1 The graph of a possible sequence

There are variants of this construction: one can assign the first k values and then assign a rule that allows to build a_{n+k} from the previous k values $a_n, a_{n+1}, \ldots a_{n+k-1}$. A typical example of this sort is the famous Fibonacci sequence $1, 1, 2, 3, 5, 8, 13, \ldots$, that is defined by:

$$a_0 = a_1 = 1, \qquad a_{n+2} = a_n + a_{n+1}.$$

Sequences are particular kinds of functions $f : I \subset \mathbb{R} \to \mathbb{R}$ and, as such, can be visualized by means of their graphs. Since the domain is (essentially) $\mathbb{N}$, the graph of a sequence is as in Fig. 4.1.

4.2 Convergence Notions

The most important notion relative to sequences is the following one.

Definition 4.2 (*Converging sequence*) The sequence $(a_n)_{n\geq 0}$ is said to *converge* to $\ell \in \mathbb{R}$, or that it *tends* to ℓ, or that the *limit* of $(a_n)_{n\geq 0}$ is ℓ, if for every $\varepsilon > 0$ there exists an integer N_ε such that if $n \geq N_\varepsilon$, then $|a_n - \ell| < \varepsilon$. If this situation occurs, then one writes

$$\lim_n a_n = \ell \qquad \text{or} \qquad a_n \to \ell.$$

The limit, if it exists, is unique.

In the definition of limit, it is stated that for any given $\varepsilon > 0$ all the terms a_n of the sequence satisfy $|a_n - \ell| < \varepsilon$ provided that n is large enough. In general, if there exists an integer N_ε such that if $n \geq N_\varepsilon$, then a_n satisfies a certain property, then for brevity it is said that the sequence $(a_n)_{n\geq 0}$ *eventually* has that property.

In most of the current literature, a sequence that does not converge is said to *diverge*. However, in this book a slightly different choice[1] is adopted, and the word "divergent" is reserved to a special subclass of sequences that do not converge, namely it is used to refer to those sequences that become arbitrarily large, either positive or negative, in the sense clarified in the next definition.

Definition 4.3 (*Diverging sequence*) The sequence $(a_n)_{n \geq 0}$ is said to *diverge* to $+\infty$, or that it tends to $+\infty$, or that the limit of $(a_n)_{n \geq 0}$ is $+\infty$ if for every $K > 0$ eventually $a_n > K$. If this situation occurs, then one writes

$$\lim_n a_n = +\infty \qquad \text{or} \qquad a_n \to +\infty.$$

Similarly, if for every $K > 0$ eventually $a_n < -K$, then $(a_n)_{n \geq 0}$ is said to diverge (or tend) to $-\infty$ and one writes

$$\lim_n a_n = -\infty \qquad \text{or} \qquad a_n \to -\infty.$$

If the sequence $(a_n)_{n \geq 0}$ is neither convergent nor divergent, then it is said that $(a_n)_{n \geq 0}$ *does not have a limit*. Thus, according to standard terminology the sequence $a_n = (-1)^n$, the values of which are $1, -1, 1, -1, \ldots$, would be called divergent, whereas according to the present convention it does not have a limit.

4.3 Results on Limits

A sequence $(a_n)_{n \geq 0}$ is said to be *bounded* if the set of its values, namely $\{a_n : n \geq 0\}$, is a bounded subset of $\mathbb{R}$. Equivalently, $(a_n)_{n \geq 0}$ is bounded if and only if there exists $M > 0$ such that for all $n \geq 0$

$$|a_n| \leq M.$$

A sequence is *unbounded* if it is not bounded, namely, if the set of its values is an unbounded subset of $\mathbb{R}$.

Proposition 4.1 *A convergent sequence is bounded. A (positively or negatively) divergent sequence is unbounded.*

It should be noticed that the previous result cannot be reversed: a bounded sequence needs not be convergent and an unbounded sequence needs not be divergent.

The four results that follow concern convergent sequences and are of basic importance.

---

[1] The primary motivation for this choice is to comply with the current terminology used in Italy.

Theorem 4.1 (Comparison, I) *Let* $a, b \in \mathbb{R}$ *and suppose that* $a_n \to a$ *and* $b_n \to b$. *Then:*

(i) *if eventually either* $a_n < b_n$ *or* $a_n \leq b_n$, *then* $a \leq b$;
(ii) *if* $a < b$, *then eventually* $a_n < b_n$;
(iii) *if* $a < \lambda \in \mathbb{R}$, *then eventually* $a_n < \lambda$; *if* $a > \mu \in \mathbb{R}$, *then eventually* $a_n > \mu$.

Corollary 4.1 (Permanence of sign) *If the sequence* $(a_n)_{n \geq 0}$ *converges to* $\ell \neq 0$, *then* $(a_n)_{n \geq 0}$ *eventually has the sign of* ℓ.

Theorem 4.2 (Comparison, II, or Squeeze Theorem) *Let* $\ell \in \mathbb{R}$ *and suppose that the sequences* $(a_n)_{n \geq 0}$, $(b_n)_{n \geq 0}$ *and* $(c_n)_{n \geq 0}$ *satisfy:*

(i) $a_n \to \ell$ *and* $c_n \to \ell$;
(ii) *eventually* $a_n \leq b_n \leq c_n$.

Then $(b_n)_{n \geq 0}$ *converges and* $\lim_{n} b_n = \ell$.

In many practical situations, a sequence is given as a sum, product or quotient of elementary sequences, the limiting behaviour of which is known. It is then of basic importance to be able to infer how algebraic operations affect the process of taking limits. The answer for convergent sequences is in the result that follows.

Theorem 4.3 (Algebra of Limits) *Suppose that* $(a_n)_{n \geq 0}$ *and* $(b_n)_{n \geq 0}$ *converge to* a *and* b, *respectively. Then:*

(i) $|a_n| \to |a|$;
(ii) $a_n + b_n \to a + b$;
(iii) $a_n b_n \to ab$;
(iv) *for every* $\lambda \in \mathbb{R}$, $\lambda a_n \to \lambda a$;
(v) *if* $b \neq 0$, *then* $a_n / b_n \to a/b$.

Notice that in item (v) of the previous theorem it is implicitly stated that the ratio a_n/b_n eventually makes sense, under the sole assumption that $b \neq 0$. Indeed, by permanence of sign, eventually $b_n \neq 0$. The preceeding results on convergent sequences admit, to some extent, a formulation to situations where some of the sequences at hand diverge, either to $+\infty$ or to $-\infty$.

Theorem 4.4 (Comparison, III) *Suppose that the sequences* $(a_n)_{n \geq 0}$ *and* $(b_n)_{n \geq 0}$ *eventually satisfy* $a_n \leq b_n$. *Then:*

(i) $a_n \to +\infty$ *implies* $b_n \to +\infty$;
(ii) $b_n \to -\infty$ *implies* $a_n \to -\infty$.

As for the algebra of limits, the possible conclusions that can be drawn for sums and products in which one of the two sequences involved is divergent is summarized in Table 4.1 below.

Question marks appear when no general conclusion may be drawn. In all of the three indecided situations it is possible, and the reader is urged to try this, to produce

Table 4.1 Extended algebra of limits of sequences

$\lim_{n} a_n$	$\lim_{n} b_n$	$\lim_{n} (a_n + b_n)$	$\lim_{n} (a_n b_n)$
$a > 0$	$+\infty$	$+\infty$	$+\infty$
$a > 0$	$-\infty$	$-\infty$	$-\infty$
$a < 0$	$+\infty$	$+\infty$	$-\infty$
$a < 0$	$-\infty$	$-\infty$	$+\infty$
0	$+\infty$	$+\infty$	?
0	$-\infty$	$-\infty$	?
$+\infty$	$+\infty$	$+\infty$	$+\infty$
$-\infty$	$-\infty$	$-\infty$	$+\infty$
$+\infty$	$-\infty$	?	$-\infty$

examples in which the resulting sequence either converges, or diverges to $+\infty$ or to $-\infty$, or has no limit. These cases are referred to as situations in which a sequence takes on an *indeterminate form*.

Much can be said about sequences that are reciprocal of sequences that are known to either diverge or to converge to 0, making precise the rule of thumb according to which the reciprocal of something large is small and, conversely, the reciprocal of something small is large.

Proposition 4.2 *Consider the sequence $(a_n)_{n \geq 0}$.*

(i) If $(a_n)_{n \geq 0}$ diverges, then $a_n^{-1} \to 0$;
(ii) if $(a_n)_{n \geq 0}$ is eventually positive and $a_n \to 0$, then $a_n^{-1} \to +\infty$;
(iii) if $(a_n)_{n \geq 0}$ is eventually negative and $a_n \to 0$, then $a_n^{-1} \to -\infty$.

A seemingly intractable sequence like $\sin n / n$ is easily seen to converge to 0 because although the factor $\sin n$ can be shown to admit no limit, it does have the redeeming feature of being bounded. Since the factor $1/n \to 0$, then the product $\sin n / n$ also tends to 0, because

$$\left| \frac{\sin n}{n} \right| \leq \frac{1}{n},$$

so that $1/n$ being eventually small implies that $\sin n / n$ is eventually small. This fact is very general, and very useful.

Proposition 4.3 *The product of a bounded sequence with a sequence that tends to 0, tends to 0.*

Finally, some results on monotonic sequences. The notion of monotonic map applies to sequences because, as already remarked, a sequence is a map defined on a subset of $\mathbb{R}$, namely $\mathbb{N}$. The so-called inductive nature of $\mathbb{N}$, that is, the fact that, unlike what happens for real numbers, for every $n \in \mathbb{N}$ there is "the next" element, namely the successive integer $n + 1$, implies the following simple test for monotonic sequences.

Proposition 4.4 *Consider the sequence* $(a_n)_{n\geq 0}$.

  (i) $(a_n)_{n\geq 0}$ *is strictly increasing if and only if* $a_n < a_{n+1}$ *for every* $n \in \mathbb{N}$;
  (ii) $(a_n)_{n\geq 0}$ *is strictly decreasing if and only if* $a_n > a_{n+1}$ *for every* $n \in \mathbb{N}$;
 (iii) $(a_n)_{n\geq 0}$ *is non decreasing if and only if* $a_n \leq a_{n+1}$ *for every* $n \in \mathbb{N}$;
 (iv) $(a_n)_{n\geq 0}$ *is non increasing if and only if* $a_n \geq a_{n+1}$ *for every* $n \in \mathbb{N}$.

The result that follows is of fundamental nature, and states that monotonic sequences always have limits, either finite, that is, real numbers, or infinite, that is, either $+\infty$ or $-\infty$. Spelled in a different way, a monotonic sequence either converges or diverges.

Theorem 4.5 *Consider the sequence* $(a_n)_{n\geq 0}$ *and denote by* $A = \{a_n : n \geq 0\}$ *the set of its values.*

  (i) *If* $(a_n)_{n\geq 0}$ *is either strictly increasing or non decreasing, then* $\lim_n a_n = \sup A$;
  (ii) *if* $(a_n)_{n\geq 0}$ *is either strictly decreasing or non increasing, then* $\lim_n a_n = \inf A$.

4.4   Guided Exercises on Sequences

4.1 Put $a_n = \sqrt{n^2 + 1} - n$.

(a) Prove that the sequence $(a_n)_{n\geq 0}$ is strictly monotonic.
(b) Compute $\lim_n a_n$.

Answer.
   (a) The sequence $(a_n)_{n\geq 0}$ is strictly decreasing. Indeed:

$$a_{n+1} < a_n \iff \sqrt{(n+1)^2 + 1} - (n+1) < \sqrt{n^2 + 1} - n$$
$$\iff \sqrt{(n+1)^2 + 1} < 1 + \sqrt{n^2 + 1}$$
$$\iff (n+1)^2 + 1 < 1 + (n^2 + 1) + 2\sqrt{n^2 + 1}$$
$$\iff n^2 + 2n + 2 < 2 + n^2 + 2\sqrt{n^2 + 1}$$
$$\iff n < \sqrt{n^2 + 1}$$
$$\iff n^2 < n^2 + 1.$$

(b) Since

$$a_n = \sqrt{n^2 + 1} - n = \frac{(\sqrt{n^2 + 1} - n)(\sqrt{n^2 + 1} + n)}{\sqrt{n^2 + 1} + n} = \frac{(n^2 + 1) - n^2}{\sqrt{n^2 + 1} + n} = \frac{1}{\sqrt{n^2 + 1} + n}$$

and since the denominator diverges to $+\infty$, it follows that $a_n \to 0$.

This is a very basic exercise on sequences. In part (a) the very definition of decreasing sequence may be applied, for it amounts to solving a very simple inequality among positive integers. In part (b) a simple rationalization technique is used. Upon multiplying and dividing by the non zero sum $\sqrt{n^2+1}+n$, the term a_n, which is the difference of two terms each of which tends to $+\infty$, takes the form of the reciprocal of something which tends to $+\infty$. This is an example of standard way to turn a sequence which takes on the indeterminate form "$\infty - \infty$" into a ratio, which is usually easier to handle. In the case at hand, the ratio is elementary.

4.2 Consider $a_n = \sin(n\pi/2)$. Establish the limiting behaviour of each of the following sequences.

(a) $(b_n)_{n\geq1}$, with $b_n = na_n$.
(b) $(c_n)_{n\geq1}$, with $c_n = a_n/n$.
(c) $(d_n)_{n\geq1}$, with $d_n = a_n \cos(n\pi/2)$.

Answer. (a) Consider the subsequence of b_n corresponding to the odd integers, namely $b_{2n+1} = (-1)^n(2n+1)$. Since this does not have a limit, neither does b_n.
(b) The sequence c_n is the product of the bounded sequence $(a_n)_{n\geq1}$ with the sequence $(1/n)_{n\geq1}$ which tends to 0. Therefore $\lim_n c_n = 0$.
(c) Observe that $d_n = a_n \cos(n\pi/2) = \sin(n\pi)/2 = 0$ for all $n \in \mathbb{N}$. This means that the sequence $(d_n)_{n\geq1}$ is constant and equal to 0, hence it converges to 0.

This is a simple exercise in which the basic ingredient is the oscillating, periodic sequence $(a_n)_{n\geq1}$, which takes the values $0, 1, 0, -1$. In (a) it is multiplied by the positively divergent sequence $(n)_{n\geq1}$, hence no limit exists; in (b) it is multiplied by a sequence that tends to 0, hence tends to 0 owing to the boundedness of $(a_n)_{n\geq1}$. Finally, in (c), $(a_n)_{n\geq1}$ is multiplied by a sequence that attains the same values periodically, but with a shift in n, so that one of the two factors is always 0.

4.3 Compute $\displaystyle\lim_n \sqrt{\frac{n^2}{3} + \frac{1}{n^2} \log(1 + 2\sin\frac{1}{n})}$.

Answer. Observe first that

$$\sqrt{\frac{n^2}{3} + \frac{1}{n^2}\log(1+2\sin\frac{1}{n})} = n\sqrt{\frac{1}{3} + \frac{1}{n^4}\frac{\log(1+2\sin\frac{1}{n})}{2\sin\frac{1}{n}}2\sin\frac{1}{n}}$$

$$= \sqrt{\frac{1}{3} + \frac{1}{n^4}\frac{\log(1+2\sin\frac{1}{n})}{2\sin\frac{1}{n}}\frac{2\sin\frac{1}{n}}{\frac{1}{n}}}.$$

Now, since

$$\lim_n \frac{\log(1+a_n)}{a_n} = \lim_n \frac{\sin a_n}{a_n} = 1 \qquad (4.2)$$

for every sequence $(a_n)_{n\geq1}$ that tends to 0, the conclusion is that the limit is $2/\sqrt{3}$.

The main task here is twofold. First, one has to realize that the first factor, the square root, is really of the form nb_n where $b_n \to 1/\sqrt{3}$. Thus, up to constants, the study of the given sequence can easily be reduced to that of the well known sequences (4.2).

4.4 Study $\lim\limits_n \dfrac{\sqrt{kn+1}-\sqrt{n}}{\sqrt{n+1}-\sqrt{n}}$ as k ranges in $(0, +\infty)$.

Answer. Put $a_n = (\sqrt{kn+1}-\sqrt{n})/(\sqrt{n+1}-\sqrt{n})$. Rationalizing, that is, multiplying and dividing by $\sqrt{kn+1}+\sqrt{n}$ and also by $\sqrt{n+1}-\sqrt{n}$, after collecting a factor $\sqrt{n}$ both at the numerator and at the denominator, it turns out that

$$a_n = ((k-1)n+1) \cdot \frac{\sqrt{1+\frac{1}{n}}+1}{1+\sqrt{k+\frac{1}{n}}}.$$

Now,

$$\lim_n \frac{\sqrt{1+\frac{1}{n}}+1}{1+\sqrt{k+\frac{1}{n}}} = \frac{2}{1+\sqrt{k}}$$

and

$$\lim_n (k-1)n+1 = \begin{cases} +\infty & k > 1 \\ 1 & k = 1 \\ -\infty & 0 < k < 1. \end{cases}$$

It follows that

$$\lim_n a_n = \begin{cases} +\infty & k > 1 \\ 1 & k = 1 \\ -\infty & 0 < k < 1. \end{cases}$$

Here some algebraic manipulation is required, basically the rationalization technique that has already been used in part (b) of Exercise 4.1, namely

$$\sqrt{a}-\sqrt{b} = \frac{a-b}{\sqrt{a}+\sqrt{b}},$$

combined with the idea of writing things in such a way that emphasizes the role of n, or, equivalently, of $1/n$.

4.5 Study $\lim\limits_n (kn^2 + n\sin(kn!))$, as k ranges in $\mathbb{R}$.

Answer. Put $a_n = kn^2 + n\sin(kn!)$. Observe that $-1 \le \sin(kn!) \le 1$, so that for every positive integer n the inequalities $-n \le n\sin(kn!) \le n$ hold. Now, if $k > 0$, then since

$$kn^2 + n\sin(kn!) \ge kn^2 - n \to +\infty,$$

it follows by comparison that $a_n \to +\infty$. Analogously, if $k < 0$ then since

$$kn^2 + n\sin(kn!) \le kn^2 + n \to -\infty$$

it follows by comparison that $a_n \to -\infty$. Finally, if $k = 0$ then $a_n = 0$ and evidently $a_n \to 0$.

This is a basic exercise on the use of comparison tests. The scary looking argument of the sine function has no effect in terms of the qualitative behaviour of a_n because the sine is bounded and because it multiplies n, and this, in turn, is dominated by n^2. In practice, the comparison criteria say that what counts is kn^2, whose limiting properties depend on the sign of k.

4.6 Compute, if existing, $\displaystyle\lim_n \frac{(n!+1)^{\frac{1}{n^2}}}{n}\sin n.$

Answer. Observe that

$$0 \le \frac{(n!+1)^{\frac{1}{n^2}}}{n} = \frac{e^{\frac{1}{n^2}\log(1+n!)}}{n} \le \frac{e^{\frac{1}{n^2}\log(n^n)}}{n}$$

for every $n \ge 2$. Indeed, it is easy to show by induction that for such values of n the inequality $1 + n! \le n^n$ holds. Therefore

$$0 \le \frac{(n!+1)^{\frac{1}{n^2}}}{n} \le \frac{e^{\frac{1}{n^2}\log(n^n)}}{n} = \frac{e^{\frac{\log n}{n}}}{n} \to 0,$$

because the exponent tends to 0. By comparison, it follows that $(n!+1)^{\frac{1}{n^2}}/n \to 0$. Since the sequence $(\sin n)_{n\ge 0}$ is bounded, the conclusion is that

$$\lim_n \frac{(n!+1)^{\frac{1}{n^2}}}{n}\sin n = 0.$$

This exercise requires a combination of simple ideas and results. First of all, the form of the sequence calls for a careful analysis of the first factor, say b_n, because the second factor, $\sin n$, does not have a limit but is bounded. It is actually reasonable to guess that b_n tends to 0, because it is a ratio in which the denominator is n, and hence tends to $+\infty$, and the numerator might be bounded, a fact that can be established writing it in exponential form. In doing so, the term $\log(1+n!)$ appears. It is natural to compare $1 + n!$ with n^n because the logarithm of a power is easy to handle. Finally, the guess $1 + n! < n^n$ is supported by the consideration that a factorial is a product of decreasing factors whereas in $n^n = n\cdots n$ the factors are always equal to n. Thus, first the comparison criterion applies to infer that $b_n \to 0$, and, finally, the product of the bounded sequence $(\sin n)_{n\ge 1}$ does not affect the limiting properties of b_n.

4.7 Compute, if existing, $\lim_n a_n$, where a_n is defined by the recurrence relation

$$a_1 = 2, \qquad a_{n+1} = \frac{a_n^2 + 3}{2a_n}.$$

Answer. First of all, observe that $a_n > 0$ for every $n \in \mathbb{N}$. Indeed $a_1 = 2 > 0$ and, assuming inductively that $a_n > 0$, it follows that

$$a_{n+1} = \frac{a_n^2 + 3}{2a_n} > 0.$$

Furthermore,

$$a_{n+1} < a_n \iff \frac{a_n^2 + 3}{2a_n} < a_n \iff 3 < a_n^2,$$

which is equivalent to $a_n > \sqrt{3}$ because $a_n > 0$. Therefore the sequence is strictly decreasing if and only if $a_n > \sqrt{3}$ for every $n \in \mathbb{N}$. This latter fact can be established by induction. Indeed, $a_1 = 2 > \sqrt{3}$ and assuming that $a_n > \sqrt{3}$, it follows that

$$a_{n+1} - \sqrt{3} = \frac{a_n^2 + 3}{2a_n} - \sqrt{3} = \frac{a_n^2 + 3 - 2\sqrt{3}a_n}{2a_n} = \frac{(a_n - \sqrt{3})^2}{2a_n} > 0,$$

whence $a_{n+1} > \sqrt{3}$. Since the given sequence is monotonic, it has a limit, and since it is decreasing and positive it converges to a non negative real number. Put $\lim_n a_n = \ell$. Evidently $\ell \geq \sqrt{3}$ and, by uniqueness of the limit

$$\ell = \lim_n a_{n+1} = \lim_n \frac{a_n^2 + 3}{2a_n} = \frac{\ell^2 + 3}{2\ell}.$$

It follows that ℓ satisfies the equation

$$\frac{\ell^2 + 3}{2\ell} = \ell, \tag{4.3}$$

which is equivalent to $\ell^2 = 3$, and since $\ell > 0$, this implies $\ell = \sqrt{3}$.

This is a prototypical exercise on the limit of a sequence defined via a recurrence relation. The recurrence relation itself suggests, taking the limit, an equation. In the case at hand, the equation is (4.3). A priori, this equation has actually the two solutions $\ell = \pm\sqrt{3}$, one of which should hopefully be ruled out. The point is that the sequence is initialized at $a_1 = 2$, and the recurrence relation preserves the sign. This motivates the procedure: first show that $a_n > 0$, then show that it decreases. This is enough to conclude that a_n actually converges, so that ℓ exists and is positive. At this point the equation does make sense and forces $\ell = \sqrt{3}$.

4.8 Take $x_0 > 0$. Compute, if existing, $\lim_n a_n$, where a_n is defined by the recurrence relation

$$a_0 = x_0, \qquad a_{n+1} = \frac{a_n}{1 + a_n}.$$

Answer. First of all, observe that the sequence is well defined because all of its terms are positive, so that in particular $1 + a_n \neq 0$. This is easily established by induction, because $a_0 = x_0 > 0$ and then a_{n+1} is a ratio of positive numbers by inductive assumption. Furthermore,

$$a_{n+1} - a_n = -\frac{a_n^2}{1 + a_n},$$

so that the sequence is strictly decreasing and bounded below by 0. Hence it converges, and if $\ell = \lim_n a_n$, then, by uniqueness of the limit,

$$\ell = \lim_n a_{n+1} = \frac{\ell}{1 + \ell}.$$

The equation $\ell = \ell/(1 + \ell)$ must hold and hence $\ell = 0$.

This is a variant of Exercise 4.7. This time the limit, the existence of which is established using the same monotonicity argument, is independent of the initial value x_0, as long as $x_0 > 0$.

4.5 Problems on Sequences

4.9 Compute, if existing, $\lim_n \dfrac{n^2 - 4n + 2}{n^2 + 1}$.

4.10 Compute, if existing, $\lim_n (\sin \sqrt{n+1} - \sin \sqrt{n})$.

4.11 Compute, if existing, $\lim_n (\sqrt{n+1} - \sqrt[3]{n^3 + 1})$.

4.12 Compute, if existing, $\lim_n n \sin(\sin(1/n))$.

4.13 Compute, if existing, $\lim_n n^k \log(1 + \sin(1/n))$, as k ranges in $\mathbb{R}$.

4.14 Compute, if existing, $\lim_n (\sin^4 n + \cos^4 n) \log(1 + n^k)$, as k ranges in $\mathbb{R} \setminus \{0\}$.

4.15 Compute, if existing, $\lim_n \dfrac{\sqrt{n} - \sin n}{\sqrt{n+1} - \cos n}$.

4.16 Compute, if existing, $\lim_n a_n$ where a_n is defined by the recurrence relation

$$a_1 = 1, \qquad a_{n+1} = \sin(a_n).$$

4.17 Compute, if existing,

$$\lim_n \frac{\sin^\alpha(\frac{1}{n}) - \cos(\frac{1}{n}) + 1}{\sin(\frac{1}{n})},$$

as α ranges in $(0, +\infty)$.

4.18 Compute, if existing, $\lim_n \sqrt{n+1}\log\left(\frac{\sqrt{n}+1}{\sqrt{n}}\right)$.

4.19 Consider the sequence $(a_n)_{n\geq 1}$ defined by the recurrence relation

$$a_1 = 1, \qquad a_{n+1} = \sqrt{k + 2a_n},$$

where k is a positive parameter.

(a) Establish if $(a_n)_{n\geq 1}$ is bounded.
(b) Establish if $(a_n)_{n\geq 1}$ is monotonic.
(c) Compute, if existing, $\lim_n a_n$.

4.20 Consider the sequence $(a_n)_{n\geq 1}$ defined by the recurrence relation

$$a_1 = x_0, \qquad a_{n+1} = \frac{n}{n+1}a_n,$$

where $x_0 \in \mathbb{R}$.

(a) Study when $(a_n)_{n\geq 1}$ is monotonic, as x_0 ranges in $\mathbb{R}$.
(b) Find explicitly $(a_n)_{n\geq 1}$.

4.21 Consider the sequence $(a_n)_{n\geq 1}$ defined by the recurrence relation

$$a_1 = \frac{1}{2}, \qquad a_{n+1} = \frac{1}{2 - a_n}.$$

(a) Establish if $(a_n)_{n\geq 1}$ is monotonic.
(b) Compute, if existing, $\lim_n a_n$.

4.22 Consider the sequence $(a_n)_{n\geq 1}$ defined by the recurrence relation

$$a_1 = 2, \qquad a_{n+1} = \frac{1 + a_n}{1 + a_n^2}.$$

(a) Establish if $(a_n)_{n\geq 1}$ is bounded.
(b) Establish if $(a_n)_{n\geq 1}$ is, at least eventually, monotonic.

4.23 Consider the sequence defined by $a_n = n^{|\sin n| + |\cos n|}$. Prove the inequality $1 \leq |\sin n| + |\cos n| \leq \sqrt{2}$ and then compute, if existing, the limits $\lim_n a_n/\sqrt{n}$ and $\lim_n a_n/\sqrt{n^\pi}$.

Chapter 5
Limits of Functions

5.1 Convergence Notions

The notion of limit of a function is one of the cornerstones in Analysis. It has to do with the behaviour of a function as the variable approaches a given point, which may or may not belong to the domain of the function, as long as it is a *limit point* for it, also known as *accumulation point*.

Definition 5.1 (*Finite limit point.*) Take a non-empty subset I of $\mathbb{R}$. The point $x_0 \in \mathbb{R}$ is called a limit point, or an accumulation point, of I if every punctured neighborhood of x_0 intersects I in at least one element.

The above definition may be rephrased by saying that x_0 is a limit point of I if

$$\text{for every } \delta > 0 \text{ there exists } x \in I \text{ such that } 0 < |x - x_0| < \delta.$$

Thus, a limit point of I cannot be separated from I by selecting a small enough radius δ in such a way that in the ball $B(x_0, \delta)$ there are no points of I other than (possibly) x_0 itself. For example, if $I = (0, 1]$ then both 0 and 1 are limit points for I although $1 \in I$ but $0 \notin I$. All the other points of $(0, 1]$ are limit points of $(0, 1]$ as well. The set $\{2, 3\}$, as any other finite set, has no limit points.

Definition 5.2 (*Isolated point.*) Given a non-empty set $I \subset \mathbb{R}$, the point $x_0 \in I$ is said to be an *isolated point* for I if it is not a limit point of I, namely, if there is a punctured neighborhood of x_0 that does not contain points of I.

For many purposes it is useful to extend Definition 5.1 to the case when the limit point is either $+\infty$ or $-\infty$.

Definition 5.3 ($\pm\infty$ *as limit points.*) Take a non-empty subset I of $\mathbb{R}$. It is said that $+\infty$ is a limit point of I if any interval $(K, +\infty)$, $K \in \mathbb{R}$, intersects I in at least one point. Otherwise, $+\infty$ is isolated from I. Similarly, it is said that $-\infty$ is a limit point of I if any interval $(-\infty, K)$ intersects I in at least one point.

© Springer International Publishing Switzerland 2016
M. Baronti et al., *Calculus Problems*, UNITEXT - La Matematica per il 3+2 101,
DOI 10.1007/978-3-319-15428-2_5

In the above definition, the sets $(K, +\infty)$ play the role of punctured neighborhoods of $+\infty$, and the sets $(-\infty, K)$ play the role of punctured neighborhoods of $-\infty$.

Below, whenever $f(x)$ appears, it is tacitly assumed that x belongs to the domain of f. The definitions are designed in such a way that in all the circumstances in which the value $f(x)$ is considered, the set of real numbers x in the domain of f for which the additional assumptions are met is non-empty.

Definition 5.4 (*Finite limit of a function.*) Suppose that $x_0 \in \mathbb{R}$ is a limit point of $I \subset \mathbb{R}$ and take $f : I \to \mathbb{R}$. The real number $\ell \in \mathbb{R}$ is called the *limit* of f as x tends to x_0 if for any neighborhood $\mathscr{U}$ of ℓ there exists a punctured neighborhood of x_0 the image of which, under f, is contained in $\mathscr{U}$. In this case, one writes

$$\lim_{x \to x_0} f(x) = \ell,$$

or $f(x) \to \ell$ as $x \to x_0$. Equivalently, the limit of f as $x \to x_0$ is ℓ if and only if

for every $\varepsilon > 0$ there exists $\delta > 0$ s.t. $0 < |x - x_0| < \delta \implies |f(x) - \ell| < \varepsilon$.

Clearly, here $\mathscr{U} = B(\ell, \varepsilon)$, and the punctured neighborhood of x_0 is $B(x_0, \delta) \setminus \{x_0\}$. Since x_0 is an accumulation point of I, there always exist elements $x \in I$ that belong to $B(x_0, \delta) \setminus \{x_0\}$, those for which it is asked that $|f(x) - \ell| < \varepsilon$. An other way of saying that the limit is ℓ is to say that f *converges* to ℓ or that it *tends* to ℓ as $x \to x_0$.

One advantage of formulating Definition 5.4 in this abstract way is that it adapts perfectly to the case where x_0 is replaced by either $+\infty$ or $-\infty$. Thus, taking into account that the punctured neighborhoods of $+\infty$ are the sets $(K, +\infty)$ and that the punctured neighborhoods of $-\infty$ are the sets $(-\infty, K)$, the definition becomes

$$\lim_{x \to +\infty} f(x) = \ell \iff \text{for every } \varepsilon > 0 \text{ there exists } K \text{ s.t. } x > K \implies |f(x) - \ell| < \varepsilon$$

$$\lim_{x \to -\infty} f(x) = \ell \iff \text{for every } \varepsilon > 0 \text{ there exists } K \text{ s.t. } x < K \implies |f(x) - \ell| < \varepsilon.$$

It is not hard to see that the limit, if it exists, is unique. With little effort, Definition 5.4 can be modified to the case where ℓ is replaced by either $+\infty$ or $-\infty$.

Definition 5.5 (*Diverging function.*) Suppose that $x_0 \in \mathbb{R}$ is a limit point of $I \subset \mathbb{R}$ and take $f : I \to \mathbb{R}$. It is said that the *limit* of f, as x tends to x_0, is $+\infty$ (or $-\infty$, respectively) if for any punctured neighborhood $\mathscr{U}$ of $+\infty$ ($-\infty$, respectively) there exists a punctured neighborhood of x_0 the image of which, under f, is contained in $\mathscr{U}$. In this case, one writes

$$\lim_{x \to x_0} f(x) = +\infty, \qquad \lim_{x \to x_0} f(x) = -\infty,$$

or $f(x) \to \pm\infty$ as $x \to x_0$. Equivalently, the limit of f as $x \to x_0$ is $\pm\infty$ if and only if

for every K there exists $\delta > 0$ s.t. $0 < |x - x_0| < \delta \implies f(x) > K$ $(+\infty)$

for every K there exists $\delta > 0$ s.t. $0 < |x - x_0| < \delta \implies f(x) < K$ $(-\infty)$.

An alternative way of saying that the limit is $\pm\infty$ is to say that f *diverges* to $\pm\infty$ or that it *tends* to $\pm\infty$ as $x \to x_0$. Finally, the definition of a diverging function at $\pm\infty$ is that for any punctured neighborhood $\mathscr{U}$ of $\pm\infty$ there exists a punctured neighborhood of $\pm\infty$ the image of which, under f, is contained in $\mathscr{U}$, or, explicitly:

$$\lim_{x \to +\infty} f(x) = +\infty \iff \text{for every } H \text{ there exists } K \text{ s.t. } x > K \implies f(x) > H$$

$$\lim_{x \to +\infty} f(x) = -\infty \iff \text{for every } H \text{ there exists } K \text{ s.t. } x > K \implies f(x) < H$$

$$\lim_{x \to -\infty} f(x) = +\infty \iff \text{for every } H \text{ there exists } K \text{ s.t. } x < K \implies f(x) > H$$

$$\lim_{x \to -\infty} f(x) = -\infty \iff \text{for every } H \text{ there exists } K \text{ s.t. } x < K \implies f(x) < H.$$

The various concepts introduced above can be expressed in a single definition by introducing the appropriate general topological notion of limit. In this book, the standard case-by-case definitions have been spelled out using the classical ε-δ language and the H-K language. The various statements are most meaningful when ε and δ are small and positive, and when H and K are large in absolute value.

Given a function $f : I \to \mathbb{R}$ and an accumulation point p of I, be it finite or not, the limit of f as $x \to p$ may or may not exist, but if it does, then it is unique. Examples of non existing limits are

$$\lim_{x \to 0} \sin \frac{1}{x}, \quad \lim_{x \to 0} \frac{x}{|x|}, \quad \lim_{x \to 0} \frac{1}{x}, \quad \lim_{x \to +\infty} \sin x, \quad \lim_{x \to -\infty} \cos x,$$

as the reader is urged to verify.

In what follows, it is often stated that $p \in \overline{\mathbb{R}}$ to mean that either $p \in \mathbb{R}$ or $p = \pm\infty$. This is primarily in order to avoid cumbersome repetitions, stressing the fact that many results hold independently of the fact that $x \to x_0 \in \mathbb{R}$ or $x \to +\infty$ or else $x \to -\infty$. Thus, one speaks of $\overline{\mathbb{R}}$ as the *extended real line*.

Finally, the notion of limit as $x \to x_0$ with $x_0 \in \mathbb{R}$ can be given in one-sided versions, by considering a single direction of approach, either from the right or from the left. To be precise, the notion of *left limit point* or *right limit point* should be given. Below is a shortcut definition.

Definition 5.6 (*One-sided limits.*) Suppose that $f : I \to \mathbb{R}$. The real number $\ell \in \mathbb{R}$ is called the limit of f as x tends to x_0 *from the left*, or *from the right*, and in this case, one writes

$$\lim_{x \to x_0^-} f(x) = \ell, \quad \text{or} \quad \lim_{x \to x_0^+} f(x) = \ell$$

according as

$$\text{for every } \varepsilon > 0 \text{ there exists } \delta > 0 \text{ s.t. } 0 < x_0 - x < \delta \implies |f(x) - \ell| < \varepsilon$$
$$\text{or, for every } \varepsilon > 0 \text{ there exists } \delta > 0 \text{ s.t. } 0 < x - x_0 < \delta \implies |f(x) - \ell| < \varepsilon.$$

What is tacitly assumed in the definition above is that, for example in the case when $x \to x_0^-$, given any $\delta > 0$ there are always points $x \in I$ that satisfy $0 < x_0 - x < \delta$. This is the notion of left limit point, and similarly for the case $x \to x_0^+$.

5.2 Results on Limits

In this Section, all the results hold for x that tends to an arbitrary point p in the extended real line $\overline{\mathbb{R}}$. They are analogous formulated almost *verbatim* for sequences. This is no coincidence, and is in fact a consequence of the theorem that follows.

Theorem 5.1 (Sequential limits) *Suppose that $p, q \in \overline{\mathbb{R}}$, that p is a limit point for $I \subset \mathbb{R}$, and take $f : I \to \mathbb{R}$. The following are equivalent:*

(i) $\lim_{x \to p} f(x) = q$;

(ii) $\lim_n f(x_n) = q$ *for every sequence $(x_n)_{n \geq 0}$ of points in $I \setminus \{p\}$ for which $x_n \to p$.*

The above result is typically used to show that a function does not have a limit: it is enough to exhibit two sequences $(x_n)_{n \geq 0}$ and $(y_n)_{n \geq 0}$ both tending to p such that $(f(x_n))_{n \geq 0}$ and $(f(y_n))_{n \geq 0}$ tend to different limits. A good example for this kind of argument is $f(x) = \sin(1/x)$ which has no limit for $x \to 0$, as the choices

$$x_n = \frac{1}{\frac{\pi}{2} + 2n\pi}, \qquad y_n = \frac{1}{\frac{3\pi}{2} + 2n\pi}.$$

show. Indeed, $f(x_n) = 1$ while $f(y_n) = -1$ for every n.

In many cases, what is needed, or infered, is a local property of a function around a limit point p of its domain. Therefore, it is said that a certain property holds *locally* at p for the function $f : I \to \mathbb{R}$ if p is a limit point of I and if there is a punctured neighborhood $\mathscr{U}$ of p such that $f(x)$ satisfies the property for every $x \in \mathscr{U} \cap I$.

Theorem 5.2 (Comparison, I) *Suppose that $p \in \overline{\mathbb{R}}$ is a limit point for I and that $f, g : I \to \mathbb{R}$ are such that $f(x) \to \ell$ and $g(x) \to m$ as $x \to p$, where $\ell, m \in \mathbb{R}$. Then:*

(i) *if $f(x) \leq g(x)$ locally at p, then $\ell \leq m$;*
(ii) *if $\ell < m$, then $f(x) < g(x)$ locally at p;*
(iii) *if $\ell < \lambda \in \mathbb{R}$, then $f(x) < \lambda$ locally at p, and similarly if $\ell > \mu \in \mathbb{R}$.*

Corollary 5.1 (Permanence of sign) *If* $\lim\limits_{x \to p} f(x) = q \in \overline{\mathbb{R}}\backslash\{0\}$, *then locally at* p *the function* f *has the sign*[1] *of* q.

Theorem 5.3 (Comparison, II, or Squeeze theorem) *Suppose that* $p \in \overline{\mathbb{R}}$ *is a limit point for* I *and suppose that the functions* f, g *and* h, *defined on* $I \subseteq \mathbb{R}$, *satisfy*

(i) $f(x) \to \ell$ *and* $h(x) \to \ell$ *as* $x \to p$, *where* $\ell \in \mathbb{R}$;
(ii) $f(x) \le g(x) \le h(x)$ *locally at* p.

  Then $\lim\limits_{x \to p} g(x)$ *exists and* $\lim\limits_{x \to p} g(x) = \ell$.

Theorem 5.4 (Algebra of limits) *Suppose that* $p \in \overline{\mathbb{R}}$ *is a limit point for* I *and that* $f(x) \to \ell$ *and* $g(x) \to m$ *as* $x \to p$, *where* $\ell, m \in \mathbb{R}$. *Then, as* $x \to p$, *the following holds:*

(i) $|f(x)| \to |\ell|$;
(ii) $f(x) + g(x) \to \ell + m$;
(iii) $f(x)g(x) \to \ell m$;
(iv) *for every* $\lambda \in \mathbb{R}$, $\lambda f(x) \to \lambda \ell$;
(v) *if* $m \ne 0$, *then* $f(x)/g(x) \to \ell/m$.

As in the analogous result for sequences, in item (v) of the above theorem the assumption $m \ne 0$ actually guarantees that the ratio $f(x)/g(x)$ makes sense close to p, by permanence of sign. Below is an extended version of the squeeze theorem.

Theorem 5.5 (Comparison, III) *Take* $f, g : I \to \mathbb{R}$ *and suppose that* $p \in \overline{\mathbb{R}}$ *is a limit point for* I. *Assume that* $f(x) \le g(x)$ *locally at* p. *Then, as* $x \to p$, *the following holds:*

(i) $f(x) \to +\infty$ *implies* $g(x) \to +\infty$;
(ii) $g(x) \to -\infty$ *implies* $f(x) \to -\infty$.

The algebra of limits extends to cover most cases of sums or products of functions one of which diverges as $x \to p$, as it is summarized in Table 5.1 below.

The meaning of question marks is that there are examples in which f and g satisfy the requirements, and the resulting sum, or product, either converges, or diverges to $+\infty$, or diverges to $-\infty$, or has no limit. Again, these cases are referred to as situations in which a function takes on an *indeterminate form*.

It is possible to infer the limiting behaviour of the reciprocal of a function which is known to either diverge or to converge to 0.

Proposition 5.1 *Suppose that* $p \in \overline{\mathbb{R}}$ *is a limit point of* I, *and take* $f : I \to \mathbb{R}$. *Then, as* $x \to p$:

---

[1] Here it is understood that the sign of $+\infty$ is positive and that the sign of $-\infty$ is negative.

Table 5.1 Extended algebra of limits for functions

$\lim\limits_{x\to p} f(x)$	$\lim\limits_{x\to p} g(x)$	$\lim\limits_{x\to p} (f(x)+g(x))$	$\lim\limits_{x\to p} f(x)g(x)$
$\ell > 0$	$+\infty$	$+\infty$	$+\infty$
$\ell > 0$	$-\infty$	$-\infty$	$-\infty$
$\ell < 0$	$+\infty$	$+\infty$	$-\infty$
$\ell < 0$	$-\infty$	$-\infty$	$+\infty$
0	$+\infty$	$+\infty$	?
0	$-\infty$	$-\infty$	?
$+\infty$	$+\infty$	$+\infty$	$+\infty$
$-\infty$	$-\infty$	$-\infty$	$+\infty$
$+\infty$	$-\infty$	?	$-\infty$

(i)   if $f(x)$ diverges, then $1/f(x) \to 0$;
(ii) if $f(x) > 0$ locally at p and $f(x) \to 0$, then $1/f(x) \to +\infty$;
(iii) if $f(x) < 0$ locally at p and $f(x) \to 0$, then $1/f(x) \to -\infty$.

Products of bounded functions and functions that tend to 0, also tend to 0, as formally stated in the next proposition.

Proposition 5.2 *Suppose that $p \in \overline{\mathbb{R}}$ is a limit point of I and take $f, g : I \to \mathbb{R}$. If f is bounded locally at p and $g(x) \to 0$ as $x \to p$, then also $f(x)g(x) \to 0$ as $x \to p$.*

Just like sequences, monotonic functions always have limits.

Theorem 5.6 *Suppose that $p \in \overline{\mathbb{R}}$ is a limit point of I and suppose that $f : I \to \mathbb{R}$ is non decreasing in I. Then:*

(i)   *if $p \in \mathbb{R}$ is a left limit point for I, then $\lim\limits_{x\to p^-} f(x) = \sup\{f(x) : x \in I,\ x < p\}$;*
(ii) *if $p \in \mathbb{R}$ is a right limit point for I, then $\lim\limits_{x\to p^+} f(x) = \inf\{f(x) : x \in I,\ x > p\}$;*
(iii) *if $p = +\infty$ is a limit point for I, then $\lim\limits_{x\to +\infty} f(x) = \sup\{f(x) : x \in I\}$;*
(iv) *if $p = -\infty$ is a limit point for I, then $\lim\limits_{x\to -\infty} f(x) = \inf\{f(x) : x \in I\}$.*

Quite obviously, the above theorem has a counterpart for non increasing functions. The next result is of high practical use.

Theorem 5.7 (Limit of Compositions) *Take $f : I \to \mathbb{R}$ and $g : J \to \mathbb{R}$ and $p, q \in \overline{\mathbb{R}}$ limit points for I and J, respectively. Suppose that $f(I) \subset J$, that $\lim_{x\to p} f(x) = q$ and that $\lim_{y\to q} g(y) = \ell$. Suppose further that one of the following assumptions is satisfied:*

(i)   *$q \in J$ and $\ell = g(q)$;*
(ii) *$q \notin J$;*
(iii) *$f(x) \neq q$ locally at p.*

Then the limit as $x \to p$ of the composition $g \circ f$ exists and

$$\lim_{x \to p} g(f(x)) = \lim_{y \to q} g(y) = \ell.$$

The above theorem is also known as the theorem on the *change of variable* for limits, because it is interpreted as follows: if one the hypotheses listed above is satisfied, then one changes the variable by putting $y = f(x)$, so that the limit of $g(f(x))$ as $x \to p$ is reduced to the limit of $g(y)$ as $y \to q$.

5.3 Local Comparison of Functions

In this Subsection, $p \in \overline{\mathbb{R}}$ and $\mathscr{F}_p$ denotes the set of functions that are defined in some punctured neighborhood of p.

Definition 5.7 Take $f, g \in \mathscr{F}_p$.

(i) The function f is said to be *negligible* with respect to g as $x \to p$ if there exists $\omega \in \mathscr{F}_p$ such that $\omega(x) \to 0$ as $x \to p$ and such that $f(x) = g(x)\omega(x)$ locally at p. In this case one writes $f \prec g$. Equivalently one says that g *dominates* f.

(ii) The function f is said to be *asymptotically equivalent* to g as $x \to p$ if there exists $h \in \mathscr{F}_p$ such that $h(x) \to 1$ as $x \to p$ and such that $f(x) = g(x)h(x)$ locally at p. In this case one writes $f \sim g$.

Clearly, if $g(x) \neq 0$ locally at p, then

$$f \prec g \quad \Longleftrightarrow \quad \lim_{x \to p} \frac{f(x)}{g(x)} = 0$$

$$f \sim g \quad \Longleftrightarrow \quad \lim_{x \to p} \frac{f(x)}{g(x)} = 1.$$

Theorem 5.8 *Take $f, g, f_1, g_1 \in \mathscr{F}_p$ and suppose that $g(x) \neq 0$ locally at p. Then:*

(i) if $f_1 \prec f$ and $g_1 \prec g$, then

$$\lim_{x \to p} \frac{f(x) + f_1(x)}{g(x) + g_1(x)} = \lim_{x \to p} \frac{f(x)}{g(x)};$$

(ii) if $f_1 \sim f$ and $g_1 \sim g$, then

$$\lim_{x \to p} \frac{f(x)}{g(x)} = \lim_{x \to p} \frac{f_1(x)}{g_1(x)}.$$

Item (i) of Theorem 5.8 is sometimes referred to as the principle of elimination of negligible terms, whereas item (ii) is also known as the principle of substitution of asymptotically equivalent terms.

The relation $f \prec g$ is transitive, and $f \sim g$ is a *bona fide* equivalence relation. Also, they satisfy the following properties

$$
\begin{aligned}
f_1 \prec g \text{ and } f_2 \prec g & \implies & f_1 + f_2 \prec g \\
f_1 \prec \alpha g \text{ and } f_2 \prec \beta g \text{ with } \alpha + \beta \neq 0 & \implies & f_1 + f_2 \prec (\alpha + \beta)g \\
f_1 \prec g_1 \text{ and } f_2 \prec g_2 & \implies & f_1 f_2 \prec g_1 g_2 \\
f_1 \sim g_1 \text{ and } f_2 \sim g_2 & \implies & f_1 f_2 \sim g_1 g_2 \\
f \prec g \text{ and } g \sim h & \implies & f \prec h \\
f \sim g \text{ and } g \prec h & \implies & f \prec h.
\end{aligned}
$$

The reader should beware of possible mistakes in handling $\prec$ and $\sim$. Indeed:

$$
\begin{aligned}
f_1 \prec g_1 \text{ and } f_2 \prec g_2 & \centernot\implies & f_1 + f_2 \prec g_1 + g_2 \\
f_1 \sim g_1 \text{ and } f_2 \sim g_2 & \centernot\implies & f_1 + f_2 \sim g_1 + g_2.
\end{aligned}
$$

Examples are easy to manufacture.

The notation $f \prec g$ is not universal. As discussed thoroughly in Chap. 8, whenever $g \in \mathscr{F}_p$, it is customary to write

$$
o(g) = \left\{ f \in \mathscr{F}_p : f \prec g \right\}
$$

and then replace the precise writing $f \in o(g)$ with the sloppier but very efficient $f = o(g)$. This notation is known as *Landau's little-o* notation. Informally speaking, if one assumes that g does not vanish locally at p, then $f = o(g)$ is meant to suggest that when dividing f by g, the result is not zero but it is something very small, that tends to zero as $x \to x_0$. The letter that best resembles the number 0 is of course o. Notice that $f = o(1)$ means that $f(x) \to 0$. The reader is referred to Chap. 8 for a detailed discussion on the use of Landau's little-o notation.

As for the notion of asymptotic equivalence, notice that if $f \sim g$, then $f = gh$ with $h \to 1$ as $x \to p$. Hence $h = 1 + \omega$ with $\omega = h - 1 \to 0$. In conclusion,

$$
\begin{aligned}
f \sim g & \iff & f = g(1 + \omega) \text{ with } \omega \to 0 \\
& \iff & f = g + o(g).
\end{aligned}
$$

The use of asymptotic equivalence and the analysis of negligible terms is best carried out by reducing the computations, whenever possible, to known limits involving elementary functions. A list of basic limits that will be used throughout is contained in Appendix B.

5.4 Orders

The notions introduced below are designed to understand the "speed" with which a function $f \in \mathscr{F}_p$ either tends to 0 or diverges as $x \to p$. This means that a given scale of reference functions is selected, typically the scale of powers, and then f is compared with these, as explained next.

Definition 5.8 (*Order, I*) Fix $p \in \mathbb{R}$. A function $f \in \mathscr{F}_p$ is said to *vanish of order* $\alpha > 0$ as $x \to p^+$ if there exists $\lambda_+ \in \mathbb{R} \backslash \{0\}$ such that $f \sim \lambda_+ |x - p|^\alpha$ as $x \to p^+$, namely if

$$\lim_{x \to p^+} \frac{f(x)}{|x - p|^\alpha} = \lambda_+.$$

Similarly, it is said to vanish of order $\alpha > 0$ as $x \to p^-$ if there exists $\lambda_- \in \mathbb{R} \backslash \{0\}$ such that $f \sim \lambda_- |x - p|^\alpha$ as $x \to p^-$. Finally, if $p \in \mathbb{R}$ and f vanishes of order $\alpha > 0$ both as $x \to p^+$ and as $x \to p^-$, then f is said to vanish of order α as $x \to p$. In this case, the *principal part* of f at p is

$$\begin{cases} \lambda_+ |x - p|^\alpha & x > p \\ \lambda_- |x - p|^\alpha & x < p. \end{cases}$$

For example, $f(x) = (\sin x)^3$ vanishes of order 3 as $x \to 0$. Indeed $\lambda_\pm = \pm 1$ and its principal part is actually equal to x^3. The above definition can be extended to the case where $p = \pm \infty$.

Definition 5.9 (*Order, II*) A function $f \in \mathscr{F}_{\pm\infty}$ is said to vanish of order $\alpha > 0$ as $x \to \pm\infty$ if there exists $\lambda_\pm \in \mathbb{R} \backslash \{0\}$ such that $f \sim \lambda_\pm |x|^{-\alpha}$ as $x \to \pm\infty$, namely

$$\lim_{x \to \pm\infty} f(x)|x|^\alpha = \lambda_\pm.$$

In this case, the principal part of f at $\pm\infty$ is $\lambda_\pm |x|^{-\alpha}$.

For example, the function $f(x) = \log(1 + x^{-n})$, with n a positive integer, vanishes of order n at $+\infty$, because, under the change of variable $y = 1/x^n$,

$$\lim_{x \to +\infty} \log(1 + x^{-n})|x|^n = \lim_{y \to 0^+} \frac{\log(1 + y)}{y} = 1.$$

Much in the same spirit, it is desirable to measure the speed with which a given function diverges at the point p, in comparison to a given scale of diverging functions, for example $1/|x - p|^\alpha$ with $\alpha > 0$.

Definition 5.10 (*Order, III*) Fix $p \in \mathbb{R}$. A function $f \in \mathscr{F}_p$ is said to *diverge of order* $\alpha > 0$ as $x \to p^+$ if there exists $\lambda_+ \in \mathbb{R}\backslash\{0\}$ such that $f \sim \lambda_+ |x - p|^{-\alpha}$ as $x \to p^+$, namely

$$\lim_{x \to p^+} f(x)|x - p|^\alpha = \lambda_+.$$

Similarly, it is said to diverge of order $\alpha > 0$ as $x \to p^-$ if there exists $\lambda_- \in \mathbb{R}\backslash\{0\}$ such that $f \sim \lambda_- |x - p|^{-\alpha}$ as $x \to p^-$. Finally, if $p \in \mathbb{R}$ and f diverges of order α both as $x \to p^+$ and as $x \to p^-$, then f is said to diverge of order α as $x \to p$. In this case, the principal part of f at p is

$$\begin{cases} \lambda_+ |x - p|^{-\alpha} & x > p \\ \lambda_- |x - p|^{-\alpha} & x < p. \end{cases}$$

For example, $f(x) = 2/\sin x$, which is defined in the punctured neighborhood $(-\pi, \pi)\backslash\{0\}$ of the origin, diverges of order 1 at the origin with $\lambda_\pm = \pm 2$ and principal part $2/x$. Finally, here is how to measure functions that diverge at $\pm\infty$.

Definition 5.11 (*Order, IV*) A function $f \in \mathscr{F}_{\pm\infty}$ is said to diverge of order $\alpha > 0$ as $x \to \pm\infty$ if there exists $\lambda_\pm \in \mathbb{R}\backslash\{0\}$ such that $f \sim \lambda_\pm |x|^\alpha$ as $x \to \pm\infty$, namely

$$\lim_{x \to \pm\infty} f(x)|x|^{-\alpha} = \lambda_\pm.$$

In this case, the principal part of f at $\pm\infty$ is $\lambda_\pm |x|^\alpha$.

For example, any polynomial $P(x) = a_0 + \cdots + a_n x^n$ of degree n diverges both at $+\infty$ and at $-\infty$ of order n. Its principal part at $+\infty$ is $a_n x^n$ and at $-\infty$ is $a_n x^n$.

It should be clear that asymptotically equivalent functions have the same order, in all the cases considered in Definitions 5.8–5.11.

Definition 5.12 Suppose that $f, g \in \mathscr{F}_p$.

(i) If both f and g tend to 0 as $x \to p$ and $f \prec g$, then one says that f vanishes with *order greater than* the order of g.
(ii) If both f and g diverge as $x \to p$ and $f \prec g$, then one says that f diverges with *order smaller than* the order of g.

If two functions have a specific order at a point, then the above definition is almost tautological, in the sense that if, say, f vanishes with order α and g with order β, and $\alpha > \beta$, then f vanishes with order greater than the order of g according to Definition 5.12. It is important to notice, however, that it may well be that f vanishes with order greater than the order of g but that one of either f or g, or maybe both, has no order in the scale of powers. This is a subtle but important issue. Indeed, not every function that tends to 0 at a given point vanishes with some order α at that point (in the scale of powers), and, analogously, not every diverging function diverges with some order.

For example, the function $f(x) = x \log x$, defined in $(0, +\infty)$, tends to 0 as $x \to 0^+$ but there is no $\alpha > 0$ such that $f(x)$ vanishes of order α. Indeed,

$$\lim_{x \to 0^+} \frac{x \log x}{|x|^\alpha} = \begin{cases} 0 & \alpha < 1 \\ +\infty & \alpha \geq 1. \end{cases}$$

Furthermore, this shows that f vanishes with order less than 1 (the order of $|x|$) as $x \to 0^+$, but greater than any $\alpha > 0$ with $\alpha < 1$ (the order of $|x|^\alpha$).

The exponential function e^x diverges at $+\infty$ of order greater than any $\alpha > 0$ because $e^x x^{-\alpha} \to +\infty$ as $x \to +\infty$ for any positive α. The limits listed in Appendix B show that there are many functions that either diverge or tend to 0 without a definite order.

One of the main applications of the notion of order is the following result on computation of limits, which is a direct consequence of the principle of substitution of asymptotically equivalent functions.

Theorem 5.9 (Orders and Limits of Ratios) *Take $p \in \overline{\mathbb{R}}$ and $f, g \in \mathscr{F}_p$, and suppose that $g(x) \neq 0$ locally at p.*

(i) If f vanishes with order α and g vanishes with order β as $x \to p^+$, then

$$\lim_{x \to p^+} \frac{f(x)}{g(x)} = \begin{cases} 0 & \alpha > \beta \\ \ell \in \mathbb{R} \backslash \{0\} & \alpha = \beta \\ \pm\infty & \alpha < \beta. \end{cases}$$

(ii) If f diverges with order α and g diverges with order β as $x \to p^+$, then

$$\lim_{x \to p^+} \frac{f(x)}{g(x)} = \begin{cases} \pm\infty & \alpha > \beta \\ \ell \in \mathbb{R} \backslash \{0\} & \alpha = \beta \\ 0 & \alpha < \beta. \end{cases}$$

Analogous conclusions are drawn as $x \to p^-$.

It follows in particular that in case (i), $\alpha > \beta$ implies $f \prec g$ and, in case (ii), this happens if $\alpha < \beta$. In both cases, if $\alpha = \beta$, then $f \sim \ell g$. Also, the sign of the diverging ratios is established by looking at the principal parts. Observe that in the case where $x \to p$, then it is quite possible that the ratio diverges with one sign to the left of p and with the opposite sign to the right of p.

Finally, it is not hard to see that in all the above cases in which the ratio either converges to 0 or diverges, then it is actually possible to establish the order with which this happens: it is just the positive difference of the orders.

5.5 Guided Exercises on Limits of Functions

5.1 Consider the function

$$f(x) = \begin{cases} \frac{x-\sqrt{x}}{x} & x > 0 \\ 0 & x = 0. \end{cases}$$

Use the definition of limit to prove that $\lim\limits_{x \to +\infty} f(x) = 1$, and $\lim\limits_{x \to 0+} f(x) \neq 0$.

Answer. By definition, $\lim_{x \to +\infty} f(x) = 1$ if and only if for every $\varepsilon > 0$ there exists K such that if $x > K$, then $1 - \varepsilon < f(x) < 1 + \varepsilon$. Fix now $\varepsilon > 0$ and consider the inequalities

$$-\varepsilon < \frac{x - \sqrt{x}}{x} - 1 < \varepsilon \quad \Longrightarrow \quad \begin{cases} -\varepsilon < -\frac{\sqrt{x}}{x} \\ -\frac{\sqrt{x}}{x} < \varepsilon. \end{cases}$$

The second is true for every $x > 0$, whereas the first gives $1/\sqrt{x} < \varepsilon$, namely, taking squares, $x > 1/\varepsilon^2 = K$. Hence, if $x > 1/\varepsilon^2$ then $1 - \varepsilon < f(x) < 1 + \varepsilon$, as desired.

Suppose by contradiction that $\lim_{x \to 0+} f(x) = 0$. Then, for every $\varepsilon > 0$ there would exist $\delta > 0$ such that if $0 < x < \delta$, then $-\varepsilon < f(x) < \varepsilon$. Take $\varepsilon < 1$ and consider the system of inequalities

$$\begin{cases} -\varepsilon < \frac{x-\sqrt{x}}{x} \\ \frac{x-\sqrt{x}}{x} < \varepsilon. \end{cases}$$

Any $x > 0$ satisfying the system, satisfies also $(1 + \varepsilon)^{-2} < x < (1 - \varepsilon)^2$ which entails that x is bounded away from 0. Thus, there is no $\delta > 0$ for which $0 < x < \delta$ implies $-\varepsilon < f(x) < \varepsilon$. Hence the limit cannot be 0.

This exercise focuses on the definition of limit, when $x \to +\infty$ and when $x \to 0^+$. In the first case, the procedure is straightforward: asking for $f(x)$ to be close to the limit 1 is tantamount to solving a system of simple inequalities, which give immediately a condition on x, depending on ε, of the correct form, namely $x > \varepsilon^{-2}$. Hence, for x large, to wit $x > \varepsilon^{-2}$, $f(x)$ is close to 1. As for the second limit, one must show that the limit is not 0, and this calls for a contradiction-type argument. Informally speaking, the idea is to prove that if the function is to be close to the impossible limit (0 in the case at hand), then the variable cannot be close to the limit point (0 in the case at hand).

5.2 Compute the limits:

$$\lim_{x \to 0^+} \frac{e^x - \cos x}{\sqrt{1 - \cos x}}, \qquad \lim_{x \to 0^-} \frac{e^x - \cos x}{\sqrt{1 - \cos x}}.$$

Answer. For x in a punctured neighborhood of the origin

$$\frac{e^x - \cos x}{\sqrt{1 - \cos x}} = \frac{e^x - 1 + 1 - \cos x}{\sqrt{1 - \cos x}}$$

$$= \frac{e^x - 1}{x} \frac{x}{\sqrt{1 - \cos x}} + \frac{1 - \cos x}{\sqrt{1 - \cos x}}$$

$$= \frac{e^x - 1}{x} \frac{x}{\sqrt{1 - \cos x}} + \sqrt{1 - \cos x}.$$

Now, $(e^x - 1)/x \to 1$ and $\sqrt{1 - \cos x} \to 0$ for $x \to 0^{\pm}$, while

$$\lim_{x \to 0^+} \frac{x}{\sqrt{1 - \cos x}} = \lim_{x \to 0^+} \frac{\sqrt{x^2}}{\sqrt{1 - \cos x}} = \lim_{x \to 0^+} \sqrt{\frac{x^2}{1 - \cos x}} = \sqrt{2},$$

$$\lim_{x \to 0^-} \frac{x}{\sqrt{1 - \cos x}} = \lim_{x \to 0^-} \frac{-\sqrt{x^2}}{\sqrt{1 - \cos x}} = \lim_{x \to 0^-} -\sqrt{\frac{x^2}{1 - \cos x}} = -\sqrt{2}.$$

Therefore the limits are $\sqrt{2}$ and $-\sqrt{2}$, respectively.

This is a simple exercise on the use of elementary limits. Hence one makes them appear by adding and subtracting, or multiplying and dividing by what is missing. The only slight subtlety arises from the presence of the square root of $1 - \cos x$, which behaves like $|x|/2$, hence differently according as $x \to 0^+$ or $x \to 0^-$.

5.3 Compute, if existing, $\displaystyle\lim_{x \to 0^+} \frac{(1 - \cos x)^2 \sin x}{x^4 (e^x - 1)^2}$.

Answer. The product $(1 - \cos x)^2 \sin x$ tends to 0 with order 5, while the product $x^4(e^x - 1)^2$ tends to 0 with order 6. As both numerator and denominator are positive in a right neighborhood of 0, the limit diverges to $+\infty$.

This is an immediate application of orders criteria, using the known orders at the origin of $1 - \cos x$, $\sin x$ and $e^x - 1$.

5.4 Compute, if existing, $\displaystyle\lim_{x \to 0^+} \frac{\sqrt{1 + x^2} - 1 - x}{x^k}$, as k ranges in $(0, +\infty)$.

Answer. A simple rationalization gives

$$f(x) = \sqrt{1 + x^2} - 1 - x = \frac{-2x}{\sqrt{1 + x^2} + 1 + x} \to 0$$

as $x \to 0^+$ with order 1. It follows that for $0 < k < 1$ the limit is 0, whereas for $k > 1$ the limit is $-\infty$. Take now $k = 1$. By the previous computation

$$\lim_{x \to 0^+} \frac{\sqrt{1 + x^2} - 1 - x}{x} = \lim_{x \to 0^+} \frac{-2}{\sqrt{1 + x^2} + 1 + x} = -1$$

This is a basic exercise on orders: upon rationalizing, the function under consideration is actually of the form

$$-\frac{2}{g(x)}x^{1-k}$$

with $g(x) \to 2$ as $x \to 0$, so the answer is straightforward.

5.5 Compute the limit: $\displaystyle\lim_{x \to \pi^+} \frac{\sin^2 x}{(e^x - e^\pi)^k}$, as k ranges in $(0, +\infty)$.

Answer. First of all observe that:

$$\lim_{x \to \pi^+} \frac{\sin^2 x}{(e^x - e^\pi)^k} = \lim_{x \to 0^+} \frac{\sin^2 x}{e^{k\pi}(e^x - 1)^k}.$$

Now, as $x \to 0^+$, the funcion $x \mapsto \sin^2 x$ vanishes with order 2 and $x \mapsto (e^x - 1)^k$ with order k. Therefore, for $k < 2$ the limit is 0, and for $k > 2$ the limit is $+\infty$. Finally, if $k = 2$, then using elementary limits

$$\lim_{x \to 0^+} \frac{\sin^2 x}{e^{k\pi}(e^x - 1)^k} = \lim_{x \to 0^+} \frac{(\sin^2 x)/x^2}{e^{2\pi}(e^x - 1)^2/x^2} = 1/e^{2\pi}.$$

This is an exercise on the use of order criteria. Both numerator and denominator are elementary functions that are asymptotically equivalent to x^2 and x^k, respectively.

5.6 Compute, provided that they make sense, the limits $\displaystyle\lim_{x \to 0^+} (x^{\sqrt{x}} - (\sqrt{x})^x)$ and $\displaystyle\lim_{x \to +\infty} (x^{\sqrt{x}} - (\sqrt{x})^x)$.

Answer. First of all, observe that the function $f(x) = x^{\sqrt{x}} - (\sqrt{x})^x$ is defined for $x > 0$ and hence that the indicated limits do make sense. Observe furthermore that

$$f(x) = e^{\sqrt{x}\log x} - e^{x\log(x^{1/2})} = e^{\sqrt{x}\log x} - e^{\frac{x}{2}\log x}.$$

The limit $\lim_{x \to 0^+} x^\alpha \log x = 0$ for every $\alpha > 0$ implies at once that both the exponents $\sqrt{x}\log x$ and $\frac{x}{2}\log x$ tend to 0 as $x \to 0^+$. Therefore, from the theorem on the limit of composition of functions it follows that

$$\lim_{x \to 0^+} f(x) = \lim_{x \to 0^+} (e^{\sqrt{x}\log x} - e^{\frac{x}{2}\log x}) = 1 - 1 = 0.$$

As for the behaviour for $x \to +\infty$, write

$$f(x) = e^{\sqrt{x}\log x} - e^{\frac{x}{2}\log x} = e^{\frac{x}{2}\log x}\left(e^{\sqrt{x}\log x - \frac{x}{2}\log x} - 1\right) = e^{\frac{x}{2}\log x}\left(e^{(\sqrt{x} - \frac{x}{2})\log x} - 1\right).$$

Now, $\left(\sqrt{x} - \frac{x}{2}\right) \log x \to -\infty$ as $x \to +\infty$, so that the factor in round brackets converges to -1. Since $e^{\frac{x}{2} \log x} \to +\infty$ as $x \to +\infty$, the conclusion is that $f(x) \to -\infty$.

As it is often the case, the behaviour of functions like the one at hand requires first to be written using the exponential form $e^{\varphi(x)}$ as much as possible. Technically, this is because the exponential function $y \mapsto e^y$ is very well understood and also because the theorem on the limit of composition of functions applies. In the present case, $f(x) = e^{g(x)} - e^{h(x)}$. As $x \to 0^+$, both g and h vanish, and hence so does f. On the contrary, when $x \to +\infty$ it is important to understand which between g and h grows faster, and the answer is clearly h. The strategy is thus to collect $e^{h(x)}$.

5.7 Compute, if existing,

$$\lim_{x \to +\infty} (\sqrt[3]{4x^a + x^2} - \sqrt[3]{x^a + x^2})^{(x-[x])},$$

as the real parameter a ranges in $(0, 2)$. Here $[x]$ is the *integral part* of x.

Answer. First of all, it is important to compute the limit of $\sqrt[3]{4x^a + x^2} - \sqrt[3]{x^a + x^2}$ as $x \to +\infty$, because it takes on the indeterminate form "$\infty - \infty$". Put $P(x) = 4x^a + x^2$ and $Q(x) = x^a + x^2$ and observe that, quite generally,

$$\sqrt[3]{P(x)} - \sqrt[3]{Q(x)} = \frac{(\sqrt[3]{P(x)} - \sqrt[3]{Q(x)}) \cdot (\sqrt[3]{P^2(x)} + \sqrt[3]{Q^2(x)} + \sqrt[3]{P(x) \cdot Q(x)})}{\sqrt[3]{P^2(x)} + \sqrt[3]{P(x)Q(x)} + \sqrt[3]{Q^2(x)}}$$

$$= \frac{P(x) - Q(x)}{\sqrt[3]{P^2(x)} + \sqrt[3]{P(x)Q(x)} + \sqrt[3]{Q^2(x)}}.$$

Substituting the explicit expressions of $P(x)$ and $Q(x)$, and by collecting the power $x^{4/3}$ at the denominator, it follows that

$$\sqrt[3]{P(x)} - \sqrt[3]{Q(x)} = \frac{3x^{a-\frac{4}{3}}}{h(x)},$$

where

$$h(x) = \sqrt[3]{(1 + 4x^{a-2})^2} + \sqrt[3]{1 + 5x^{a-2} + 4x^{2a-4}} + \sqrt[3]{(1 + x^{a-2})^2}.$$

Since $a < 2$, the limit of $h(x)$ as $x \to +\infty$ is 3, whence

$$\lim_{x \to +\infty} (\sqrt[3]{4x^a + x^2} - \sqrt[3]{x^a + x^2}) = \lim_{x \to \infty} \frac{3x^{a-\frac{4}{3}}}{h(x)} = \begin{cases} +\infty & \frac{4}{3} < a < 2 \\ 1 & a = \frac{4}{3} \\ 0 & 0 < a < \frac{4}{3}. \end{cases}$$

With this information available, it is possible to proceed, using the exponential form

$$\lim_{x\to+\infty} (\sqrt[3]{4x^a + x^2} - \sqrt[3]{x^a + x^2})^{x-[x]} = \lim_{x\to+\infty} e^{(x-[x])\log\left(\frac{3x^{a-\frac{4}{3}}}{h(x)}\right)}.$$

Observe next that if $a \neq 4/3$ then the exponent, denoted $g(x)$, has no limit, and hence its exponential has no limit either. Indeed:

- if $4/3 < a < 2$, then select the sequences $(x_k)_{k>0}$ and $(y_k)_{k>0}$ with $x_k = k$ and $y_k = k + 1/2$, so that $g(x_k) = 0$ and

$$g(y_k) = \frac{1}{2}\log\left(\frac{3y_k^{a-\frac{4}{3}}}{h(y_k)}\right) \to +\infty$$

because $y_k \to +\infty$ whereas $h(y_k) \to 3$;
- the case $0 < a < 4/3$ is analogous.

Finally, if $a = \frac{4}{3}$ then

$$g(x) = (x - [x])\log\left(\frac{3}{h(x)}\right) \to 0$$

as $x \to +\infty$ because it is the product of a bounded function ($0 \le x - [x] < 1$) with a function that tends to 0. In conclusion, if $a \neq 4/3$ the limit does not exist, and if $a = 4/3$ the limit is 1.

The function that needs to be analyzed is a generalized exponential, where the basis function has the form $\sqrt[3]{P(x)} - \sqrt[3]{Q(x)}$ and can thus be rationalized. After doing so, it is clear that the critical value of a is 4/3, because below this value the basis tends to 0, above 4/3 it diverges positively and at $a = 4/3$ it tends to 1. Writing the given function in exponential form, in the exponent $g(x)$ the logarithm of the basis function is multiplied by the *mantissa* function $x - [x]$, which is bounded. This fact can be exploited to show that when the logarithm diverges (that is, for all $a \neq 4/3$) then one can select mantissa values that make g either constantly equal to 0 or diverging. When the logarithm tends to 0, that is for $a = 4/3$, the mantissa stays bounded.

5.8 Compute the limit $\displaystyle\lim_{x\to+\infty} \frac{\log x}{\sin^4 x + \cos^4 x}$.

Answer. Observe that $\lim_{x\to+\infty}(\sin^4 x + \cos^4 x)$ does not exist, so that no immediate use of the extended algebra of limits applies. Consider thus the denominator, namely the function $x \mapsto \sin^4 x + \cos^4 x$, which is defined on $\mathbb{R}$, positive and periodic. Now,

$$\sin^4 x + \cos^4 x = (\sin^2 x + \cos^2 x)^2 - 2\sin^2 x \cos^2 x = 1 - \frac{1}{2}\sin^2(2x).$$

It follows that

$$\frac{1}{2} \leq \sin^4 x + \cos^4 x \quad \Longrightarrow \quad \frac{\log x}{2} \leq \frac{\log x}{\sin^4 x + \cos^4 x}.$$

By comparison, since $\log x \to +\infty$, as $x \to +\infty$, the indicated limit is $+\infty$.

This is an application of comparison criteria. The point is that although the denominator has no limit, it is bounded below and above, and positive. So the same holds to its reciprocal. In particular, the indicated function is the product of a positively diverging function and a function that is bounded below by a positive constant. Anything greater than something diverging to $+\infty$ diverges to $+\infty$.

5.9 Compute the limit $\lim\limits_{x \to 0}(\cos x)^{|\sin x|^{-\alpha}}$ as α ranges in $(0, +\infty)$.

Answer. Write

$$(\cos x)^{|\sin x|^{-\alpha}} = \exp\left(\frac{1}{|\sin x|^\alpha} \log(\cos x)\right),$$

where it is understood that $\exp y = e^y$. Look first at the behaviour of the exponent as $x \to 0$. Multiplying and dividing by x^2 and by $\cos x - 1$, it follows that

$$\frac{1}{|\sin x|^\alpha} \log(\cos x) = \frac{x^2}{|\sin x|^\alpha} \frac{\cos x - 1}{x^2} \frac{\log(\cos x)}{\cos x - 1}.$$

The last two factors in the right hand side converge to $-1/2$ and 1, respectively. The limit of the exponent thus depends on the limit of $x^2/|\sin x|^\alpha$ as α varies. Since

$$\lim_{x \to 0} \frac{x^2}{|\sin x|^\alpha} = \begin{cases} 0 & 0 < \alpha < 2 \\ 1 & \alpha = 2 \\ +\infty & \alpha > 2, \end{cases}$$

it follows that

$$\lim_{x \to 0} \frac{1}{|\sin x|^\alpha} \log(\cos x) = \begin{cases} 0 & 0 < \alpha < 2 \\ -\frac{1}{2} & \alpha = 2 \\ -\infty & \alpha > 2 \end{cases}$$

and, by the theorem on the limit of the composition of functions,

$$\lim_{x \to 0}(\cos x)^{|\sin x|^{-\alpha}} = \lim_{x \to 0} \exp\left[\frac{1}{|\sin x|^\alpha} \log(\cos x)\right] = \begin{cases} 1 & 0 < \alpha < 2 \\ e^{-\frac{1}{2}} & \alpha = 2 \\ 0 & \alpha > 2. \end{cases}$$

This is a typical exercise on the use of asymptotic behaviours, after writing the given function in standard exponential form. Indeed, this preliminary step permits to reduce the study of the given function to the study of the exponent. This, in turn, is a ratio, where the numerator is asymptotically equivalent to x^2 and the denominator is asymptotically equivalent to $|x|^\alpha$, facts that are actually spelled out by multiplying and dividing by the correct factors, so as to make the appropriate elementary limits appear. Thus, in the end, one must compare the fixed order 2 of the numerator with the varying order α of the denominator.

5.10 Compute the limits of $f(x) = \dfrac{\sin x}{|e^x - 1| - 1}$ at the boundary points of its domain.

Answer. The domain of f is $(-\infty, \log 2) \cup (\log 2, +\infty)$. First, it is clear that

$$\lim_{x \to \log 2^+} f(x) = +\infty \qquad \lim_{x \to \log 2^-} f(x) = -\infty.$$

as the denominator vanishes as $x \to \log 2$ and is positive when $x > \log 2$ and negative when $x < \log 2$. Furthermore, for $x < 0$ the denominator is actually $-e^x$, hence

$$\lim_{x \to -\infty} f(x) = \lim_{x \to -\infty} -\frac{\sin x}{e^x}.$$

Consider the sequences $(a_n)_{n \geq 1}$ and $(b_n)_{n \geq 1}$ defined by

$$a_n = \frac{\pi}{2} - 2n\pi, \quad b_n = \pi - 2\pi n.$$

Evidently, $\sin a_n = 1$ and $\sin b_n = 0$ and obviously $a_n, b_n \to -\infty$. Therefore

$$\lim_n f(a_n) = \lim_n -\frac{1}{e^{a_n}} \sin a_n = -\infty$$

$$\lim_n f(b_n) = \lim_n -\frac{1}{e^{b_n}} \sin b_n = 0$$

so that the limit of f as $x \to -\infty$ does not exist. Finally, consider

$$\lim_{x \to +\infty} f(x) = \lim_{x \to +\infty} \frac{\sin x}{e^x - 2}.$$

Observe that for $x > \log 2$

$$-\frac{1}{e^x - 2} \leq \frac{\sin x}{e^x - 2} \leq \frac{1}{e^x - 2}$$

so that by the squeeze theorem $\lim_{x \to +\infty} f(x) = 0$.

In this exercise, the behaviour of the given function is very easy to analyze when $x \to \log 2^\pm$ because the denominator clearly vanishes whereas the numerator tends to a positive number. As $x \to \mp\infty$, the function has the two different forms

$$-\frac{\sin x}{e^x}, \qquad \sin x \cdot (e^x - 2)^{-1}.$$

The first, when $x \to -\infty$ is clearly very sensitive to the behaviour of the sine function because the denominator tends to 0. Thus, it is natural to select two sequences where the sine is either 0 or 1, thereby exhibiting two different limits. The second, when $x \to +\infty$, can be seen as the product of a bounded function with a function that tends to 0, or by directly appealing to the squeeze theorem.

5.11 Compute the order with which the function $f(x) = e^{x^2} + 3 \sin^2 x - \cos x$ tends to 0 as $x \to 0$.

Answer. By asymptotic equivalence,

$$e^x - 1 = x(1 + \omega_1(x)),$$
$$\sin x = x(1 + \omega_2(x)),$$
$$1 - \cos x = x^2 \left(\frac{1}{2} + \omega_3(x) \right),$$

where $\omega_i(x) \to 0$ as $x \to 0$. Therefore

$$e^{x^2} = 1 + x^2(1 + \omega_1(x^2)), \quad \sin^2 x = x^2((1 + \omega_2(x))^2$$

and consequently

$$f(x) = x^2 \left(\frac{9}{2} + \omega_1(x^2) + 6\omega_2(x) + 3\omega_2^2(x) + \omega_3(x) \right).$$

It follows that $f(x) \to 0$ with order 2 as $x \to 0$, because indeed

$$\lim_{x \to 0} \frac{f(x)}{x^2} = \frac{9}{2} \neq 0.$$

This is a basic exercise on asymptotic equivalence. Reading the known limits as statements about the asymptotic equivalence of the functions $e^x - 1$ and $\sin x$ with x, and of $1 - \cos x$ with $x^2/2$, permits to single out the principal part of f and therefore to establish its order as $x \to 0$.

5.12 Consider the function

$$f(x) = \frac{\sqrt{6x - 22} - \sqrt{x^2 - 4x + 3}}{x - 5}.$$

Prove that there exists $\ell \in \mathbb{R}$ such that $f(x) - \ell$ tends to 0 as $x \to \infty$, and determine the order with which this happens.

Answer. First of all, the limit of f as $x \to +\infty$ exists and it is finite. Indeed, the domain f is $[11/3, 5) \cup (5, +\infty)$, which is not bounded above, so that it makes sense to compute the limit. The numerator takes on the indeterminate form $+\infty - \infty$ but the function $x \mapsto \sqrt{x^2 - 4x + 3}$ dominates $x \mapsto \sqrt{6x - 22}$ and has order 1. The denominator diverges to $+\infty$ with order 1 and hence $f(x) \to -1 = \ell$. It follows that $h(x) = f(x) + 1$ tends to 0 as $x \to +\infty$. As for the order,

$$h(x) = \frac{\sqrt{6x - 22} - \sqrt{x^2 - 4x + 3} + x - 5}{x - 5} = \frac{\sqrt{6x - 22} + k(x) - 5}{x - 5},$$

where $k(x) = x - \sqrt{x^2 - 4x + 3}$.

Now, multiplying and dividing by $x + \sqrt{x^2 - 4x + 3}$

$$k(x) = \frac{4x - 3}{x + \sqrt{x^2 - 4x + 3}} = \frac{4 - \frac{3}{x}}{1 + \sqrt{1 + \frac{3}{x^2} - \frac{4}{x}}} \to 2$$

as $x \to +\infty$. Therefore h vanishes with order $1/2$ as $x \to +\infty$. Indeed,

$$\lim_{x \to +\infty} \sqrt{x} h(x) = \sqrt{6}.$$

This is an exercise on the computation of orders at $+\infty$. The bottom line is that the first summand in the numerator has order $1/2$ and the second has order 1 and thus dominates. Since the denominator has clearly order 1, the main result on orders implies that the ratio tends to a non zero limit ℓ. Then one computes $h(x) = f(x) - \ell$, which is again a ratio. The numerator is a sum of three terms, the first of which diverges with order $1/2$; the second and third term give rise to a difference of diverging functions which actually converges to a finite number. Since the denominator of h clearly diverges with order 1, the ratio that defines h tends to 0 with order $1/2$.

5.13 Determine the order with which $f(x) = \pi/2 - \arccos(\log(x^x - x^2))$ vanishes as $x \to 0^+$.

Answer. The domain of the function contains $(0, 1)$, and since $x^x - x^2 \to 1$ as $x \to 0^+$, by permanence of sign there exists a right neighborhood of $x_0 = 0$ contained in $\text{Dom}(f)$. By the theorem on the limit of compositions, if $g(x) \to 0$ as $x \to 0$ and $g(x) \neq 0$ locally at 0, then $\sin(g(x))/g(x) \to 1$, and hence $g \sim \sin(g)$. Thus, it is enough to determine the order of

$$\sin f(x) = \sin(\pi/2 - \arccos(\log(x^x - x^2))) = \log(x^x - x^2).$$

By a similar argument, using that $\log(1 + y)/y \to 1$ as $y \to 0$, the order of $\sin f(x)$ is the same as that of

$$x \mapsto x^x - x^2 - 1 = e^{x \log x} - 1 - x^2.$$

Now $e^{x \log x} - 1 \sim x \log x$, and $x \log x$ has order less than that of x^2, so in conclusion f has the same order as that of $x \log x$. As seen in Sect. 5.4, this implies that f vanishes with order less than 1 as $x \to 0^+$, but greater than any $\alpha > 0$ with $\alpha < 1$.

In this exercise on orders, the theorem on the limit of compositions of functions, namely Theorem 5.7, is used together with basic elementary limits (see Appendix B). The point is that these lead to the formulae

$$g \sim \sin(g), \qquad \log(1 + h) \sim h, \qquad e^k - 1 \sim k,$$

provided that g, h and k tend to 0 in a way that is compatible with the assumptions of Theorem 5.7. The rest of the exercise is just using the rules.

5.14 Compute, if existing, $\displaystyle\lim_{x \to \pi} \frac{\log(2 + \cos x) \log |\sin x| - \sin^2 x}{|\cos x|^{\log |x - \pi|} - 1}$.

Answer. First of all, observe that $\cos x = -\cos(\pi - x)$ and $\sin x = \sin(\pi - x)$, so that by the theorem on the limit of compositions of functions

$$\lim_{x \to \pi} \frac{\log(2 + \cos x) \log |\sin x| - \sin^2 x}{|\cos x|^{\log |x - \pi|} - 1}$$

$$= \lim_{x \to \pi} \frac{\log(2 - \cos(\pi - x)) \log |\sin(\pi - x)| - \sin^2(\pi - x)}{|\cos(\pi - x)|^{\log |x - \pi|} - 1}$$

$$= \lim_{t \to 0} \frac{\log(2 - \cos t) \cdot \log |\sin t| - \sin^2 t}{|\cos t|^{\log |t|} - 1}.$$

Consider the numerator and denominator separately. Since as $t \to 0$

$$\log(2 - \cos t) \sim 1 - \cos t \sim \frac{t^2}{2}, \qquad \log |\sin t| \sim \log |t|, \qquad \sin^2 t \sim t^2$$

and since t^2 is negligible with respect to $t^2 \log |t|/2$ (the ratio of the former by the latter tends to 0), in view of Theorem 5.8, that is, by eliminating negligible terms, the numerator $N(t)$ can be replaced by $t^2 \log |t|/2$. In other words,

$$N(t) \sim \frac{t^2}{2} \log |t|.$$

Similarly, taking into account that $\cos t > 0$ near 0, the denominator $D(t)$ satisfies

$$D(t) = |\cos t|^{\log |t|} - 1 = e^{\log |t| \log |\cos t|} - 1$$

$$\sim \log |t| \log(\cos t) \sim \log |t|(\cos t - 1) \sim -\frac{t^2}{2} \log |t|.$$

Therefore, by the principle of substitution with asymptotically equivalent terms

$$\lim_{t \to 0} \frac{\log(2 - \cos t) \cdot \log|\sin t| - \sin^2 t}{(\cos t)^{\log|t|} - 1} = \lim_{t \to 0} \frac{\frac{t^2}{2} \log|t|}{-\frac{t^2}{2} \log|t|} = -1.$$

After a preliminary manipulation that puts the function in the form of a ratio of elementary functions, and shifts the limit point to the origin, the exercise calls for a careful analysis of numerator and denominator separately, in the attempt of using the idea that negligible terms should be neglected and dominant terms should be substituted by asymptotically equivalent terms of a simpler form. One could avoid the use of this technique by multiplying and dividing out by the correct factors so as to make the elementary limits appear and then appeal to the limit of composite functions to conclude. The outcoming expressions, however, are far more complicated, so that this exercise highlights the power and simplicity of asymptotic analysis.

5.6 Problems on Limits

5.15 Compute, if existing, the following limits:

(a) $\displaystyle\lim_{x \to +\infty} \sqrt{x^2 + 1} - \sqrt{x - 1}$

(b) $\displaystyle\lim_{x \to +\infty} \sqrt{2x + 1} - \sqrt{x^2 - 1}$

(c) $\displaystyle\lim_{x \to +\infty} \frac{\sqrt{x^2 + x^3}}{\sin x + \sqrt{x}}$

(d) $\displaystyle\lim_{x \to 0^+} \frac{1 + \cos(2x) - 2e^x}{\sin \sqrt{x}}$

(e) $\displaystyle\lim_{x \to 0} \frac{2x^3 - x^2|x|}{x}$.

5.16 Compute, if existing, the following limits:

(a) $\displaystyle\lim_{x \to +\infty} \sqrt{x^3 - x^2} - \sqrt{x^3 - 2x}$

(b) $\displaystyle\lim_{x \to \frac{\pi}{4}} \frac{\cos 2x}{2\sin^2 x - 1}$

(c) $\displaystyle\lim_{x \to \frac{\pi}{4}} \frac{\sin x - \cos x}{\log(\sin x) - \log(\cos x)}$

(d) $\displaystyle\lim_{x \to \pi} \frac{\sqrt{1 + \sin x} - \sqrt{1 - \sin x}}{\sin x}$

(e) $\displaystyle\lim_{x \to +\infty} \cos\sqrt{x^2 + x + 2} - \cos\sqrt{x^2 + x + 1}$

(f) $\displaystyle\lim_{x \to 0^+} \sin(x^{-4}) - \log(4x)$.

5.17 Compute, as k ranges in $\mathbb{R}$, the limits:

(a) $\displaystyle\lim_{x \to 0^+} \frac{e^x + ke^{-x}}{x}$

(b) $\displaystyle\lim_{x \to +\infty} \sin^k \frac{1}{\sqrt{x}}$

(c) $\displaystyle\lim_{x\to 0^+}\frac{\log(x+1)-k\arctan x+\frac{x^2}{2}}{x}$.

5.18 Compute, if existing, the following limits, as a and b range in $[0,+\infty)$.

(a) $\displaystyle\lim_{x\to+\infty}\left(ax+\frac{b}{x}\right)\cos x$

(b) $\displaystyle\lim_{x\to+\infty}\frac{ax+\frac{b}{x}}{|\sin x|+|\cos x|}$

(c) $\displaystyle\lim_{x\to+\infty}a^x\cos\frac{1}{b^x}$.

5.19 Compute the limits of $\dfrac{e^x-e^{\pi/2}}{\sqrt{1-\sin^2 x}}$, as x tends to $(\pi/2)^+$ and to $(\pi/2)^-$.

5.20 Compute the limits of $\log(1+e^{\lambda x})-x$, as x tends to $\pm\infty$ and λ ranges in $\mathbb{R}$.

5.21 Determine the order with which the following functions tend to 0 as $x\to 0$:

(a) $\sin x-\sin(x/2)$
(b) $\sqrt[3]{e^x}-\sqrt{x+1}$
(c) $\log^3(1+\sin(x^2))$
(d) $\sqrt[3]{x}-2\sqrt[3]{\sin x}$.

5.22 Compute the limit

$$\lim_{x\to 0^+}\frac{\log(1+1/x)}{x^\alpha-x\log(1+1/x)},$$

as α ranges in $(0,+\infty)$, specifying, if applicable, the order with which it either tends to 0 or diverges.

5.23 Compute $\displaystyle\lim_{x\to 0}\frac{\sin x(\log(\cos x)-x)}{x^2}$.

5.24 Compute the limit

$$\lim_{x\to+\infty}\frac{x^2 e^{\sin x}+\log(1+e^x)}{\sqrt{x^4+x^3}-\sqrt{x^4-x^3}}$$

specifying, if applicable, the order with which it either tends to 0 or diverges.

5.25 Consider the function $f(x)=\dfrac{kx-\sqrt{x^2+x}-\sqrt{x}}{2+\sqrt{x}}$.

(a) Compute the limit as $x\to+\infty$ as k varies in $\mathbb{R}$, specifying, if applicable, the order with which it either tends to 0 or diverges.
(b) For which k does there exists $\alpha\in\mathbb{R}$ such that f is positive in $(\alpha,+\infty)$?

5.26 Compute $\displaystyle\lim_{x\to 1^+}\left(3-2^x\right)^{(x-1)^\alpha}$ as α varies in $\mathbb{R}$.

Chapter 6
Continuous Functions

6.1 Basic Properties of Continuous Functions

In this chapter, I denotes a non empty subset of $\mathbb{R}$. Typically, I is an interval but this is not actually necessary.

Definition 6.1 Let $f : I \to \mathbb{R}$ and suppose that $x_0 \in I$ is a limit point of I. We say that f is *continuous* at x_0 if

$$\lim_{x \to x_0} f(x) = f(x_0).$$

If $x_0 \in I$ is not a limit point, we say that f is continuous at x_0. If $J \subset I$ and if f is continuous at every point in J, we say that f is continuous on J and if f is continuous on I, we simply say that it is continuous.

By definition, a function is automatically continuous at the isolated points of I. This fact is required in order to comply with the general definition of continuity, according to which a function is continuous if the inverse image of an open set is open.

Proposition 6.1 (Sequential continuity) *Suppose that $x_0 \in I$ is a limit point for $I \subseteq \mathbb{R}$ and take $f : I \to \mathbb{R}$. The following are equivalent:*

(i) f is continuous at x_0;
(ii) for every sequence $(x_n)_{n \geq 0}$ of elements of I different from x_0 such that $x_n \to x_0$ it holds that $\lim_n f(x_n) = f(x_0)$.

Proposition 6.2 (Permanence of sign) *Suppose that $x_0 \in I$ is a limit point for I, take $f : I \to \mathbb{R}$ and assume $f(x_0) \neq 0$. Then there exists a punctured neighborhood of x_0 in the points of which f has the same sign as $f(x_0)$.*

© Springer International Publishing Switzerland 2016
M. Baronti et al., *Calculus Problems*, UNITEXT - La Matematica per il 3+2 101,
DOI 10.1007/978-3-319-15428-2_6

Proposition 6.3 (Algebra of continuous functions) *Suppose that $f, g : I \to \mathbb{R}$ are continuous at $x_0 \in I$. Then also the functions $|f|$, $f + g$, fg, λf with $\lambda \in \mathbb{R}$ and f/g, if $g(x_0) \neq 0$, are continuous at x_0.*

From Proposition 6.3 we infer that linear combinations of continuous functions at a point are continuous at the same point, so that the set

$$C(I) = \big\{ f : I \to \mathbb{R} : f \text{ is continuous in } I \big\}$$

is a real *vector space* for any subset I of $\mathbb{R}$, for example an interval.

In the next proposition, the usual hypotheses under which the composition $g \circ f$ makes sense are tacitly assumed, as are those under which the various continuities are meaningful.

Proposition 6.4 (Continuity of compositions) *If f is continuous at x_0 and g is continuous at $f(x_0)$, then $g \circ f$ is continuous at x_0.*

Large families of elementary functions are continuous. All polynomials are continuous on $\mathbb{R}$ and every rational function (that is, every ratio of polynomials) is continuous on its domain. For example, $f(x) = 1/x$ is continuous on $\mathbb{R}\backslash\{0\}$. The powers $x \mapsto x^r$ with $r \in \mathbb{R}$ are continuous on $(0, +\infty)$. Also, whenever $a > 0$, the exponential function $x \mapsto a^x$ is continuous on $\mathbb{R}$. Consequently, by the algebra of continuous functions, also the hyperbolic sine and cosine are continuous on $\mathbb{R}$. Also, all elementary trigonometric functions and their inverses are continuous on their domains.

6.2 Discontinuities, Continuous Extensions

Take a function $f : I \subseteq \mathbb{R} \to \mathbb{R}$ and suppose that $x_0 \in I$ is a limit point for I. We say that f has a *removable discontinuity* at x_0 if

$$\lim_{x \to x_0} f(x) = \ell \in \mathbb{R}, \qquad \ell \neq f(x_0).$$

In the case in which $x_0 \notin I$ but again $f(x) \to \ell \in \mathbb{R}$ as $x \to x_0$, then we say that f admits a *continuous extension* at x_0. Indeed, in both cases, the function

$$\tilde{f}(x) = \begin{cases} f(x) & x \in I \\ \ell & x = x_0 \end{cases}$$

is continuous at x_0. In the first case, when $x_0 \in I$, the discontinuity of f has been removed by simply changing the value at x_0, while, in the second, $\tilde{f}$ extends f which was not originally defined at x_0, and the resulting function is continuous at x_0. For example, $\sin x/x$, defined in $\mathbb{R}\backslash\{0\}$, has a continuous extension at the origin and

$$\text{sinc}(x) = \begin{cases} \dfrac{\sin x}{x} & x \neq 0 \\[2mm] 1 & x = 0 \end{cases}$$

is the continuous extension of it. This function is called the *cardinal sine*.

Continuous extensions may be defined more generally. A function $g : J \subseteq \mathbb{R} \to \mathbb{R}$ is called a *continuous extension* of the continuous function $f : I \subseteq \mathbb{R} \to \mathbb{R}$ if $I \subseteq J$, if $g\big|_I = f$, and if g is continuous on J, that is, if g is an extension of f and is itself a continuous function.

The function $f : I \subseteq \mathbb{R} \to \mathbb{R}$ is said to have a *jump discontinuity* at $x_0 \in I$, or a *discontinuity of the first kind at* x_0, if x_0 is a limit point for I, if the limits of f exist in $\mathbb{R}$ (that is, if they are finite), both as $x \to x_0^+$ and as $x \to x_0^-$, but

$$[f]_{x_0} := \lim_{x \to x_0^+} f(x) - \lim_{x \to x_0^-} f(x) \neq 0.$$

The quantity $[f]_{x_0}$ is called the *jump* of f at x_0. A discontinuity of f, that is, a point $x_0 \in I$ at which f is not continuous, which is neither removable nor a jump discontinuity is called a *a discontinuity of the second kind at* x_0. For example, the function

$$f(x) = \begin{cases} x/|x| & x \neq 0 \\ 0 & x = 0 \end{cases}$$

has a jump discontinuity at the origin, whereas the function

$$f(x) = \begin{cases} \sin(1/x) & x \neq 0 \\ 0 & x = 0 \end{cases}$$

has a discontinuity of the second kind at the origin because neither the limit from the left nor from the right exist at the origin, hence the discontinuity is neither removable nor is it a jump discontinuity.

6.3 Global Properties of Continuous Functions

One important task of Calculus is the computation of local or global maxima (or minima) of a function. Many useful results are available for continuous functions, specially when the domain is a closed and bounded set, for example an interval of the form $[a, b]$. The exact notions of *extreme point* are formalized next.

Definition 6.2 (*Global and local extreme points*) A point x_0 in the domain I of the function $f : I \to \mathbb{R}$ is called:

(i)   a *global maximum* for f if $f(x_0) \geq f(x)$ for every $x \in I$;
(ii) a *global minimum* for f if $f(x_0) \leq f(x)$ for every $x \in I$;
(iii) *local maximum* for f if there exists $\delta > 0$ such that $f(x_0) \geq f(x)$ for every
      $x \in (x_0 - \delta, x_0 + \delta) \cap I$;
(iv) *local minimum* for f if there exists $\delta > 0$ such that $f(x_0) \leq f(x)$ for every
      $x \in (x_0 - \delta, x_0 + \delta) \cap I$.

If x_0 satisfies either (i) or (ii), then it is called a *global extreme point* for f,
whereas if it satisfies either (iii) or (iv), then it is called a *local extreme point* for f.
If furthermore the inequalities hold in the strong sense for $x \neq x_0$, then, say, a local
minimum is called a *strong local minimum*, and similarly in the other cases.

Observe that any global extreme point is also a local extreme point, but the converse is false, as the reader can verify with simple examples. The next result is of
fundamental importance.

Theorem 6.1 (Weierstrass) *If $f : [a, b] \to \mathbb{R}$ is continuous, then it admits global
maxima and minima, that is, there exist $\xi, \eta \in [a, b]$ such that $f(\xi) \leq f(x) \leq f(\eta)$
for every $x \in [a, b]$.*

The above theorem guarantees the existence of at least one global maximum and
at least one global minimum. They are in general not unique. A more general version
of Weierstrass' theorem involves *compact* subsets of $\mathbb{R}$, that is, closed and bounded
sets, for example, finite disjoint unions like

$$K = [a_0, b_0] \cup [a_1, b_1] \cup \cdots \cup [a_n, b_n].$$

Thus, a continuous function on a compact set has always extrema. The next result,
which is again formulated in its simplest version where the domain I of the function
is an interval $[a, b]$, relies on the fact that I is an interval, not on the property of
being closed or bounded.

Theorem 6.2 (Intermediate zero theorem) *Suppose $f : [a, b] \to \mathbb{R}$ continuous. If
$f(a)$ and $f(b)$ have opposite signs, then f has at least one zero, that is, there exists
$\xi \in [a, b]$ such that $f(\xi) = 0$.*

In view of a general formulation, we say that $s_1, s_2 \in \overline{\mathbb{R}}$ have opposite signs if
one of the following circumstances occur:

- $s_1 < 0 < s_2$;
- $s_1 = -\infty$ and $s_2 > 0$;
- $s_1 < 0$ and $s_2 = +\infty$;
- $s_1 = -\infty$ and $s_2 = +\infty$.

Corollary 6.1 *Assume that f is continuous on the interval I. Suppose further that
f admits limits (finite or not) at the boundary of I, and that they have opposite signs.
Then f has at least one zero in I.*

The next corollary is quite useful in order to show that an equation of the form $f(x) = g(x)$ admits a solution, when f and g are continuous functions.

Corollary 6.2 *Suppose that $f, g : [a, b] \to \mathbb{R}$ are continuous functions and assume that $f(a) < g(a)$ whereas $f(b) > g(b)$. Then there exists a point $\xi \in (a, b)$ such that $f(\xi) = g(\xi)$.*

The next two results generalize the intermediate zero theorem.

Theorem 6.3 (Intermediate value theorem) *Suppose that $f : [a, b] \to \mathbb{R}$ is continuous. Then f attains all the values between $f(a)$ and $f(b)$.*

Corollary 6.3 *Suppose that $f : [a, b] \to \mathbb{R}$ is continuous. Then $f([a, b])$ is a closed and bounded interval, namely $f([a, b]) = [m, M]$, where $m = \min f$ and $M = \max f$. More generally, the image of an interval under a continuous function is an interval.*

6.4 Continuous Monotonic Functions

Proposition 6.5 *Suppose $f : I \to \mathbb{R}$ monotonic. If $f(I)$ is an interval, then f is continuous.*

Corollary 6.4 (Continuity of monotonic functions) *Let I be an interval and suppose $f : I \to \mathbb{R}$ monotonic. Then f is continuous if and only if $f(I)$ is an interval.*

In general, the continuity of f^{-1} does not follow from the continuity of f. Consider the function defined on $I = (-\infty, -1) \cup [1, +\infty)$ by

$$f(x) = \begin{cases} x + 1 & x \leq -1 \\ x - 1 & x \geq 1. \end{cases}$$

It is clearly continuous, it is invertibile and its inverse $f^{-1} : \mathbb{R} \to I$ is

$$f^{-1}(x) = \begin{cases} x - 1 & x < 0 \\ x + 1 & x \geq 0. \end{cases}$$

The graphs of f and f^{-1} are in Fig. 6.1. Even though f is continuous, f^{-1} is not. This can happen because the domain is not an interval.

Theorem 6.4 (Continuity of the inverse function) *Let I be an interval and assume $f : I \to \mathbb{R}$ continuous. Then:*

(i) *f is injective on I if and only if f is strictly monotonic on I;*
(ii) *if f is increasing (decreasing) on I, then $f(I) = J$ is an interval and $f^{-1} : J \to I$ is continuous and increasing (decreasing).*

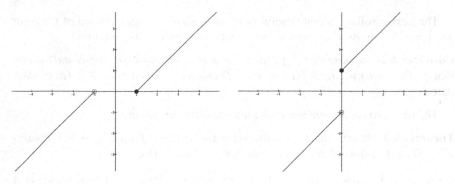

Fig. 6.1 Graphs of f and f^{-1}

6.5 Guided Exercises on Continuous Functions

6.1 Consider the function:

$$f(x) = \begin{cases} \dfrac{\log(1 - 2x)}{x} & x < 0 \\ a & x = 0 \\ x^2 + b\cos x & x > 0. \end{cases}$$

Find for which values of $a, b \in \mathbb{R}$, if any, f is continuous at the origin.

Answer. Clearly, $f(x) \to b$ as $x \to 0^+$, while

$$\lim_{x \to 0^-} \frac{\log(1 - 2x)}{x} = \lim_{x \to 0^-} \frac{\log(1 - 2x)}{-2x}(-2) = -2.$$

Therefore the limit of f as $x \to 0$ exists if $b = -2$. By definition, f is continuous at the origin if the limit is equal to $f(0) = a$. Hence f is continuous at the origin if and only if $a = b = -2$.

This is a basic exercise, in which the request is to apply the definition of continuity in one specific point: one has to tune the parameters in such a way that the limit from the left is equal to the limit from the right and that both are equal to the value of the function at the point.

6.2 Show that $e^x + \log x = 0$ admits a unique real solution.

Answer. The function $f(x) = e^x + \log x$ is continuous and strictly increasing on $(0, +\infty)$ because it is the sum of e^x and $\log x$, which are both continuous and strictly increasing. Furthermore, by the extended version of the intermediate zero theorem, since $f(x) \to -\infty$ as $x \to 0^+$ and $f(x) \to +\infty$ as $x \to +\infty$, there exists exactly one zero of f in its domain.

This is an exercise on the extended intermediate zero theorem. One has to show that at the boundary of the interval the function attains (or tends to) values that have opposite signs and, of course, that the function is continuous. If uniqueness is also to be shown, then monotonicity provides the answer.

6.3 Determine the number of zeroes of $f(x) = x^5 - x^2 + 3$ and, for each of them, find an interval of length one that contains it.

Answer. The function f is certainly continuous and increasing in $(-\infty, 0)$ because such are both $x \mapsto x^5 + 3$ and $x \mapsto -x^2$. Since $f(-2) = -2^5 - 4 + 3 = -33$ and $f(-1) = 1$, the intermediate zero theorem implies that f has exactly one zero in $(-\infty, 0)$, which actually belongs to the interval $(-2, -1)$. Further, if $x \in [0, 1]$

$$-1 \leq x^3 - 1 \leq 0 \implies |x^2(x^3 - 1)| \leq x^2 \leq 1 \implies x^2(x^3 - 1) + 3 \geq -1 + 3 = 2,$$

and hence $f(x) = x^2(x^3 - 1) + 3 > 0$ for any $x \in [0, 1]$. Finally, if $x \in (1, +\infty)$, then the inequalities $f(x) = x^2(x^3 - 1) + 3 > 3 > 0$ hold, so that f is positive for every $x \geq 0$. Therefore f has exactly one zero which belongs to $(-2, -1)$.

Even simple third degree polynomials may be hard to handle when it comes to finding roots. Here one argues by first breaking the real line into three intervals. In one of them the intermediate zero theorem is used to find one zero, and in the other two one sees that the function is always positive by simple estimates.

6.4 Consider the function $h(x) = 1/\log(4x^2 - 3x)$. Determine if there exits a continuous function $q : (-1/8, 4/5) \to \mathbb{R}$ such that $h(x) = q(x)$ for every $x \in (-1/8, 4/5) \cap \text{Dom}(h)$.

Answer. It is easy to see that $\text{Dom}(h) = (-\infty, -1/4) \cup (-1/4, 0) \cup (3/4, 1) \cup (1, +\infty)$, and consequently $(-1/8, 4/5) \cap \text{Dom}(h) = (-1/8, 0) \cup (3/4, 4/5)$. Observe that h is continuous on its domain because it is the ratio and composition of functions that are each continuous on its own domain. Any function q with the required properties is then a continuous extension of h to the interval $(-1/8, 4/5)$. If such an extension exists, then the following limits must all exist and the following equalities must hold true:

$$\lim_{x \to 0} q(x) = q(0) = \lim_{x \to 0^-} q(x) = \lim_{x \to 0^-} h(x) = \lim_{x \to 0^-} \frac{1}{\log(4x^2 - 3x)} = 0.$$

Similarly, it must be

$$\lim_{x \to \frac{3}{4}} q(x) = q\left(\frac{3}{4}\right) = \lim_{x \to \frac{3}{4}^+} h(x) = \lim_{x \to \frac{3}{4}^+} \frac{1}{\log(4x^2 - 3x)} = 0.$$

A continuous extension of h to the interval $(-1/8, 4/5)$ is then for instance

$$q(x) = \begin{cases} 0 & x \in [0, \frac{3}{4}] \\ h(x) & x \in (-\frac{1}{8}, 0) \cup (\frac{3}{4}, \frac{4}{5}). \end{cases}$$

This is an exercise about understanding what a continuous extension really is. First, one must check what the original domain of the given function is, and then one must compare it to the domain on which the required extension should be defined. The problem is then to compute the various limits of the given function at the appropriate boundary points and, finally, to define the extension in such a way that the two functions, the old and the new, meet nicely, that is, in a continuous fashion.

6.5 Prove that $f(x) = e^{1/x} \arctan x - e^x \arctan(1/x)$ has exactly one zero in the negative half-line, namely $x_0 = -1$.

Answer. Since

$$\lim_{x \to 0^-} e^{\frac{1}{x}} \arctan x = 0, \qquad \lim_{x \to 0^-} e^x \arctan(\frac{1}{x}) = -\frac{\pi}{2},$$

it follows that $f(x) \to \pi/2$ as $x \to 0^-$. Furthermore,

$$\lim_{x \to -\infty} e^{\frac{1}{x}} \arctan x = -\frac{\pi}{2}, \qquad \lim_{x \to -\infty} e^x \arctan(1/x) = 0$$

implies that $f(x) \to -\pi/2$ as $x \to -\infty$. We next show that f is (strictly) increasing, which then implies that f has exactly one zero in the negative half-line because f is continuous on $(-\infty, 0)$. Indeed, $x \mapsto e^{1/x}$ is decreasing and positive, while $x \mapsto \arctan x$ is increasing and negative, so that $x \mapsto e^{1/x} \arctan x$ is increasing. Similarly, $x \mapsto e^x$ is increasing and positive, $x \mapsto \arctan(1/x)$ is decreasing and negative, so that $x \mapsto e^x \arctan(1/x)$ is decreasing. Consequently, $x \mapsto -e^x \arctan(1/x)$ is increasing. Being the sum of two increasing functions, f is increasing.

This is a standard exercise on the use of the extended intermediate zero theorem. This time the main effort is in showing that the function is increasing, which can be inferred from Table 2.2. The reader should notice that, in this case, no use of derivatives is necessary to this end.

6.6 Consider the function $g(x) = \dfrac{x}{\log(x - \frac{6}{x})}$.

(a) Wherever possible, extend g by continuity.
(b) Find $h > 0$ such that for every $k > h$ the equation $g(x) = k$ has at least two solutions.

Answer. (a) Observe that $\text{Dom}(g) = (-\sqrt{6}, -2) \cup (-2, 0) \cup (\sqrt{6}, 3) \cup (3, +\infty)$ and that g is continuous on $\text{Dom}(g)$ because it is obtained as sum, composition and quotient of functions that are continuous on $\text{Dom}(g)$. We first compute the limits of g at the boundary of $\text{Dom}(g)$:

$$\lim_{x \to -\sqrt{6}^+} \frac{x}{\log(x - \frac{6}{x})} = 0 \qquad\qquad \lim_{x \to -2^\pm} \frac{x}{\log(x - \frac{6}{x})} = \mp\infty$$

$$\lim_{x \to 0^-} \frac{x}{\log(x - \frac{6}{x})} = 0 \qquad\qquad \lim_{x \to \sqrt{6}^+} \frac{x}{\log(x - \frac{6}{x})} = 0$$

$$\lim_{x \to 3^\pm} \frac{x}{\log(x - \frac{6}{x})} = \pm\infty.$$

Hence g admits a continuous extension at the points $-\sqrt{6}, 0$ and $\sqrt{6}$. Such extension is actually unique, and is given by

$$h(x) = \begin{cases} g(x) & x \in \mathrm{Dom}(g) \\ 0 & x \in \{-\sqrt{6}, 0, \sqrt{6}\}. \end{cases}$$

A maximal (non unique) extension is

$$q(x) = \begin{cases} g(x) & x \in \mathrm{Dom}(g) \\ 0 & x \in (-\infty, -\sqrt{6}] \cup [0, \sqrt{6}\}. \end{cases}$$

(b) Consider g in the interval $(3, +\infty)$. Evidently,

$$\lim_{x \to 3^+} g(x) = \lim_{x \to 3^+} \frac{x}{\log(x - \frac{6}{x})} = +\infty$$

$$\lim_{x \to +\infty} g(x) = \lim_{x \to +\infty} \frac{x}{\log(x - \frac{6}{x})} = +\infty.$$

Take now any value of g in $(3, +\infty)$, for example $g(6) = 6/\log 5$. Since g is continuous in $(3, 6]$, by the intermediate value theorem g takes all the values in $[6/\log 5, +\infty)$. The same happens in the interval $[6, +\infty)$. Hence, for every $k > 6/\log 5$ the equation $g(x) = k$ has at least two solutions in $(3, +\infty)$: one in $(3, 6)$, the other in $(6, +\infty)$.

This exercise poses two rather different tasks. The first is again about continuous extensions, and displays the fact that when the original domain has punctured intervals, then such an extension, if existing, is necessarily unique at those points, whereas if a whole interval is missing then if the behaviour at the boundary points is good, namely the limits are finite, then continuous extensions are possible in infinitely many ways. As it is often the case, the simpler the better, so that the selected extension is the so-called *extension by zero*, which is the extension obtained by defining the function to be equal to 0 in the points that have been added to the domain. The second task is some reasoning on the fact that the intermediate value theorem is really a theorem about solving equations.

6.7 Prove that

$$f(x) = \begin{cases} xe^{-x} & x \neq 0 \\ 3 & x = 0 \end{cases}$$

admits a global maximum in the interval $[-1, 1]$.

Answer. The function f is not continuous in $[-1, 1]$ because $f(x) \to 0$ as $x \to 0$ but $f(0) = 3$. Hence we cannot argue by using directly Weierstrass' theorem on extreme values. Consider now $g(x) = xe^{-x}$ with $x \in [-1, 1]$. Since g is continuous in a closed and bounded interval, by Weierstrass' theorem there exists $c \in [-1, 1]$ such that $g(x) \leq g(c)$ for every $x \in [-1, 1]$. In particular, $f(x) = g(x) \leq g(c)$ for every $x \in [-1, 1]\backslash\{0\}$. Thus $f(x) \leq \max(g(c), f(0)) = \max(g(c), 3)$. Observe that $g(c) = ce^{-c} \leq |c|e^{-c} \leq e^{-c} \leq e \leq 3$. Therefore $f(x) \leq \max(g(c), 3) = 3 = f(0)$ and f has in $x_0 = 0$ a global maximum.

This is an exercise on Weierstrass' theorem, where, though, a little extra argument is required because, as it stands, the given function is not a continuous function on a compact interval. In some sense, one selects among the maxima of two functions, both of which are continuous in the given interval: g and a constant function.

6.8 Consider the function

$$f(x) = \begin{cases} x - (x+1)\log(x+1) & x > -1 \\ -1 & x \leq -1 \end{cases}$$

and prove that $g(x) = \sqrt{-f(x)}$ is well defined in a neighborhood of $x_0 = -1$.

Answer. Clearly, f is continuous at least on $\mathbb{R}\backslash\{-1\}$. Further, $f(x) \to -1$ as $x \to -1^-$ and $f(x) \to -1$ as $x \to -1^+$ (recall the limit $\lim_{t \to 0^+} t^\alpha \log t = 0$ for every $\alpha > 0$) and both are equal to (-1). Hence f is continuous on $\mathbb{R}$. Since $f(-1) = -1 < 0$, and since f is continuous on $\mathbb{R}$, by permanence of sign there exists a neighborhood of $x_0 = -1$ on which $f(x) < 0$. In such neighborhood, $g(x) = \sqrt{-f(x)}$ is then well defined (and continuous).

This exercise asks to evaluate the sign of $-f$, hence of f, when x is close to $x_0 = -1$. The key technique is to use permanence of sign, so that it is sufficient to compute the limit as $x \to -1$.

6.9 Prove that $f(x) = x + \arctan x$ is invertible on its domain and denote by g its inverse. Define next

$$h(x) = \begin{cases} f(x) & x \geq c \\ g(x) & x < c \end{cases}$$

and find all the values of $c \in \mathbb{R}$ for which h is continuous on $\mathbb{R}$.

Answer. Since f is the sum of strictly increasing functions on $\mathbb{R}$, it is itself strictly increasing on $\mathbb{R}$. Further, $f(x) \to \pm\infty$ as $x \to \pm\infty$ and $f \in C^0(\mathbb{R})$. By the intermediate value theorem, $f(\mathbb{R}) = \mathbb{R}$ and by the theorem on the continuity of the inverse function, $g : \mathbb{R} \to \mathbb{R}$ is itself continuous on $\mathbb{R}$. Hence h is continuous at least on $\mathbb{R}\backslash\{c\}$ and it will be continuous at $x_0 = c$ if

$$\lim_{x \to c^+} f(x) = \lim_{x \to c^-} g(x) = h(c) = f(c).$$

Recall that $f(c) = \lim_{x \to c^+} f(x)$ and $\lim_{x \to c^-} g(x) = g(c)$ has already been established, because both f and g are continuous on $\mathbb{R}$.

Observe that $f(0) = 0$, and hence $g(0) = 0$. Therefore $c = 0$ is a value of c for which h is continuous on $\mathbb{R}$. We next show that this happens for no other value of c. Suppose on the contrary that there exists $c \neq 0$, for example $c > 0$, such that $f(c) = g(c)$. Since $f(x) > x$ if $x > 0$, it follows that $g(c) = f(c) > c$ and since f is strictly increasing on $\mathbb{R}$, it follows $c = f(g(c)) > f(c)$, a contradiction. If one assumes that $c < 0$ the argument is similar. Hence zero is the only value of c for which h is continuous.

The first part of this exercise is standard, and asks to show that a function is invertible by appealing to the fact that it is strictly monotonic (increasing, in this case). The second part contains a subtlety, and has a geometric flavour. First, the reader is urged to check for which values of c the function h makes sense, and discovers that since both f and g are defined everywhere, then all values of c are to be considered. Finally, one must realize that the continuity of h is equivalent to asking $f(c) = g(c)$. The symmetry of the graphs of f and g with respect to the bisector line $y = x$ then says that this may happen only at the points on the bisector. Since f crosses it only (it is increasing!) at the origin, the answer is found.

6.10 Given the function $f(x) = \sin x + \log(e^{x^2} - \sin x)$, prove that for every $r > 0$ the equation $f(x) = r$ has at least two solutions.

Answer. Clearly, $\mathrm{Dom}(f) = \{x \in \mathbb{R} : e^{x^2} - \sin x > 0\}$. Now, $e^{x^2} > 1$ for every $x \in \mathbb{R}\backslash\{0\}$. Therefore, $e^{x^2} > 1 \geq \sin x$ for every $x \in \mathbb{R}\backslash\{0\}$. For $x = 0$ the argument of the logarithm is 1, hence $\mathrm{Dom}(f) = \mathbb{R}$. Further, $f \in C^0(\mathbb{R})$ because f is a sum and composition of continuous functions. Consider the interval $[0, +\infty)$. Now, $f(0) = 0$ and $f(x) \to +\infty$ as $x \to +\infty$. Hence for every $r > 0$ there exits $\overline{x} \in (0, +\infty)$ such that $f(\overline{x}) > r$. Since $f \in C^0([0, +\infty))$, by the intermediate value theorem f takes all the values between 0 and $f(\overline{x})$, in particular it attains the value r. Therefore the equation $f(x) = r$ has at least one solution in $(0, +\infty)$. An analogous argument holds in the interval $(-\infty, 0)$ because $f(x) \to +\infty$ as $x \to -\infty$. The intermediate value theorem applies again and for every $r > 0$ yields a solution in $(-\infty, 0)$ of the equation $f(x) = r$. Hence the equation $f(x) = r$ has at least two solutions for every $r > 0$.

This exercise is a plain application of the intermediate value theorem.

6.11 Consider the function $g : [0, \pi] \to \mathbb{R}$ defined by $g(x) = \sqrt{\sin x} - \sin \sqrt{x}$. Establish whether g has at least one zero in $(0, \pi]$ or not. If yes, locate it with an error not greater than 8×10^{-1}.

Answer. Clearly, $g \in C^0([0, \pi])$. Moreover, $g(\pi) = -\sin\sqrt{\pi} < 0$ and $g(\pi/2) = 1 - \sin\sqrt{\pi/2} > 0$. By the intermediate zero theorem there exists $\overline{x} \in (\pi/2, \pi)$ such that $g(\overline{x}) = 0$. For $x_0 = 3\pi/4$ it holds $|\overline{x} - x_0| \le \pi/4 < 8/10$. Therefore $x_0 = 3\pi/4$ is an approximation of $\overline{x}$ with an error not greater than 8×10^{-1}.

This is an application of the intermediate zero theorem, with an additional remark: the mid point of the interval to which it is applied provides an approximation of the searched zero with an error that is less than half of the length of the interval. Alternatively, one could have chosen to approximate the intermediate zero with either $\pi/2$ or with π, the endpoints of the interval. In this case one would have obtained a larger a priori error but would have gained some extra information: in the former case the approximation is from below in the latter from above.

6.12 Consider the function $g(x) = x \log\left(\dfrac{x + \cos x}{x^2}\right)$.

(a) Find the domain of g.
(b) Prove that g has at least two zeroes, and locate one of them with an error not greater than 3×10^{-1}.

Answer. (a) Clearly, $\mathrm{Dom}(g) = \{x \in \mathbb{R}\backslash\{0\} : x + \cos x > 0\}$. The function $h(x) = x + \cos x$ is strictly increasing on $[-\pi, 0]$ because h is a sum of strictly increasing functions on $[-\pi, 0]$. Furthermore, $h(-\pi) = -\pi - 1 < 0$ and $h(0) = 1 > 0$. Since $h \in C^0([-\pi, 0])$, by the intermediate zero theorem there exists $\overline{x} \in (-\pi, 0)$ such that $h(\overline{x}) = 0$. Since h is strictly increasing, $\overline{x}$ is actually the only zero of h in $(-\pi, 0)$ and $h(x) < 0$ for $x \in [-\pi, \overline{x})$ whereas $h(x) > 0$ if $x \in (\overline{x}, 0)$. Further, if $x < -\pi$, then $h(x) < -\pi + \cos x < 0$. If $x \in (0, \pi/2)$, then $h(x) > 0$ because it is a sum of two positive functions. Finally, if $x \ge \pi/2$, then $h(x) \ge \pi/2 + \cos x > 0$. Therefore $h(x) > 0$ if and only if $x \in (\overline{x}, +\infty)$. It follows that $\mathrm{Dom}(g) = (\overline{x}, 0) \cup (0, +\infty)$.

(b) Notice that

$$\{x \in \mathrm{Dom}(g) : g(x) = 0\} = \left\{x \in \mathrm{Dom}(g) : \frac{x + \cos x}{x^2} = 1\right\}$$
$$= \{x \in \mathrm{Dom}(g) : x + \cos x - x^2 = 0\}.$$

The function $b(x) = x + \cos x - x^2$ belongs to $C^0(\mathbb{R})$ and $b(\overline{x}) = -\overline{x}^2 < 0$ because $h(\overline{x}) = 0$. Moreover $b(0) = 1$. By the intermediate zero theorem there exists $x_1 \in (\overline{x}, 0)$ such that $b(x_1) = 0$. Hence x_1 is a zero of g. Further, $b(1) = \cos 1 > 0$ and $b(\pi/2) = \pi/2 - \pi^2/4 < 0$. Again by the intermediate zero theorem there exists $x_2 \in (1, \pi/2)$ such that $b(x_2) = 0$, so that x_2 is a zero of g. Taking the mid point of the interval, that is $x_0 = (\pi + 2)/4$, it holds that

$$|x_2 - x_0| \le \frac{1}{2}\left(\frac{\pi}{2} - 1\right) = \frac{\pi - 2}{4} < 3 \times 10^{-1},$$

so that x_0 is an approximation of x_2 with an error less than 3×10^{-1}.

This is again a repeated application of the intermediate zero theorem, and similar comments to those made for Exercise 6.11 apply.

6.6 Problems on Continuity

6.13 Determine the number of zeroes of the function $f(x) = \dfrac{x}{2^x} - \dfrac{2^x}{x}$ in the interval $(-\infty, 0)$ and locate each of them in an interval of length $1/2$.

6.14 Denote by $\alpha, \beta \in \mathbb{R}$ two real parameters and consider the function

$$\varphi(x) = \begin{cases} \alpha \log(1 + x) & x \geq 0 \\ \beta x^2 \sin\left(\frac{1}{x}\right) & x < 0. \end{cases}$$

Establish for which values of α e β, if any, φ is continuous at the origin.

6.15 Denote by $a \in \mathbb{R}$ a real parameter and consider the function

$$g(x) = \begin{cases} e^{ax} - 1 & x > 0 \\ 0 & x = 0 \\ \dfrac{1 - \cos x}{2 \arctan x} & x < 0. \end{cases}$$

Establish for which values of $a \in \mathbb{R}$, if any, g is continuous at the origin.

6.16 Denote by $\alpha, \beta \in \mathbb{R}$ two real parameters and consider the function

$$\varphi(x) = \begin{cases} x^{\alpha x} & x > 1 \\ \beta x + 1 - \beta & x \leq 1. \end{cases}$$

Establish for which values of α e β, if any, φ is continuous at the point $x_0 = 1$.

6.17 Study the possible continuous extensions to the real line, as k ranges in $\mathbb{R}$, of the function $f(x) = (e^{-x} - \cos x + kx)/x$.

6.18 Consider the function:

$$f(x) = \begin{cases} xe^{-\frac{1}{x^2}} & x < 0 \\ a & x = 0 \\ b + \dfrac{1 - \cos x}{x} & x > 0. \end{cases}$$

(a) For which values of the real parameters a and b, if any, the function is continuous in its domain?

(b) Put $a = b = 1$. Is the function bounded in $[1, +\infty)$?

6.19 Consider the function:

$$f(x) = \begin{cases} x(1-x) & x \in \mathbb{Q} \\ 0 & x \in \mathbb{R} - \mathbb{Q}. \end{cases}$$

At which points is f continuous?

6.20 Determine the values of the parameters $a, b \in \mathbb{R}$ for which the following functions are continuous on their domains with:

$$f(x) = \begin{cases} \dfrac{\sin(x^2 + x)}{x \log x} & x > 0 \\ a & x = 0 \\ bx + c & x < 0, \end{cases} \qquad g(x) = \begin{cases} \sin(x + a) & x \geq 0 \\ \sin(\sin(x + a)) & x < 0. \end{cases}$$

6.21 Consider the function $f(x) = e^x + kx$.

(a) Find the number of zeroes of f as k ranges in $[0, +\infty)$.

(b) Put $k = 1$. Locate the only zero of f.

6.22 Consider the function:

$$f(x) = \begin{cases} \dfrac{\arctan \log(1 + x^2)}{|x|^a} & x \neq 0 \\ b & x = 0. \end{cases}$$

For which values of the real parameters a and b the function is continuous on $\mathbb{R}$?

6.23 For which values of the real parameter k is the function $f(x) = \dfrac{e^x}{e^{2x} + e^x + k}$ continuous on $\mathbb{R}$?

6.24 Suppose that f is a continuous function on $[0, 1]$ with image in $[0, 1]$. Prove that there exists a point c such that $f(c) = c$.

6.25 Establish whether the function $f(x) = (\sin x^2 - \sin^2 x)/x$ admits a continuous extension at $x_0 = 0$.

6.26 Consider the equation $\arctan x = 1 - x^3$. Prove that it admits one and only one solution, and compute an approximation of it with an error not larger than 10^{-1}.

6.27 Prove that for every $y < 0$ the equation $\log_2(1 + x) - \sqrt{x} = y$ has at least one solution, and that there exists $a > 0$ such that for every $y \in [0, a)$ the equation has at least two solutions.

6.28 Consider the function $f(x) = (\arctan x)(\sin x)$.

(a) Determine the image of f.
(b) Establish if f admits global extrema on $\mathbb{R}$.

6.29 Consider the function:

$$f(x) = \frac{1}{\log_2 \left(x\sqrt{x^2 + x} \right)}.$$

(a) Find the domain of f.
(b) Does there exist a continuous function $g : \mathbb{R} \to \mathbb{R}$ such that $f(x) = g(x)$ for every $x \in \mathrm{Dom}(f)$?

6.30 Consider the function $f(x) = (\log x)^{\log x}$.

(a) Find the domain of f.
(b) Does there exist a continuous function $g : \mathbb{R} \to \mathbb{R}$ such that $f(x) = g(x)$ for every $x \in \mathrm{Dom}(f)$?

6.31 Suppose that $f : \mathbb{R} \to \mathbb{R}$ is a continuous function such that $f(x) > 0$ for every $x \in \mathbb{R}$ and $f(x) \to 0$ as $x \to \pm\infty$. Prove that f has at least one global maximum and that it does not have any global minimum.

6.32 Consider the function:

$$f(x) = \begin{cases} a \cos x + b \sin x & -\frac{\pi}{2} < x < 0 \\ \sqrt{x^2 + 1} & 0 \le x \le 1. \end{cases}$$

Find $a, b > 0$ such that f has a global maximum, a global minimum, or both.

Chapter 7
Differentiable Functions

7.1 The Derivative of a Function

Definition 7.1 (*Linearization*) The function $f : (a, b) \to \mathbb{R}$ admits a *linearization* at the point $x_0 \in (a, b)$ if there exists a first order polynomial $p_1(x)$ that satisfies $p_1(x_0) = f(x_0)$ and is such that

$$\lim_{x \to x_0} \frac{f(x) - p_1(x)}{x - x_0} = 0.$$

In this case p_1 is called the linearization of f at x_0.

Evidently, any first order polynomial that coincides with f at x_0 can be written in the form $p(x) = f(x_0) + d(x - x_0)$ for some $d \in \mathbb{R}$. This implies at once that the linearization of f at x_0, if existing, is unique. It is also immediate to see that if f is itself a first order polynomial, then it is its own linearization at every point. Notice that a first order polynomial may well be of degree zero, that is, constant.

Geometric intuition suggests that a function that exhibits a "corner" does not admit a linearization at that point. Indeed, the prototype of such a behaviour is $x \mapsto |x|$ at the origin. This is because any polynomial $p(x) = ax$ vanishing at $x_0 = 0$ satisfies

$$\frac{|x| - ax}{x} = \frac{|x|}{x} - a = \begin{cases} 1 - a & x > 0 \\ -1 - a & x < 0, \end{cases}$$

so that the limit for $x \to 0$ does not exist for any a. Thus, $|x|$ does not admit a linearizaition at the origin.

Closely related to the notion of linearization at a point is the concept that follows, one of the cornerstones of Mathematical Analysis.

Definition 7.2 (*Derivative*) Take $f : (a, b) \to \mathbb{R}$ and $x_0 \in (a, b)$. The ratio

$$\frac{f(x) - f(x_0)}{x - x_0},$$

© Springer International Publishing Switzerland 2016
M. Baronti et al., *Calculus Problems*, UNITEXT - La Matematica per il 3+2 101,
DOI 10.1007/978-3-319-15428-2_7

which is well defined for $x \in (a, b)\setminus\{x_0\}$, is called the *difference quotient* of f at x_0. The function f is said to be *differentiable* at x_0 if the difference quotient converges to a finite real number as $x \to x_0$, that is, if the limit

$$\lim_{x \to x_0} \frac{f(x) - f(x_0)}{x - x_0} = \ell,$$

exists and is a real number. In this case, the real number ℓ is called the *derivative* of f at x_0 and it denoted in one of the following ways

$$f'(x_0), \quad \frac{df}{dx}(x_0), \quad Df(x_0).$$

It is also customary to write the difference quotient in the form

$$\frac{f(x_0 + h) - f(x_0)}{h},$$

with the understanding that h is small enough so that $x_0 + h \in (a, b)$. Thus, when existing, the derivative of f at x_0 can also be expressed as

$$f'(x_0) = \lim_{h \to 0} \frac{f(x_0 + h) - f(x_0)}{h}.$$

The existence of a linearization and of the derivative at a point turn out to be equivalent, as stated in the fundamental result that follows

Theorem 7.1 *Let* $f : (a, b) \to \mathbb{R}$ *and take* $x_0 \in (a, b)$. *The following are equivalent:*

(i) *f admits a linearization at x_0;*
(ii) *f is differentiable at x_0;*
(iii) *there exist a real number d and a function $R_1 : (a, b) \to \mathbb{R}$ that satisfy:*

$$f(x) = f(x_0) + d(x - x_0) + R_1(x)$$

for every $x \in (a, b)$ and

$$\lim_{x \to x_0} \frac{R_1(x)}{x - x_0} = 0. \tag{7.1}$$

By the above theorem, the differentiability of f is thus equivalent to the formula

$$f(x) = p_1(x) + R_1(x) = f(x_0) + f'(x_0)(x - x_0) + R_1(x), \tag{7.2}$$

Table 7.1 Tangent lines

$f(x)$	$f'(0)$	Tangent line at $(0, f(0))$
$\sin x$	1	$y = x$
$\cos x$	0	$y = 1$
e^x	1	$y = x + 1$
$\log(1 + x)$	1	$y = x$

provided that (7.1) holds. Formula (7.2) is called the *first order Taylor expansion* of f at x_0 and R_1 is called the *first order remainder*. As explained below in Chap. 8, (7.2) is the first of a family of formulae, known as (higher order) Taylor expansions. Evidently,

$$p_1(x) = f(x_0) + f'(x_0)(x - x_0)$$

is nothing but the linearization of f at x_0. Geometrically, the graph of p_1 represents the *tangent* to the graph of f at the point $(x_0, f(x_0))$, as formalized in the definition that follows.

Definition 7.3 Let $f : (a, b) \to \mathbb{R}$ and take $x_0 \in (a, b)$. If f is differentiable at x_0, then the tangent to the graph of f at the point $(x_0, f(x_0))$ is the straight line with equation

$$y = f(x_0) + f'(x_0)(x - x_0). \tag{7.3}$$

If f is not differentiable at x_0, then the function does not admit a tangent to its graph f at the point $(x_0, f(x_0))$.

Important examples of tangent lines are in Table 7.1.

Proposition 7.1 *If $f : (a, b) \to \mathbb{R}$ is differentiable at $x_0 \in (a, b)$, then it is also continuous at x_0.*

The converse of Proposition 7.1 does not hold. A simple example is $x \mapsto |x|$, which is continuous but not differentiable at the origin.

Proposition 7.2 (Derivative of sums and products) *Suppose that $f, g : (a, b) \to \mathbb{R}$ are differentiable at $x_0 \in (a, b)$. Then so are also the functions $f + g$, fg and αf for any $\alpha \in \mathbb{R}$, and the following formulae hold*

(i) $(f + g)'(x_0) = f'(x_0) + g'(x_0)$;
(ii) $(fg)'(x_0) = f'(x_0)g(x_0) + f(x_0)g'(x_0)$;
(iii) $(\alpha f)'(x_0) = \alpha f'(x_0)$.

The formula for the derivative of the product is also known as *Leibnitz rule*. The rule for the differentiation of the composition of two differentiable functions, described next, is called the *chain rule* ad it says that, under differentiation, composition becomes multiplication.

Proposition 7.3 (Chain rule) *Suppose that* $f : (a, b) \to \mathbb{R}$ *and* $g : (c, d) \to \mathbb{R}$ *are differentiable at* $x_0 \in (a, b)$ *and at* $f(x_0) \in (c, d)$, *respectively, and suppose further that* $f((a, b)) \subseteq (c, d)$. *Then the composition* $g \circ f$ *is differentiable at* $x_0 \in (a, b)$ *and*

$$(g \circ f)'(x_0) = g'(f(x_0)) f'(x_0).$$

Proposition 7.4 (Derivative of the inverse function) *Suppose that* $f : (a, b) \to \mathbb{R}$ *is continuous and invertible in* (a, b) *and assume further that it is differentiable at* $x_0 \in (a, b)$ *with* $f'(x_0) \neq 0$. *Then the inverse function* f^{-1} *is differentiable at* $y_0 = f(x_0)$ *and*

$$(f^{-1})'(y_0) = \frac{1}{f'(x_0)} = \frac{1}{f'(f^{-1}(y_0))}.$$

Under the assumptions of Proposition 7.4, the inverse function has a tangent to its graph at the point $(y_0, f^{-1}(y_0)) = (f(x_0), x_0)$. Considering the inverse function as a function of x, as it is customary, the equation of the tangent to its graph at the point $(y_0, f^{-1}(y_0))$ is then

$$y = f^{-1}(y_0) + (f^{-1})'(y_0)(x - y_0) = x_0 + \frac{x - y_0}{f'(x_0)}. \qquad (7.4)$$

Proposition 7.5 (Derivative of the reciprocal) *Suppose that* $f : (a, b) \to \mathbb{R}$ *is differentiable at* $x_0 \in (a, b)$ *and assume that* $f(x_0) \neq 0$. *Then the reciprocal* $1/f$, *defined in a neighborhood of* x_0, *is differentiable at* x_0 *and*

$$\left(\frac{1}{f} \right)'(x_0) = -\frac{f'(x_0)}{(f(x_0))^2}.$$

Corollary 7.1 (Derivative of the quotient) *Suppose that* $f, g : (a, b) \to \mathbb{R}$ *are differentiable at* $x_0 \in (a, b)$ *and assume that* $g(x_0) \neq 0$. *Then the quotient* f/g, *defined in a neighborhood of* x_0, *is differentiable at* x_0 *and*

$$\left(\frac{f}{g} \right)'(x_0) = \frac{f'(x_0)g(x_0) - f(x_0)g'(x_0)}{(g(x_0))^2}.$$

Definition 7.4 (*Derivative as a function*) Given a function $f : \text{Dom}(f) \to \mathbb{R}$, the function, defined in the subset $\text{Dom}(f')$ of $\text{Dom}(f)$ consisting of the points at which f is differentiable, and that assigns to $x \in \text{Dom}(f')$ the value $f'(x)$, is called the *derivative of* f and is denoted f'.

We conclude this section with the notions of left and right derivatives.

Definition 7.5 Let $f : [a, b] \to \mathbb{R}$.

(i) Take $x_0 \in [a, b)$. If the limit

$$\lim_{x \to x_0^+} \frac{f(x) - f(x_0)}{x - x_0}$$

exists and is a real number, then it is denoted by $f'_+(x_0)$ and it is called the *right derivative* of f at x_0.

(ii) Take $x_0 \in (a, b]$. If the limit

$$\lim_{x \to x_0^-} \frac{f(x) - f(x_0)}{x - x_0}$$

exists and is a real number, then it is denoted by $f'_-(x_0)$ and it is called the *left derivative* of f at x_0.

Evidently, if $x_0 \in (a, b)$ then f is differentiable at x_0 if and only if both the left and right derivatives exist at x_0 and $f'_+(x_0) = f'_-(x_0)$. In general, f is said to be differentiable in $[a, b]$ if it is such at all points of (a, b) (in the sense of Definition 7.2) and if it has the right derivative $f'_+(a)$ and the left derivative $f'_-(b)$.

Notice that although $f(x) = |x|$ is not differentiable at the origin, nevertheless it has both left and right derivatives. Explicitly, $f'_+(0) = 1$ and $f'_-(0) = -1$.

The derivative function of a given f is usually intended in the broader sense that the notions of left and right derivatives entail. This means that if, say, $f : [a, b] \to \mathbb{R}$ is differentiable in $[a, b]$, then $f' : [a, b] \to \mathbb{R}$ is the function

$$f'(x) = \begin{cases} f'_+(a) & x = a \\ f'(x) & x \in (a, b) \\ f'_-(b) & x = b. \end{cases}$$

7.2 Derivatives of Elementary Functions

Most elementary functions are differentiable in their domains. In Appendix B the derivatives of the most common elementary functions are listed. They will be used throughout and considered as known.

Of particular importance is the formula for the derivative of power functions with basis f and exponent g, namely functions of the form $f(x)^{g(x)}$, where f and g are any two differentiable functions and where $f(x) > 0$ in the common domain of f and g. The formula reads

$$\frac{d}{dx} f(x)^{g(x)} = f(x)^{g(x)} \left[g'(x) \log f(x) + \frac{g(x)}{f(x)} f'(x) \right]$$

and follows from the various differentiation rules if one takes $f(x)^{g(x)} = e^{g(x) \log(f(x))}$ as the (only reasonable) definition of power function with basis f and exponent g.

7.3   The Classical Theorems of Differential Calculus

Recall Definition 6.2 in Chap. 6 on local and global extreme points of a function.

Theorem 7.2   (Fermat) *Let $f : (a, b) \to \mathbb{R}$ and suppose that $x_0 \in (a, b)$ is a local extreme point. If f is differentiable at x_0, then $f'(x_0) = 0$.*

Definition 7.6   A point $x_0 \in (a, b)$ at which $f : (a, b) \to \mathbb{R}$ is differentiable and $f'(x_0) = 0$ is called a *critical point* of f.

Fermat's theorem states that a local extreme point at which a function is differentiable is a critical point. The converse, however, fails to be true. For example, the origin is a critical point for the function $f(x) = x^3$ because $f'(x) = 3x^2$ vanishes at the origin, but it is neither a local maximum nor a local minimum.

The next three results are mutually equivalent and form a classical triptych in differential calculus.

Theorem 7.3   (Rolle) *Suppose that $f : [a, b] \to \mathbb{R}$ is continuous in $[a, b]$ and differentiable in (a, b). If $f(a) = f(b)$, then there exits $\xi \in (a, b)$ such that $f'(\xi) = 0$.*

Theorem 7.4   (Cauchy) *Suppose that $f, g : [a, b] \to \mathbb{R}$ are continuous in $[a, b]$ and differentiable in (a, b). Then there exists $\xi \in (a, b)$ such that*

$$\det \begin{bmatrix} f(b) - f(a) & f'(\xi) \\ g(b) - g(a) & g'(\xi) \end{bmatrix} = 0,$$

that is $(f(b) - f(a))g'(\xi) - (g(b) - g(a))f'(\xi) = 0$.

Theorem 7.5   (Lagrange, or Mean Value Theorem) *Suppose that $f : [a, b] \to \mathbb{R}$ is continuous in $[a, b]$ and differentiable in (a, b). Then there exists $\xi \in (a, b)$ such that*

$$f(b) - f(a) = f'(\xi)(b - a).$$

The three theorems have equivalent geometrical interpretation: they all claim that there is a tangent to the natural curve associated to the data (the graph of f for Rolle's and Lagrange's theorems, the planar curve $x \mapsto (f(x), g(x))$ for Cauchy's theorem) that is parallel to the segment joining the endpoints of the curve.

Corollary 7.2   *Suppose that $f : [a, b] \to \mathbb{R}$ is continuous in $[a, b]$ and differentiable in (a, b). Then f is constant in $[a, b]$ if and only if $f'(x) = 0$ for every $x \in (a, b)$.*

In the above corollary, the assumption that the domain of f is an interval is essential, and cannot be removed. The boundedness of the interval, however, is not necessary and may be removed. For example, if $f : (c, +\infty) \to \mathbb{R}$ has everywhere vanishing derivative, then the same is true on every subinterval (a, b) with $a > c$ and f is therefore constant in every $[a, b]$, hence in $(c, +\infty)$. For instance,

$$f(x) = \arctan x + \arctan \frac{1}{x}$$

is differentiable in $\mathbb{R}\backslash\{0\}$, and it is easily seen that $f'(x) = 0$ for every $x \neq 0$. Since $f(1) = \arctan 1 + \arctan 1 = \pi/2$, Corollary 7.2 yields the formula

$$\frac{\pi}{2} - \arctan x = \arctan \frac{1}{x}$$

in the interval $(0, +\infty)$. The reader is urged to find the analogous formula in $(-\infty, 0)$.

Corollary 7.3 *Suppose that $f : [a, b] \rightarrow \mathbb{R}$ is continuous in $[a, b]$ and differentiable in (a, b).*

(i) If $f'(x) > 0$ for every $x \in (a, b)$, then f is strictly increasing in $[a, b]$;
(ii) if $f'(x) < 0$ for every $x \in (a, b)$, then f is strictly decreasing in $[a, b]$.

The function $f : [-1, 1] \rightarrow \mathbb{R}$ defined by $f(x) = x^3$ serves as a counterexample to the converse of (i) in Corollary 7.3, because f is strictly increasing but $f'(0) = 0$. For a converse statement, weaker properties must be considered.

Corollary 7.4 *Suppose that $f : [a, b] \rightarrow \mathbb{R}$ is continuous in $[a, b]$ and differentiable in (a, b).*

(i) $f'(x) \geq 0$ for every $x \in (a, b)$ if and only if f is nondecreasing in $[a, b]$;
(ii) $f'(x) \leq 0$ for every $x \in (a, b)$ if and only if f is nonincreasing in $[a, b]$.

The next Corollary to Rolle's theorem, and to Corollary 7.4, is due to Darboux. It states that the image of an interval under a derivative is an interval. Here is the version that we are interested in.

Corollary 7.5 (Darboux) *Suppose that $f : (a, b) \rightarrow \mathbb{R}$ is differentiable in (a, b) and take $\xi, \eta \in (a, b)$. Then f' attains all the values between $f'(\xi)$ and $f'(\eta)$.*

It follows from Darboux's theorem that a derivative cannot have jump discontinuities.

We end this chapter with the important theorems by de l'Hôpital and their consequences. For simplicity, we state one prototypical formulation and then comment on its possible variants.

Theorem 7.6 (de l'Hôpital) *Suppose that $f, g : (a, b) \rightarrow \mathbb{R}$ are differentiable in (a, b). Assume further that $g'(x) \neq 0$ for every $x \in (a, b)$ and that the limit*

$$\lim_{x \to a^+} \frac{f'(x)}{g'(x)} = \ell,$$

exists as an element in the extended real line, that is, $\ell \in \overline{\mathbb{R}}$. If either one of the following assumptions holds

(i) both f and g tend to 0 as x → a⁺,

(ii) g diverges for x → a⁺,

then also the limit $\lim\limits_{x \to a^+} (f(x)/g(x))$ *exists and*

$$\lim_{x \to a^+} \frac{f(x)}{g(x)} = \ell.$$

In the mathematical jargon, it is often said that de l'Hôpital's theorem holds when the quotient f/g takes on one of the indeterminate forms

$$\frac{0}{0} \quad \text{or} \quad \frac{\text{"anything"}}{\pm\infty}$$

as $x \to a^+$. The theorem can be formulated for the cases when $x \to b^-$ or when $x \to +\infty$ or $x \to -\infty$, with essentially no modifications. The important issue is the non vanishing of g' in a neighborhood of the limit point. The attentive reader will notice that this assumption actually implies that then g does not vanish sufficiently close to the limit point.

Among the standard applications of the theorem are the limits

$$\lim_{x \to +\infty} \frac{\log x}{x^\alpha} = 0, \qquad \lim_{x \to +\infty} \frac{e^x}{x^\alpha} = +\infty, \qquad \lim_{x \to 0^+} x^\alpha \log x = 0, \qquad \alpha > 0.$$

An example where the theorem cannot be applied, and would indeed lead to a wrong conclusion, is

$$\lim_{x \to +\infty} \frac{x - \sin x}{x + \sin x}.$$

Corollary 7.6 (Limit of the derivative) *Suppose that* $f : [a, b) \to \mathbb{R}$ *is continuous in* $[a, b)$ *and differentiable in* (a, b) *suppose further that*

$$\lim_{x \to a^+} f'(x) = d \in \overline{\mathbb{R}}.$$

Thus, if $d \in \mathbb{R}$, *then the right derivative of* f *exists at* a *and is equal to* d, *that is*

$$\lim_{x \to a^+} \frac{f(x) - f(a)}{x - a} = d,$$

whereas if $d = \pm\infty$ *the derivative does not exist.*

Clearly, the analogous result holds when a is an internal point of the domain of f, or when it is the rightmost extreme and the limits are taken from the left.

7.4 Guided Exercises on Differentiable Functions

7.1 Find the number of zeroes of the function $f(x) = x + (x^2 - 1)\log(1+x)$ in its domain.

Answer. The function f is defined for the values of x for which the argument of the logarithm is positive, that is in $I = (-1, +\infty)$. Observe that $f(0) = 0$ and that consequently the origin is a zero of f. Next consider the limits of f at the boundary of I. Clearly, f diverges to $+\infty$ as $x \to +\infty$ because so do $x \mapsto x$, $x \mapsto (x^2 - 1)$ and $x \mapsto \log(1 + x)$. In order to compute the limit as $x \to -1^+$, consider $\psi(x) = (x^2 - 1)\log(1+x)$, which takes on the indeterminate form "$0 \cdot \infty$". However,

$$\psi(x) = (x - 1)\big[(1 + x)\log(1 + x)\big]$$

and the term inside square brackets tends to 0 (see the Appendix). Hence $f(x) \to -1$ as $x \to -1^+$. Now, f is differentiable in I and

$$f'(x) = 1 + 2x\log(1 + x) + \frac{x^2 - 1}{x + 1} = x\big[1 + 2\log(1 + x)\big].$$

Clearly, $f'(x) > 0$ for $x > 0$ and f is therefore increasing in $[0, +\infty)$. Furthermore,

$$1 + 2\log(1 + x) < 0 \iff \log(1 + x) < -1/2 \iff 1 + x < e^{-1/2} \iff x < e^{-1/2} - 1.$$

Put $x_0 = e^{-1/2} - 1$ and observe that $-1 < x_0 < 0$. Thus, $f'(x)$ is positive in $(-1, x_0)$ and negative in $(x_0, 0)$. Therefore f is increasing in $(-1, x_0]$ and decreasing in $[x_0, 0]$. Finally, since $f(0) = 0$, and f is decreasing in $[x_0, 0]$, it follows that $f(x_0) > 0$. Therefore, besides the origin, f vanishes in exactly one other point, which belongs to the interval $(-1, x_0)$.

The natural tool to use for finding zeroes is of course the intermediate zero theorem. Since the question is rather precise (find all of them), one has to look at the limits at the boundary and then to perform a careful analysis of the monotonicity of f. The first task can be achieved with known limits (the function, near -1, is a translated multiple of $x \log x$ near the origin), and the second calls for the computation of f' and a study of its sign.

7.2 Consider $g(x) = x^2 - x + \arctan x + k$, where $k \in \mathbb{R}$ is a parameter.

(a) Compute $g'(x)$ wherever it exists and study its sign.
(b) Study the monotonicity and the number of zeroes of g as k ranges in $\mathbb{R}$.
(c) Set $k = -1$. For each zero of g find an interval of width 1 which contains it.

Answer. (a) The function g is defined and differentiable on $\mathbb{R}$ and

$$g'(x) = \frac{x(2x^2 - x + 2)}{1 + x^2}.$$

Since $1 + x^2 > 0$ and $2x^2 - x + 2 > 0$ for every $x \in \mathbb{R}$, $g'(x)$ has the same sign as x and vanishes for $x = 0$.

(b) From the above observation on the sign of its derivative, g is decreasing in $(-\infty, 0]$ and increasing in $[0, +\infty)$. It follows that the origin is the global minimum of g, which implies that for $k > 0$ no zeroes of g occur. For $k = 0$ there is exactly one zero of g at the origin. Finally, since $g(x) \to +\infty$ as $x \to \pm\infty$, the function has exactly two zeroes when $k < 0$ by the intermediate zero theorem, since g is continuous on $\mathbb{R}$.

(c) As discussed above, if $k = -1$ then g has two zeroes. Now, $g(1) = \pi/4 - 1 < 0$ and $g(2) = 1 + \arctan 2 > 0$. Again by the intermediate zero theorem it follows that g has one zero in $(1, 2)$. Analogously, since $g(-1) = 1 - \pi/4 > 0$ and $g(0) = -1 < 0$ the other zero of g lies in $(-1, 0)$.

This is a routine exercise: just use the natural tools, keeping in mind (recall Chap. 1) that the parameter k moves the graph of g up and down.

7.3 Consider the function defined on $(0, +\infty)$ by

$$
f(x) = \begin{cases}
\arctan\left(\log \dfrac{1}{1-x}\right) - \dfrac{\pi}{2} & 0 < x < 1 \\
0 & x = 1 \\
\cos(\log x) - 1 & x > 1.
\end{cases}
$$

(a) Establish if f is continuous at $x = 1$.
(b) Establish if f is differentiable at $x = 1$.
(c) Determine all the zeroes of f.

Answer. (a) Since $1/(1-x) \to +\infty$ as $x \to 1^-$, it follows that $\log[1/(1-x)] \to +\infty$, so that $f(x) \to 0$ as $x \to 1^-$. Now, $\log x \to 0$ as $x \to 1^+$ and hence $\cos(\log x) \to 1$, so that $f(x) \to 0$ as $x \to 1^+$. Therefore

$$
\lim_{x \to 1^-} f(x) = f(1) = \lim_{x \to 1^+} f(x) = 0
$$

and so f is continuous at $x = 1$.

(b) Consider the left difference quotient at $x = 1$. Observe that since the arctangent is odd and $\log(a/b) = \log a - \log b$ whenever $a > 0, b > 0$, it follows that $\arctan(\log(1/(1-x))) = -\arctan(\log(1-x))$. Thus, the left difference quotient at $x = 1$ is

$$
\frac{\arctan\left(\log \frac{1}{1-x}\right) - \frac{\pi}{2}}{x - 1} = -\frac{\arctan\left(\log(1-x)\right) + \frac{\pi}{2}}{x - 1}.
$$

Consider next

$$
\frac{\frac{d}{dx}\left[\arctan(\log(1-x)) + \frac{\pi}{2}\right]}{\frac{d}{dx}(x - 1)} = -\frac{1}{1 + (\log(1-x))^2}\frac{1}{1-x},
$$

whose denominator takes on the indeterminate form $\infty \cdot 0$. Upon writing it as $[1 + \log^2(1-x)]/(1-x)^{-1}$ and taking again the ratio of derivatives,

$$\frac{\frac{d}{dx}\left[1 + (\log(1-x))^2\right]}{\frac{d}{dx}(1-x)^{-1}} = -\frac{2\log(1-x) \cdot (1-x)^{-1}}{(1-x)^{-2}} = -2(1-x)\log(1-x).$$

Thus, using that $y \log y \to 0$ for $y \to 0$, and applying de l'Hôpital's theorem, it follows that

$$\lim_{x \to 1^-} \frac{1 + (\log(1-x))^2}{(1-x)^{-1}} = 0.$$

Applying again de l'Hôpital's theorem, the left difference quotient considered above, diverges and hence the left derivative at $x = 1$ does not exist. This implies that f is not differentiable at $x = 1$.

(c) Since the arctangent takes values in $(-\pi/2, \pi/2)$, the function does not vanish in $(0, 1)$. By definition, it does vanish at $x = 1$. In the interval $(1, +\infty)$ it vanishes whenever

$$\cos(\log x) = 1 \iff \log x = 2k\pi, \ k \in \mathbb{N}\backslash\{0\} \iff x = e^{2k\pi}, \ k \in \mathbb{N}\backslash\{0\}$$

In conclusion, the zeroes of f are the points $e^{2k\pi}$ with k a positive integer.

Here the main issue is the differentiability at the point 1 because f has a different expression to the left and to the right of it. On obvious way to proceed is to look at the sided difference quotients. The left difference quotient blows up, as one sees with de l'Hôpital's theorem. The final question is easier than expected: one finds the actual zeroes.

7.4 Consider the functions $g(x) = \sqrt{1-x}\cos(\frac{\pi}{2}x)$ and $h(x) = \frac{1}{g(x)}$.

(a) Determine the points at which g is differentiable.
(b) Write the linearization of h at $x_0 = -8$.
(c) Compute, when existing, $\lim_{x \to -3} \frac{g(x)}{|x+3|^a}$ for $a > 0$.

Answer. (a) Evidently, $\mathrm{Dom}(g) = \{x \in \mathbb{R} : 1 - x \geq 0\} = (-\infty, 1]$. Regard g as the product of $p(x) = \cos(\frac{\pi}{2}x)$ and $q(x) = \sqrt{1-x}$. Now, p is the composition of functions which are differentiable on $\mathbb{R}$, hence is itself differentiable on $\mathbb{R}$, and q is the composition of a polynomial function, everywhere differentiable, and of the square root function, which is differentiable in $(0, +\infty)$. It follows that q is differentiable at least in $\{x \in \mathbb{R} : 1 - x > 0\} = (-\infty, 1)$. Being the product of p and q, g is differentiable at least in $(-\infty, 1)$. As for its differentiability at $x_0 = 1$, it is necessary to establish whether the limit

$$\lim_{x \to 1^-} \frac{g(x) - g(1)}{x - 1} = \lim_{x \to 1^-} \frac{\sqrt{1-x}\cos(\frac{\pi}{2}x)}{x - 1}$$

is finite or not. Since $p(x) = \cos(\pi/2x)$ is differentiable at $x_0 = 1$, its first order Taylor expansion at $x_0 = 1$, namely

$$p(x) = p(1) + p'(1)(x - 1) + (x - 1)\omega(x) = -\frac{\pi}{2}(x - 1) + (x - 1)\omega(x)$$

is available, where $\omega = \omega(x) \to 0$ as $x_0 \to 1$. It follows that

$$\lim_{x \to 1^-} \frac{g(x) - g(1)}{x - 1} = \lim_{x \to 1^-} \frac{\sqrt{1-x}(x-1)[-\frac{\pi}{2} + \omega(x)]}{x - 1} = 0.$$

Hence g is differentiable in $(-\infty, 1]$.

(b) Observe that $g \in C^0((-\infty, 1])$ because it is differentiable. Further, $g(-8) = 3 \neq 0$ and, by permanence of sign, there is an open neighborhood of $x_0 = -8$ where g does not vanish in which h is thus well defined. The linearization of h at $x_0 = -8$ exists provided that at that point h is differentiable. Now, h is the reciprocal of g, so that we must use the known rules for its derivative. It was shown in item (a) that g is differentiable at $x_0 = -8$, and that $g(-8) = 3 \neq 0$. Therefore h is differentiable at $x_0 = -8$ and its linearization is $P(x) = h(-8) + h'(-8)(x+8)$. Next, $h(-8) = 1/3$ and

$$h'(x) = -\frac{g'(x)}{[g(x)]^2}.$$

Now,

$$g'(x) = -\frac{1}{2}(1 - x)^{-\frac{1}{2}} \cos\left(\frac{\pi}{2}x\right) - \frac{\pi}{2}\sqrt{1 - x}\sin\left(\frac{\pi}{2}x\right).$$

Therefore $g'(-8) = -1/6$ and $h'(-8) = 1/6 \cdot 1/9 = 1/54$. Thus, the linearization of h at $x_0 = -8$ is $P(x) = 1/3 + 1/54(x + 8)$.

(c) Since $g(x) \to 0$ as $x \to -3$, the limit takes on an indeterminate form. In item (a) it was shown that g is differentiable at $x_1 = -3$. Consider then the first order Taylor expansion of g at $x_1 = -3$, namely

$$g(x) = g(-3) + g'(-3)(x + 3) + (x + 3)\omega_1(x),$$

where $\omega_1 = \omega_1(x)$ tends to 0 as $x_1 \to -3$. Now, $g(-3) = 0$ and $g'(-3) = -\pi$, so that $g(x) = (x + 3)(-\pi + \omega_1(x))$ and

$$\lim_{x \to -3} \frac{g(x)}{|x + 3|^a} = \lim_{x \to -3} \frac{x + 3}{|x + 3|^a}(-\pi + \omega_1(x)) = \begin{cases} 0 & a < 1 \\ \text{does not exist} & a \geq 1. \end{cases}$$

7.5 Consider the function $f(x) = \dfrac{x^2 + x}{\log(\frac{2\arccos x}{\pi} - \arcsin x)}$.

(a) Find the domain of f.

(b) Compute the limits at the boundary of Dom(f).

(c) Find, if existing, a function g differentiable in $(-\infty, 0)$ such that $g(x) = f(x)$
for every $x \in \text{Dom}(f) \cap (-\infty, 0)$.

Answer. (a) Evidently,

$$\text{Dom}(f) = \left\{ x \in \mathbb{R} : x \in [-1, 1], \frac{2}{\pi} \arccos x - \arcsin x > 0, \frac{2}{\pi} \arccos x - \arcsin x \neq 1 \right\}.$$

Put $h(x) = (2/\pi) \arccos x - \arcsin x$. In order to understand the sign of h, which
is required in order to find $\text{Dom}(f)$, observe first that h is strictly decreasing in
$[-1, 1]$ because it is the sum of two such functions. Further, $h(-1) = 2 + \pi/2 >
0, h(1) = -\pi/2 < 0$ and $h(0) = 1$. Since $h \in C^0([-1, 1])$, by the intermediate zero
theorem h admits exactly one zero $x_0 \in (0, 1)$, and $h(x) > 0$ for every $x \in [-1, x_0)$.
Therefore $\text{Dom}(f) = [-1, 0) \cup (0, x_0)$.

(b) Clearly, f is continuous because it is obtained by taking ratios, compositions
and sums of continuous functions. Hence

$$\lim_{x \to -1+} f(x) = f(-1) = 0.$$

Also, $f(x) \to 0$ as $x \to x_0^-$, while, as $x \to 0$, f takes on an indeterminate
form. Denote by $q(x) = \log(h(x))$ the function at the denominator, and observe that
it is differentiable in $(-1, x_0)$ because it is a composition and sum of differentiable
functions in $(-1, x_0)$. Consider next the first order McLaurin expansion of q, namely
$q(x) = q(0) + q'(0)x + x\omega(x)$, where $\omega(x)$ tends to 0 as $x \to 0$. Now, $q(0) =
\log(h(0)) = \log 1 = 0$ and

$$q'(x) = \frac{h'(x)}{h(x)} = \frac{-\frac{2}{\pi} \frac{1}{\sqrt{1-x^2}} - \frac{1}{\sqrt{1-x^2}}}{h(x)}.$$

Hence $q'(0) = -(2/\pi) - 1$ and McLaurin's expansion reads

$$q(x) = x(-\frac{2}{\pi} - 1 + \omega(x)).$$

It follows that

$$\lim_{x \to 0} f(x) = \lim_{x \to 0} \frac{x(x+1)}{x(-\frac{2}{\pi} - 1 + \omega(x))} = \frac{1}{-\frac{2}{\pi} - 1}.$$

(c) Now, $\text{Dom}(f) \cap (-\infty, 0) = [-1, 0)$ and f is differentiable in $(-1, 0)$ because it
is the ratio of two differentiable functions in that set. Suppose now that there exists a
differentiable funcion g that extends f in $(-\infty, 0)$. Then $g(-1) = f(-1) = 0$ and
the limit

$$\lim_{x \to -1} \frac{g(x) - g(-1)}{x+1} = \lim_{x \to -1+} \frac{f(x)}{x+1} = \lim_{x \to -1+} \frac{x}{\log(\frac{2 \arccos x}{\pi} - \arcsin x)} = -\frac{1}{\log(2 + \frac{\pi}{2})}$$

exists and is finite. Thus, such an extension of f to $(-\infty, 0)$ is given by

$$
g(x) = \begin{cases} f(x) & -1 \le x < 0 \\ -\dfrac{x+1}{\log(2 + \frac{\pi}{2})} & x < -1. \end{cases}
$$

Part (a) is about solving, at least in a qualitative sense, an inequality of the form $h(x) > 0$. This is tackled by a combination of monotonicity and the intermediate zero theorem, a very standard approach for the study of the sign of a function. Part (b) is the computation of a limit of a quotient. Since the numerator is elementary, the best strategy is to expand the denominator in order to detect its leading term. This is an application of a linearization technique. Part (c) is about finding a differentiable extension g of f, which entails a twofold task: first g has to be an extension (so the boundary behaviour of f is needed) and then g must be differentiable (so the boundary behaviour of f' is needed).

7.6 Consider the function $g(x) = x \log(\sin x) - \sin(2x) \log(\cos x)$ and define

$$
h(x) = \begin{cases} g(x) & x \in \mathrm{Dom}(g) \\ 0 & x \notin \mathrm{Dom}(g). \end{cases}
$$

(a) Determine the set on which h is differentiable.
(b) Prove that h' has at least one zero in $(0, \pi/2)$.

Answer. (a) For later use, it is useful to put, for $k \in \mathbb{Z}$,

$$
y_k := 2k\pi, \qquad x_k := 2k\pi + \frac{\pi}{2}, \qquad I_k := (y_k, x_k), \qquad J_k := (x_k, y_{k+1})
$$

The domain of g is the set

$$
\mathrm{Dom}(g) = \{x \in \mathbb{R} : \sin x > 0, \cos x > 0\} = \bigcup_{k \in \mathbb{Z}} \left(2k\pi, 2k\pi + \frac{\pi}{2}\right) = \bigcup_{k \in \mathbb{Z}} I_k,
$$

which is open, and g is differentiable on it because it is given as sum, product and composition of differentiable functions. Clearly, h is differentiable in every interval J_k, hence in $\bigcup_{k \in \mathbb{Z}} J_k$, because it is constant and equal to zero on each J_k.

In order to determine the differentiability at the boundary points $\{x_k, y_k : k \in \mathbb{Z}\}$, a preliminary study of continuity is required. Consider first the continuity of h at the points $\{x_k : k \in \mathbb{Z}\}$. Clearly, $h(x) \to h(0) = 0$ as $x \to x_k^+$. Next, using that $t \log t \to 0$ for $t \to 0^+$ and the limit of composite functions, it follows

$$
\lim_{x \to x_k^-} h(x) = \lim_{x \to x_k^-} \left(x \log(\sin x) - \sin(2x) \log(\cos x)\right)
$$

$$
= \lim_{x \to x_k^-} \left(x \log(\sin x) - 2 \sin x \cos x \log(\cos x)\right) = 0
$$

for every $k \in \mathbb{Z}$. Hence h is continuous at x_k for all $k \in \mathbb{Z}$. As for the y_k's, evidently $h(x) \to h(0) = 0$ for $x \to y_k^-$. Finally,

$$\lim_{x \to y_k^+} h(x) = \lim_{x \to y_k^+} \left(x \log(\sin x) - \sin(2x) \log(\cos x) \right) = \begin{cases} +\infty & k < 0 \\ 0 & k = 0 \\ -\infty & k > 0. \end{cases}$$

Hence h is continuous at $y_0 = 0$ and it is not continuous at $y_k = 2k\pi$ if $k \neq 0$. Thus, among the points $\{x_k, y_k : k \in \mathbb{Z}\}$, h is continuous only at $\{x_k : k \in \mathbb{Z}\} \cup \{0\}$.

As far as differentiability at the points of $\{x_k : k \in \mathbb{Z}\}$ is concerned, a good strategy is to look at the limit of the derivative, because h is continuous in a neighborhood of x_k and differentiable in a punctured neighborhood of x_k, for every $k \in \mathbb{Z}$. Now, writing

$$C := \bigcup_{k \in \mathbb{Z}} (x_k, y_{k+1}) = \bigcup_{k \in \mathbb{Z}} J_k$$

a simple computation gives

$$h'(x) = \begin{cases} \log(\sin x) + x \dfrac{\cos x}{\sin x} - 2 \cos 2x \log(\cos x) + 2 \sin^2 x & x \in \mathrm{Dom}(g) \\ 0 & x \in C \end{cases}$$

and $h'(x) \to -\infty$ as $x \to x_k^-$, for every $k \in \mathbb{Z}$. It follows that h is not differentiable at x_k for any $k \in \mathbb{Z}$. As for the point $y_0 = 0$, it is easier to use the definition of derivative:

$$\lim_{x \to 0^+} \frac{h(x) - h(0)}{x} = \lim_{x \to 0^+} \frac{g(x)}{x} = \lim_{x \to 0^+} \left(\log(\sin x) - 2 \frac{\sin x}{x} \cos x \log(\cos x) \right) = -\infty.$$

Therefore h is not differentiable at $y_0 = 0$. Recall that at the other points y_k it cannot be differentiable because it is not continuous. Thus, h is differentiable only away from the boundary points of $\mathrm{Dom}(g)$, that is, it is differentiable on

$$\bigcup_{k \in \mathbb{Z}} (I_k \cup J_k) = \mathbb{R} \setminus \{x_k, y_k : k \in \mathbb{Z}\}.$$

(b) Evidently, $h \in C^0([0, \pi/2])$ and it is differentiable in $(0, \pi/2)$. By Rolle's theorem, since $h(\pi/2) = h(0) = 0$ there exists $x_0 \in (0, \pi/2)$ such that $h'(x_0) = 0$.

The main issue of this exercise is to realize that g is defined in a disjoint union of (well spaced) open intervals $I_k = (y_k, x_k)$ and needs to be analyzed at all the boundary points y_k and x_k. In every such interval except $I_0 = (0, \pi/2)$, it behaves as follows: it tends to 0 as x approaches x_k, from the right, while it diverges as x approaches y_k from the left. The exception is at $y_0 = 0$ where, unlike the other y_k's, the function tends to 0. The extension by zero of g is thus continuous at $\{0\}$ and at all the points x_k. However, it is not differentiable at any of these points, as a study of the limiting behaviour of the derivative reveals. Question (b) is designed to highlight the exceptionality of I_0, where the assumptions of Rolle's theorem are met, including continuity, with equal values, at the boundary points.

7.7 Consider

$$f(x) = \begin{cases} e^{-\frac{1}{x^2}} \sin \dfrac{1}{x} & x \neq 0 \\ k & x = 0. \end{cases}$$

(a) For which values of $k \in \mathbb{R}$ is f differentiable on $\mathbb{R}$?
(b) For which values of $k \in \mathbb{R}$ is f of class $C^1(\mathbb{R})$?

Answer. (a) By definition, the domain of f is $\mathbb{R}$. For $x \neq 0$ f is differentiable because it is a composition and product of differentiable functions. Since $|f(x)| \leq e^{-1/x^2}$, the squeeze theorem yields $\lim_{x \to 0} f(x) = 0$. Hence, f is continuous at 0 if and only if $k = 0$. Put now $k = 0$. Since for every $y \in \mathbb{R}$

$$\left| \frac{y \sin y}{e^{y^2}} \right| \leq \frac{|y|}{e^{y^2}},$$

by the squeeze theorem and change of variable

$$\lim_{x \to 0^{\pm}} \frac{f(x) - f(0)}{x} = \lim_{x \to 0^{\pm}} \frac{e^{-\frac{1}{x^2}} \sin \frac{1}{x}}{x} = \lim_{y \to \pm\infty} \frac{e^{-y^2} \sin y}{\frac{1}{y}} = \lim_{y \to \pm\infty} \frac{y \sin y}{e^{y^2}} = 0.$$

Therefore, for $k = 0$ f is differentiable at 0 and

$$f'(x) = \begin{cases} \dfrac{1}{x^2} \left(\dfrac{2}{x} \sin \dfrac{1}{x} - \cos \dfrac{1}{x} \right) e^{-\frac{1}{x^2}} & x \neq 0 \\ 0 & x = 0. \end{cases}$$

(b) At all points $x \neq 0$ the derivative is continuous because it is the composition of continuous functions. Further, if $x \neq 0$

$$|f'(x)| \leq \frac{1}{x^2} \left(\frac{2}{|x|} + 1 \right) e^{-\frac{1}{x^2}}.$$

Under the change of variable $x = 1/y$

$$\lim_{x \to 0^{\pm}} \frac{1}{x^2} \left(\frac{2}{|x|} + 1 \right) e^{-\frac{1}{x^2}} = \lim_{y \to \pm\infty} y^2 (2|y| + 1) e^{-y^2} = \lim_{y \to \pm\infty} \frac{y^2 (2|y| + 1)}{e^{y^2}} = 0.$$

By the squeeze theorem, $\lim_{x \to 0} f'(x) = 0 = f'(0)$. Thus $f \in C^1(\mathbb{R})$ if $k = 0$.
   This exercise is a variation of the classic fact that e^{-1/x^2} can be extended to a C^∞ function at the origin. Here, the main tool is the squeeze theorem, that allows to

handle the wild oscillations of $\sin(1/x)$ near the origin using its redeeming feature, the fact that it is bounded.

7.8 Let $Q(x)$ denote the first order Taylor polynomial of $f(x) = \sin(\log x)$ at 1, and define, for $x \in \text{Dom}(f)\setminus\{1\}$

$$g(x) = \frac{f(x) - Q(x)}{(x - 1)Q(x)}.$$

(a) Does g admit a continuous extension at 1?
(b) Does g admit a differentiable extension at 1?

Answer. (a) The domain of f is $(0, +\infty)$, where it is differentiable because it is a composition of differentiable functions. A quick computation gives $f(1) = 0$, $f'(x) = 1/x \cos(\log x)$ and $f'(1) = 1$. Hence $Q(x) = f(1) + f'(1)(x - 1) = x - 1$ and

$$g(x) = \frac{\sin(\log x) - (x - 1)}{(x - 1)^2},$$

is defined in $(0, 1) \cup (1, +\infty)$. As $x \to 1$, $g(x)$ takes on the indeterminate form "0/0". An application of the de l'Hôpital theorem gives

$$\lim_{x \to 1} g(x) = \lim_{x \to 1} \frac{\cos(\log x) - x}{2x(x - 1)} = \lim_{x \to 1} \frac{-\frac{1}{x}\sin(\log x) - 1}{4x - 2} = -\frac{1}{2}.$$

Therefore g admits a (unique) continuous extension at 1.
(b) The limit as $x \to 1$ of the difference quotient at 1 of the continuous extension of g can be computed applying three times the de l'Hôpital theorem. The equalities that follow are true *a posteriori*, in the sense that once the last limit is seen to be equal to $1/3$ then all the previous ones are finite and also equal to $1/3$. Explicitly:

$$\lim_{x \to 1} \frac{\frac{\sin(\log x) - (x - 1)}{(x - 1)^2} + \frac{1}{2}}{x - 1} = \lim_{x \to 1} \frac{2(\sin(\log x) - (x - 1)) + (x - 1)^2}{(x - 1)^3}$$

$$= \lim_{x \to 1} \frac{2\frac{1}{x}\cos(\log x) - 2 + 2(x - 1)}{3(x - 1)^2}$$

$$= \lim_{x \to 1} \frac{2\cos(\log x) - 2x + 2x(x - 1)}{3x(x - 1)^2}$$

$$= \lim_{x \to 1} \frac{-\frac{2}{x}\sin(\log x) - 2 + 2(x - 1) + 2x}{3(x - 1)^2 + 6x(x - 1)}$$

$$= \lim_{x \to 1} \frac{\frac{2}{x^2}\sin(\log x) - \frac{2}{x^2}\cos(\log x) + 2 + 2}{6(x - 1) + 6(x - 1) + 6x} = \frac{1}{3}.$$

Therefore g admits a differentiable extension at 1, in the sense that its continuous extension is actually differentiable at all points, including 1.

This exercise can be successfully solved by repeatedly applying the de l'Hôpital theorem. The only laborious aspect is a careful check of all the correct hypotheses.

7.9 Consider the function $f(x) = \sqrt{\dfrac{x - x^3 + 6}{x}}$.

(a) Determine where f is continuous and where it is differentiable.
(b) Show that f is invertible and compute, if existing, $(f^{-1})'(\sqrt{6})$.

Answer. (a) The domain of f is

$$\left\{ x \in \mathbb{R} : \frac{x - x^3 + 6}{x} \geq 0 \right\} = \left\{ x \in \mathbb{R} : \frac{(x^2 + 2x + 3)(2 - x)}{x} \geq 0 \right\} = (0, 2].$$

The function f is continuous because it is a ratio of continuous functions in $\text{Dom}(f) = (0, 2]$. Further, it is differentiable in $(0, 2)$ because it is a ratio of functions that are differentiable in $(0, 2)$. Using the various derivation rules,

$$f'(x) = \frac{1}{2\sqrt{\frac{x-x^3+6}{x}}} \frac{(1 - 3x^2)x - x + x^3 - 6}{x^2}$$

$$= \frac{1}{2}\sqrt{\frac{x}{x - x^3 + 6}} \cdot \frac{-2x^3 - 6}{x^2}$$

$$= -\sqrt{\frac{x}{x - x^3 + 6}} \cdot \frac{x^3 + 3}{x^2}.$$

It remains to be seen if f admits a left derivative at 2. Looking at the limit of the derivative, it turns out that $f'(x) \to -\infty$ for $x \to 2^-$, so that

$$\lim_{x \to 2^-} \frac{f(x) - f(2)}{x - 2} = -\infty.$$

Thus, f is differentiable only in $(0, 2)$.
(b) The expression of f' reveals that it is strictly negative in $(0, 2)$. Hence f is decreasing in $(0, 2]$ and therefore invertible. Furthermore, $f(1) = \sqrt{6}$ so that $\sqrt{6}$ is in the range of f. Using again the previous computation, $f'(1) = -4/\sqrt{6} \neq 0$. The derivative of the inverse therefore exists at $\sqrt{6}$ and it is equal to

$$(f^{-1})'(\sqrt{6}) = \frac{1}{f'(1)} = -\frac{\sqrt{6}}{4}.$$

This is essentially a routine exercise, where the various rules must be applied carefully. The derivative is used to infer that f is decreasing, hence invertible. Finally, the rule for the computation of the derivative of the inverse function is used.

7.10 Suppose that $f : \mathbb{R} \to \mathbb{R}$ is continuously differentiable and that $|f'(x)| \leq 1$ for every $x \in \mathbb{R}$. Knowing that $f(0) = 0$, give an estimate of the error if f is approximated with $P(x) = x$ in the interval $[0, 1/100]$.

Answer. The quantity that must be estimated is

$$E = \sup_{x \in [0,1/100]} |f(x) - P(x)|.$$

Put $g(x) = f(x) - P(x)$, so that $g(0) = f(0) - P(0) = 0$. Keeping in mind the properties of f, the mean value theorem implies that for a suitable $c \in (0, x) \subset [0, 1/100]$

$$
\begin{aligned}
|f(x) - P(x)| = |g(x)| = |g(x) - g(0)| &= |x \cdot g'(c)| \\
&= |x| \cdot |f'(c) - p'(c)| = |x| \cdot |f'(c) - 1| \\
&\leq |x| \cdot (|f'(c)| + 1) \leq 2|x|.
\end{aligned}
$$

Therefore

$$E = \sup_{x \in [0,1/100]} |f(x) - P(x)| \leq 2 \times \frac{1}{100} = 2 \times 10^{-2}.$$

This is a basic application of the mean value theorem. The estimate of the derivative is given as a datum, because f has no specific form. Thus the result holds for a whole class of functions.

7.11 Establish if the continuous extension at $x_0 = 0$ of the function:

$$f(x) = \frac{e^x - \cos x - x\sqrt{1 - x}}{x^2}$$

is differentiable at $x_0 = 0$.

Answer. Put $h(x) = e^x - \cos x - x\sqrt{1 - x}$ and $g(x) = x^2$. As $x \to 0$, the ratio $f(x) = h(x)/g(x)$ takes on the indeterminate form "0/0". Hence

$$
\begin{aligned}
\lim_{x \to 0} \frac{h'(x)}{g'(x)} &= \lim_{x \to 0} \frac{e^x + \sin x - \sqrt{1 - x} + \frac{x}{2\sqrt{1-x}}}{2x} \\
&= \lim_{x \to 0} \frac{e^x \sqrt{1 - x} - 1}{2x\sqrt{1 - x}} + \frac{\sin x}{2x} + \frac{3}{4\sqrt{1 - x}}.
\end{aligned}
$$

Now,

$$\lim_{x \to 0} \frac{e^x \sqrt{1 - x} - 1}{x} = F'(0)$$

where $F(x) = e^x\sqrt{1-x}$. Therefore,

$$\frac{e^x\sqrt{1-x}-1}{x} \quad \frac{1}{2\sqrt{1-x}} \to \frac{1}{4}, \quad \frac{\sin x}{2x} \to \frac{1}{2}, \quad \frac{3}{4\sqrt{1-x}} \to \frac{3}{4}$$

as $x \to 0$, so that $h'(x)/g'(x) \to 3/2$. By the de l'Hôpital theorem $f(x) \to 3/2$ as $x \to 0$. The continuous extension of f at $x_0 = 0$ is therefore

$$\tilde{f}(x) = \begin{cases} \dfrac{e^x - \cos x - x\sqrt{1-x}}{x^2} & x \in (-\infty, 0) \cup (0, 1] \\ 3/2 & x = 0. \end{cases}$$

As for its differentiability at the origin, observe that

$$\lim_{x \to 0} \frac{\tilde{f}(x) - \tilde{f}(0)}{x} = \lim_{x \to 0} \frac{2e^x - 2\cos x - 2x\sqrt{1-x} - 3x^2}{2x^3}.$$

Applying three times the de l'Hôpital theorem it follows

$$\lim_{x \to 0} \frac{\tilde{f}(x) - \tilde{f}(0)}{x} = \frac{7}{24},$$

so that $\tilde{f}$ is differentiable at $x_0 = 0$.

This exercise is just a repeated application of the de l'Hôpital theorem.

7.12 Establish for which values of the parameter $\alpha > 0$ the function

$$f(x) = \begin{cases} \dfrac{x}{|x|}\left| x \sin |2x| - |\sin 2x| \right| - \log(1 + |x|^\alpha) & x \in (-10^{-1}, 0) \cup (0, 10^{-1}) \\ 0 & x = 0 \end{cases}$$

is differentiable at the origin and for these values, if any such exists, compute $f'(0)$.

Answer. In the interval $(0, 10^{-1})$ the expression of f can be simplified to

$$\begin{aligned} f(x) &= 1 \cdot \left| x \sin 2x - \sin 2x \right| - \log(1 + x^\alpha) \\ &= \left| (x - 1)\sin 2x \right| - \log(1 + x^\alpha) \\ &= |x - 1| \sin 2x - \log(1 + x^\alpha) \\ &= (1 - x)\sin 2x - \log(1 + x^\alpha). \end{aligned}$$

The right derivative at the origin is

$$f'_+(0) = \lim_{x \to 0^+} \frac{f(x) - f(0)}{x}$$

$$= \lim_{x \to 0^+} \frac{(1-x)\sin 2x - \log(1+x^\alpha)}{x}$$

$$= \lim_{x \to 0^+} 2(1-x)\frac{\sin 2x}{2x} - \frac{\log(1+x^\alpha)}{x^\alpha} x^{\alpha-1}$$

$$= \begin{cases} -\infty & \alpha < 1 \\ 1 & \alpha = 1 \\ 2 & \alpha > 1. \end{cases}$$

It follows that f is not differentiable at the origin if $\alpha < 1$. In $(-10^{-1}, 0)$ the expression of f can be simplified to:

$$f(x) = -1 \cdot \left| x(-\sin 2x) - (-\sin 2x) \right| - \log(1 + |x|^\alpha)$$

$$= -\left| (1-x)\sin 2x \right| - \log(1 + |x|^\alpha)$$

$$= -|1-x||\sin 2x| - \log(1 + |x|^\alpha)$$

$$= -(1-x)(-\sin 2x) - \log(1 + |x|^\alpha)$$

$$= (1-x)\sin 2x - \log(1 + |x|^\alpha).$$

Excluding the values $\alpha < 1$, the left derivative at the origin is thus

$$f'_-(0) = \lim_{x \to 0^-} \frac{f(x) - f(0)}{x}$$

$$= \lim_{x \to 0^-} \frac{(1-x)\sin 2x - \log(1 + |x|^\alpha)}{x}$$

$$= \lim_{x \to 0^-} 2(1-x)\frac{\sin 2x}{2x} + \frac{\log(1 + |x|^\alpha)}{|x| \cdot |x|^{\alpha-1}} |x|^{\alpha-1}$$

$$= 2 + \begin{cases} 1 & \alpha = 1 \\ 0 & \alpha > 1 \end{cases}$$

$$= \begin{cases} 3 & \alpha = 1 \\ 2 & \alpha > 1. \end{cases}$$

In conclusion, f is differentiable at the origin only if $\alpha > 1$, in which case $f'(0) = 2$.

This exercise requires a few standard steps, each of which is elementary but must be performed with care. The presence of many absolute values calls for a rewriting of f into a more manageable form according as $x < 0$ or $x > 0$. In each of these cases, the limit of the one-sided difference quotient at the origin reduces to a sum of classical limits, each depending in a simple fashion on the parameter α. The final step is to glue together the behaviours from the left and from the right, that is, to see for which values of α they are both finite and coincide.

7.13 Where is $f(x) = (\arcsin \sqrt{|x|} - \sqrt{|x|})\sqrt{\log(1 + |x|)}$ differentiable?

Answer. First of all, $\text{Dom}(f) = [-1, 1]$. Away from the origin, it is possible to write f as a sum of two composite functions, and to apply for each of them the known theorems on differentiability. Indeed, put $g(x) = \arcsin \sqrt{|x|} = (h \circ q \circ p)(x)$ where

$$h(z) = \arcsin z, \qquad q(y) = \sqrt{y}, \qquad p(x) = |x|.$$

Now, p is differentiable in $\mathbb{R} - \{0\}$, q is differentiable in $(0, +\infty)$ and h is differentiable in $(-1, 1)$. Therefore, the composition g is differentiable at least in $(-1, 0) \cup (0, 1)$. Furthermore, in $I = (-1, 0) \cup (0, 1)$, the composition $q(p(x)) = \sqrt{|x|}$ is differentiable, so that such is also the composition $\sqrt{\log(1 + |x|)}$. It follows that f is differentiable at least in $(-1, 0) \cup (0, 1)$.

It remains to be seen what happens at the origin and at the endpoints ± 1. The difference quotient at 0 is

$$\frac{f(x) - f(0)}{x} = \frac{(\arcsin \sqrt{|x|} - \sqrt{|x|})\sqrt{\log(1 + |x|)}}{x}.$$

The first order Taylor expansion of h centered at 0 of the arcsin function reads $\arcsin z = z + z\omega(z)$, with $\lim_{z \to 0} \omega(z) = 0$. Hence

$$\lim_{x \to 0} \frac{f(x) - f(0)}{x} = \lim_{x \to 0} \frac{\sqrt{|x|}\omega(\sqrt{|x|})}{\sqrt{|x|}} \frac{\sqrt{\log(1 + |x|)}}{\sqrt{|x|}} \frac{|x|}{x}$$

$$= \lim_{x \to 0} \sqrt{\frac{\log(1 + |x|)}{|x|}} \omega_1 \left(\sqrt{|x|}\right) = 0.$$

Thus f is differentiable at 0 and $f'(0) = 0$. As for the point 1, observe that $\sqrt{\log(1 + x)}$ is differentiable at 1 with non vanishing derivative. Thus, f is differentiable at 1 if and only if $\beta(x) = \arcsin \sqrt{|x|} - \sqrt{|x|}$ is such. The difference quotient at $x = 1$ of the latter is

$$\lim_{x \to 1^-} \frac{\beta(x) - \beta(1)}{x - 1} = \lim_{x \to 1^-} \frac{\arcsin \sqrt{|x|} - \sqrt{|x|} - \arcsin \sqrt{1} + 1}{x - 1}.$$

Since

$$\lim_{x \to 1^-} \left(\frac{1}{\sqrt{1 - x}} \cdot \frac{1}{2\sqrt{x}} - \frac{1}{2\sqrt{x}}\right) = +\infty$$

by the de L'Hôpital theorem

$$\lim_{x \to 1^-} \beta'(x) = \lim_{x \to 1^-} \frac{\beta(x) - \beta(1)}{x - 1} = +\infty.$$

which shows that f is not differentiable at 1. Being an even function, it is not differentiable at -1 either. In conclusion, f is differentiable in $(-1, 1)$.

This exercise requires a separate analysis inside the open set $(-1, 0) \cup (0, 1)$, where no singular points of the absolute value occur, and where f is a composition of nice functions, at $x = 0$, and at the endpoints ± 1. At the origin, some fine information about the arcsin function is needed, as provided by the Taylor expansion. At the endpoints, de l'Hôpital's theorem is the right tool.

7.5   Problems on Differentiability

7.14 Let $f(x) = e^x + x^5$. Prove that f is invertible and denote by g its inverse. Compute, if existing, the linearization of g at $y_0 = 1 + e$.

7.15 Consider the functions

$$f(x) = \sqrt{1 - x}\, \arcsin \sqrt{x}, \qquad g(x) = \frac{f(x) + \alpha x + \beta}{4x - 3},$$

with α, β real parameters.

(a) Compute, if existing, the linearization of f at $x_0 = 3/4$.
(b) Find all the values of α and β, if any, for which $g \to 0$ as $x \to 3/4$.

7.16 Consider the function $g(x) = x \arcsin \frac{1}{x}$.

(a) Compute, if existing, $\displaystyle\lim_{x \to -2} \frac{3g(x) - \pi}{|x + 2|^a}$ as a ranges in $(0, +\infty)$.
(b) Write, if existing, the linearization of $\sin(g(x))$ at $x_0 = \sqrt{2}$.

7.17 Consider the function $g(x) = \log(e + x) - e^x$.

(a) Write, if existing, the linearization of $g(e^x)$ at $x_0 = \log 2$.
(b) Compute, if existing, $\displaystyle\lim_{x \to 0} g(x)/|x|^a$ as a ranges in $(0, +\infty)$.
(c) Prove that the restriction of g to $[0, +\infty)$ is invertible.

7.18 Consider the function $f(x) = \sin(\pi x) \arcsin(1/x)$.

(a) Where is f differentiable?
(b) Find an interval in which f is invertible.
(c) Compute, if existing, $\displaystyle\lim_{x \to +\infty} x^a f(x)$ as a ranges in $(0, +\infty)$.

7.19 Consider the function $h(x) = \sqrt{\sqrt[3]{x} - \sin x}$.

(a) Where is h defined?

(b) Compute, if existing, $\lim\limits_{x \to \pi^+} \dfrac{h(x) - \sqrt[6]{\pi}}{(x - \pi)^a}$ as a ranges in $(0, +\infty)$.

7.20 Consider the function $f(x) = 1/(1 - \sqrt{2}\cos x)$.

(a) Compute, if existing, $\lim\limits_{x \to \frac{\pi}{4}^+} f(x)(4x - \pi)$.

(b) Write, if existing, the linearization of $\log(f(x))$ at $x_0 = 3\pi/4$.

7.21 (a) Determine the number of zeroes of $h(x) = \log(x + \sin x)$ and compute an approximate value of one of them with an accuracy of 2×10^{-2}.

(b) Compute, if existing,

$$\lim_{x \to \pi} \frac{\log\left(\frac{x + \sin x}{\pi}\right)}{|x - \pi|^b}$$

as b ranges in $(0, +\infty)$.

7.22 Consider the functions $g(x) = \arccos(\cos x - \sin x)$ and $h(x) = g(x^2)$

(a) Find the domain of g.

(b) Determine if h is differentiable at $x_0 = 0$ and, if so, compute $h'(0)$.

7.23 Consider the function $f(x) = \arctan\left|x - \dfrac{1}{x^2}\right|$.

(a) Find $\mathrm{Dom}(f)$ and compute the limits of f at the boundary points of $\mathrm{Dom}(f)$.

(b) Determine where f is monotonic.

(c) Find, if existing, local and global minima and maxima of f.

(d) Draw a qualitative graph of f.

(e) Prove that f is invertible in a neighborhood of $x_0 = 2$ and write the equation of the tangent to the graph of f^{-1} at the point $(f(2), 2)$.

7.24 Consider the function $f(x) = \dfrac{1 - x}{1 + x}\log(x - 1)$.

(a) Find $\mathrm{Dom}(f)$ and compute the limits of f at the boundary points of $\mathrm{Dom}(f)$.

(b) Determine where f is monotonic.

(c) Find, if existing, local and global minima and maxima of f.

(d) Draw a qualitative graph of f.

(e) Prove that f is invertible in a neighborhood of $x_0 = 2$ and write the equation of the tangent to the graph of f^{-1} at the point $(f(2), 2)$.

7.25 Consider the function $f(x) = \arctan\left(e^x + \dfrac{1}{x}\right)$.

(a) Compute the limits of f at the boundary points of $\mathrm{Dom}(f)$.

(b) Determine where f is monotonic.

(c) Locate all the local extreme points of f in intervals of length $1/2$.

(d) Draw a qualitative graph of f.

(e) Discuss the number of solutions to the equation $f(x) = k$, as k ranges in $\mathbb{R}$.

7.26 Consider the function $f(x) = \dfrac{\arctan(1 - e^x)}{x^\alpha}$, where $\alpha \in \mathbb{R}$ is a real parameter.

(a) Compute the limits $\lim_{x \to 0^+} f(x)$ e $\lim_{x \to +\infty} f(x)$ as α ranges in $\mathbb{R}$.

(b) Put $\alpha = 2$ and compute, if existing, $f'(1)$.

(c) Let $g(x) = \arctan(1 - e^x)$; determine where g is monotonic and draw a qualitative graph of g.

7.27 Consider the function $f(x) = \dfrac{\log[e(x^3 - 2x + 1)]}{e(x^3 - 2x + 1)}$.

(a) Find $\mathrm{Dom}(f)$.

(b) Compute the limits of f at the boundary points of $\mathrm{Dom}(f)$.

(c) Determine where f is monotonic and find, if existing, local and global minima and maxima of f.

(d) Find all the zeroes of f, if any exist.

(e) Draw a qualitative graph of f.

7.28 Consider the function $f(x) = (\arctan x)e^{\arctan x}$.

(a) Find $\mathrm{Dom}(f)$ and compute the limits of f at the boundary points of $\mathrm{Dom}(f)$.

(b) Compute f', where it exists.

(c) Determine where f is monotonic.

(d) Find, if existing, local and global minima and maxima of f, and draw a qualitative graph of f.

7.29 Consider the function $f(x) = \dfrac{1 - 2e^{-x}}{x}$.

(a) Find $\mathrm{Dom}(f)$ and compute the limits of f at the boundary points of $\mathrm{Dom}(f)$.

(b) Determine where f is monotonic.

(c) Find, if existing, local and global minima and maxima of f.

(d) Draw a qualitative graph of f.

(e) Denote by h the restriction of f to $(0, 1)$. Prove that h is invertible and write the equation of the tangent to the graph of h^{-1} at the point $(f(\log 2), \log 2)$.

7.30 Consider the function $f(x) = \sqrt[3]{|x \log x|} - \sqrt[3]{e}$.

(a) Compute the limits of f at the boundary points of $\mathrm{Dom}(f)$.

(b) Determine where f is monotonic.

(c) Find, if existing, local and global minima and maxima of f.

(d) Determine the number of zeroes of f.

(e) Draw a qualitative graph of f.

7.31 Consider the function $f(x) = e^{x(\log x)^2} - e^{-x(\log x)^2}$.

(a) Compute the limits of f at the boundary points of Dom(f).
(b) Determine where f is monotonic.
(c) Find, if existing, local and global minima and maxima of f.
(d) Draw a qualitative graph of f.

7.32 Consider the function

$$f(x) = \begin{cases} \sqrt{\arctan x} + a & \text{if } x > 0 \\ be^x - e^{-x} & \text{if } x \le 0. \end{cases}$$

(a) Determine for which values of the parameters $a, b \in \mathbb{R}$, if any, f is continuous.
(b) Determine for which values of the parameters $a, b \in \mathbb{R}$, if any, f is differentiable and, where it exists, compute $f'(x)$.
(c) Put $a = 1$ and $b = -e$. Determine where f is monotonic and draw a qualitative graph of f.

7.33 Consider the function $f(x) = \log(1 + x) - \sqrt{|x|} + \dfrac{1}{2}$.

(a) Compute the limits of f at the boundary points of Dom(f).
(b) Determine in which intervals f is differentiable and, in those intervals, compute the derivative.
(c) Determine where f is monotonic.
(d) Find, if existing, local and global minima and maxima of f.
(e) Determine the number of zeroes of f.
(f) Draw a qualitative graph of f.

7.34 Consider the function $g(x) = x \log(1 + |x|) - |x|^3$.

(a) Discuss the sign and the number of zeroes of $g(x)$.
(b) Find in which points of its domain the function $f(x) = |g(x)|$ is differentiable, and compute the derivative at those points.

7.35 Consider the function $f(x) = \log\left(2 + (2x)^x\right)$.

(a) Prove that f is invertible in $(1/2e, +\infty)$.
(b) Write, if meaningful, the equation of the tangent to the graph of f^{-1} at $(f(1), 1)$.

7.36 Determine where the function

$$f(x) = \begin{cases} \dfrac{x \sin x}{|x|^k} & x \ne 0 \\ 0 & x = 0 \end{cases}$$

is differentiable, as the parameter k ranges in $\mathbb{R}$.

Chapter 8
Taylor Expansions

8.1 Taylor Expansions

A *Taylor expansion* of a function $f : (a, b) \to \mathbb{R}$ is a local representation of f as a sum of a polynomial p_n and a remainder term. The expansion is centered at a given point $x_0 \in (a, b)$. The polynomial p_n is used in order to approximate f close to x_0. The higher the degree n of p_n, the better the approximation, in the sense that f and p_n have the same value and the same first n derivatives, and, as x approaches x_0, the remainder term vanishes with an order that is higher than n. It is important to stress that Taylor expansions are useful to capture local phenomena but are useless to capture global ones.

Definition 8.1 (*Contact order*) Let n be a positive integer and suppose that the functions $f, g : (a, b) \to \mathbb{R}$ have at least n derivatives at $x_0 \in (a, b)$. We say that f and g have *order of contact* greater than n at x_0 if they, together with all their derivatives of order up to n, coincide at x_0, that is, if $f^{(k)}(x_0) = g^{(k)}(x_0)$ for $0 \le k \le n$. Explicitly, f and g have order of contact n at x_0 if $f(x_0) = g(x_0)$, $f'(x_0) = g'(x_0)$, $f''(x_0) = g''(x_0), \ldots, f^{(n)}(x_0) = g^{(n)}(x_0)$.

Using the notation of Chap. 5, (see in particular Definitions 5.7 and 5.12), it turns out that

$$f \text{ and } g \text{ have order of contact greater than } n \iff f - g \prec (x - x_0)^n.$$

Theorem 8.1 (Taylor Polynomial) *Take* $f : (a, b) \to \mathbb{R}$ *and* $x_0 \in (a, b)$. *Suppose that* f *admits at least* n *derivatives at* x_0. *Then there exists exactly one polynomial of degree less than or equal to* n *which has order of contact greater than* n *with* f *at* x_0. *It is called the* Taylor polynomial *of* f *of order* n *at* x_0 *and it is given by the formula*

$$p_n(x) = \sum_{k=0}^{n} \frac{f^{(k)}(x_0)}{k!} (x - x_0)^k.$$

© Springer International Publishing Switzerland 2016

M. Baronti et al., *Calculus Problems*, UNITEXT - La Matematica per il 3+2 101,
DOI 10.1007/978-3-319-15428-2_8

If $x_0 = 0$, p_n is called the McLaurin polynomial *of order n of f.*

Notice that p_n may well be of degree less than n because some of the derivatives of f at x_0 may vanish. For example, the McLaurin polynomial of order 4 of $f(x) = x^2$ is x^2, because $f(0) = f'(0) = 0$, $f''(0) = 2$, and $f^{(3)}(0) = f^{(4)}(0) = 0$.

Proposition 8.1 *If $f, g : (a, b) \to \mathbb{R}$ are differentiable n times at $x_0 \in (a, b)$ and $\alpha, \beta \in \mathbb{R}$, then $p_n[\alpha f + \beta g] = \alpha p_n[f] + \beta p_n[g]$ where p_n denotes the Taylor polynomial of order n at x_0 of the indicated function.*

The next theorem is the central result of this chapter. It describes the accuracy of the approximation.

Theorem 8.2 (Taylor expansion) *Suppose that $f : (a, b) \to \mathbb{R}$ is differentiable n times at $x_0 \in (a, b)$, let p_n denote its Taylor polynomial of order n at x_0, and put*

$$R_n(x) = f(x) - p_n(x),$$

called the remainder term *of order n. Then*

$$\lim_{x \to x_0} \frac{R_n(x)}{(x - x_0)^n} = 0 \quad \text{(Peano's formula).} \tag{8.1}$$

If we further assume that f is differentiable $(n + 1)$ times at $(a, b) \setminus \{x_0\}$, then for every $x \in (a, b)$ there exists ξ between x_0 and x such that

$$R_n(x) = \frac{f^{(n+1)}(\xi)}{(n + 1)!}(x - x_0)^{n+1} \quad \text{(Lagrange's formula).}$$

Notice that, in Lagrange's formula, the point ξ depends on x, and hence the remainder R_n is not a constant multiple of $(x - x_0)^{n+1}$, that is, it is not a monomial. The point ξ is "between x_0 and x" in the sense that $\xi \in (x, x_0)$ when $x < x_0$ and $\xi \in (x_0, x)$ when $x > x_0$. The next proposition establishes a uniqueness result.

Proposition 8.2 *Suppose that $f : (a, b) \to \mathbb{R}$ is differentiable n times at $x_0 \in (a, b)$ and that P is a polynomial of degree less than or equal to n such that*

$$\lim_{x \to x_0} \frac{f(x) - P(x)}{(x - x_0)^n} = 0,$$

then P is the Taylor polynomial of order n of f at x_0.

There is a very common technique used to handle the remainder in Taylor expansions in an efficient yet intuitive way. It is due to Landau, and it was introduced in Sect. 5.3 of Chap. 5. Recall that $\mathscr{F}_{x_0}$ denotes the set of functions that are defined in a punctured neighborhood of x_0 and that if $g \in \mathscr{F}_{x_0}$, then

$$o(g) = \left\{ f \in \mathscr{F}_{x_0} : f \prec g \right\}.$$

As already remarked in Chap. 5, with some abuse of notation it is customary to write $f = o(g)$ instead of $f \in o(g)$. Using Landau's notation, Peano's formula (8.1) becomes

$$f(x) = p_n(x) + o\left((x - x_0)^n\right).$$

There are alternative notations. A natural one is just the explicit unraveling of the former according to Definition 5.3, namely

$$f(x) = p_n(x) + (x - x_0)^n \omega(x),$$

where $\omega : (a, b) \to \mathbb{R}$ is a function that tends to 0 as $x \to x_0$. Yet another popular notation is

$$f(x) = p_n(x) + \omega(x),$$

where this time $\omega(x)/(x - x_0)^n \to 0$ as $x \to x_0$.

Landau's notation requires a little care. Since $f = o(g)$ really means $f \in o(g)$, some equalities of common use must be clarified. For example, $o(f) = o(g)$ is a set-theoretic equality, which states that the functions that are negligible with respect to f are the same as those that are negligible with respect to g. Further, the sum $o(f) + o(g)$ is the set of sums $\varphi + \psi$ with $\varphi \in o(f)$ and $\psi \in o(g)$, and similarly for $o(f)o(g)$ or for $\lambda o(g)$, where λ is either a real number or a fixed function. Also, $o(o(g))$ is the set of functions that are negligible with respect to all the functions that are negligible with respect to g. With this understanding, one can simplify notation by means of the proposition that follows.

Proposition 8.3 (The algebra of little-o) *Take a function $g : (a, b) \to \mathbb{R}$ and assume that it does not vanish in a punctured neighborhood of $x_0 \in (a, b)$ but tends to zero as $x \to x_0$. Then, for $x \to x_0$:*

(i) $o(\lambda g) = o(g)$, for every $\lambda \in \mathbb{R} \setminus \{0\}$;
(ii) $\varphi(x)o(g) = o(g)$, for every $\varphi : (a, b) \to \mathbb{R}$ bounded in a neighborhood of x_0.

If n and m are non negative integers, the following hold:

(iii) $o\left((x - x_0)^n\right) \pm o\left((x - x_0)^m\right) = o\left((x - x_0)^p\right)$, with $p = \min\{n, m\}$;
(iv) $o\left((x - x_0)^n\right) o\left((x - x_0)^m\right) = o\left((x - x_0)^{n+m}\right)$;
(v) $\left[o\left((x - x_0)^n\right)\right]^m = o\left((x - x_0)^{mn}\right)$;
(vi) $(x - x_0)^n o\left((x - x_0)^m\right) = o\left((x - x_0)^{n+m}\right)$;
(vii) $o\left[o\left((x - x_0)^n\right)\right] = o\left((x - x_0)^n\right)$;
(viii) if $f = o\left((x - x_0)^n\right)$, with $n > 0$, then $f = o\left((x - x_0)^m\right)$ for every $0 \le m < n$.

One of the most useful properties of Taylor expansions is that they behave well under composition: if we know the Taylor expansions of two functions, we may compute a Taylor expansion of their composition, in the precise sense described next.

Theorem 8.3 *Suppose that:*

(i) $f : (a, b) \to (c, d)$ is n times differentiable at x_0;
(ii) $g : (c, d) \to \mathbb{R}$ is m times differentiable at $y_0 = f(x_0)$.

Denote by P the Taylor polynomial of order n of f at x_0 and by Q the Taylor polynomial of order m of g at $y_0 = f(x_0)$ and write

$$Q \circ P(x) = \sum_{k=0}^{nm} c_k (x - x_0)^k.$$

Then, the polynomial

$$T(x) = \sum_{k=0}^{\mu} c_k (x - x_0)^k,$$

where $\mu = \min\{n, m\}$, is the Taylor polynomial of order μ of $g \circ f$ at x_0.

For the reader's convenience, the most common McLaurin expansions are listed in Appendix B and taken for granted in what follows.

8.2 Guided Exercises on Taylor Expansions

8.1 Compute the second order Taylor polynomial of $f(x) = 2(\log x)^2 - \log x$ at the point $x_0 = 1$.

Answer. Taking derivatives, we have that

$$f'(x) = \frac{4\log x}{x} - \frac{1}{x} = \frac{4\log x - 1}{x}, \qquad f''(x) = \frac{\frac{4}{x}x - (4\log x - 1)}{x^2} = \frac{5 - 4\log x}{x^2}.$$

Hence, $f(1) = 0$, $f'(1) = -1$ and $f''(1) = 5$, so that the second order Taylor polynomial of f at $x_0 = 1$ is $-(x - 1) + 5(x - 1)^2/2$.

This is a most basic exercise, where a Taylor expansion is computed directly by evaluating the necessary number of derivatives at the assigned point.

8.2 Consider the function $f(x) = x \sin^2 x - \lambda x^3$.

(a) Determine the order with which f vanishes at $x_0 = 0$ as the parameter λ ranges in $\mathbb{R}$.
(b) Compute $f^{(3)}(0)$ and $f^{(5)}(0)$.

Answer. (a) From the expansion $\sin x = x - x^3/6 + \omega_1(x)$ with $\omega_1(x)/x^4 \to 0$, we get

$$\sin^2 x = x^2 - \frac{1}{3}x^4 + \omega_2(x)$$

with $\omega_2(x)/x^4 \to 0$. It follows that

$$x \sin^2 x - \lambda x^3 = (1 - \lambda)x^3 - \frac{1}{3}x^5 + \omega_3(x)$$

with $\omega_3(x)/x^5 \to 0$. Therefore, if $\lambda \neq 1$, then f vanishes with order 3, whereas if $\lambda = 1$, then f vanishes with order 5.

(b) The previous computation shows that $(1 - \lambda)x^3 - (1/3)x^5$ is the fifth order McLaurin polynomial of f, so that

$$f^{(3)}(0) = 3!(1 - \lambda) = 6(1 - \lambda), \qquad f^{(5)}(0) = 5!(-\frac{1}{3}) = -40.$$

This is a standard exercise in which Taylor expansions are used to determine the order with which a function vanishes: it is just the degree of the first non zero term of the polynomial. One observation: when computing the expansion of $\sin^2 x$, one simply takes the square of the expansion of $\sin x$ and then keeps only the first terms, collecting the higher order ones into the remainder ω_2. Finally, one easy but crucial remark: Taylor expansions centered at x_0 contain informations about the derivatives at x_0: up to factorials, they are just the coefficients of the polynomial.

8.3 Determine the order with which $f(x) = \sin x^2 - \sin^2 x$ vanishes at $x_0 = 0$.

Answer. The bisection trigonometric formula and the McLaurin expansions of $\sin x$ and $\cos x$ imply

$$
\begin{aligned}
f(x) &= \sin x^2 - \sin^2 x \\
&= \sin x^2 - \frac{1 - \cos 2x}{2} \\
&= x^2 - \frac{x^6}{6} + x^8 \omega_1(x^2) - \frac{1}{2}[1 - (1 - 2x^2 + \frac{2}{3}x^4 + x^5 \omega_2(x))] \\
&= x^2 - \frac{x^6}{6} + x^8 \omega_1(x^2) - x^2 - \frac{1}{3}x^4 - \frac{x^5}{2}\omega_2(x) \\
&= -\frac{1}{3}x^4 + x^5 \omega_3(x)
\end{aligned}
$$

with $\lim_{x \to 0} \omega_i(x) = 0$, $i = 1, 2, 3$. Hence f tends to zero of order 4.

This very standard exercise is similar to the previous one. Notice that, in order to expand $\sin^2 x$, this time trigonometry is used instead of taking squares and selecting powers.

8.4 Compute the order with which $(x/2 - \sin x)^2 - \tan(\lambda x^2) + 2\lambda x^4/3$ tends to zero as $x \to 0$ as the parameter λ ranges in $\mathbb{R}$.

Answer. Start from the expansions:

$$\sin x = x - \frac{1}{6}x^3 + \frac{1}{120}x^5 + o(x^6), \qquad \tan y = y + \frac{1}{3}y^3 + o(y^4).$$

Taking squares and summing, it follows that

$$\left(\frac{x}{2} - \sin x\right)^2 - \tan(\lambda x^2) + \frac{2}{3}\lambda x^4$$

$$= \left(\frac{x}{2} - \left(x - \frac{1}{6}x^3 + \frac{1}{120}x^5 + o\left(x^6\right)\right)\right)^2 - \left(\lambda x^2 + \frac{1}{3}\lambda^3 x^6 + o\left(x^8\right)\right) + \frac{2}{3}\lambda x^4$$

$$= \left(-\frac{x}{2} + \frac{1}{6}x^3 - \frac{1}{120}x^5 + o\left(x^6\right)\right)^2 - \lambda x^2 + \frac{2}{3}\lambda x^4 - \frac{1}{3}\lambda^3 x^6 + o\left(x^8\right)$$

$$= \frac{1}{4}x^2 + \frac{1}{36}x^6 - \frac{1}{6}x^4 + \frac{1}{120}x^6 - \lambda x^2 + \frac{2}{3}\lambda x^4 - \frac{1}{3}\lambda^3 x^6 + o\left(x^7\right)$$

$$= \left(\frac{1}{4} - \lambda\right)x^2 + \left(\frac{4\lambda - 1}{6}\right)x^4 + \left(\frac{13 - 120\lambda^3}{360}\right)x^6 + o\left(x^7\right).$$

Hence, for $\lambda \neq 1/4$, the order is 2, whereas for $\lambda = 1/4$ the order is 6.

Again, this is a routine exercise, where the parameter affects the value of the coefficients of a local expansion. One has to detect which is the first, that is, the lowest, non vanishing coefficient as the parameter varies. This time Landau's notation is faster.

8.5 Compute the order with which

$$g(x) = \log(\cos x) + \alpha x^2$$

tends to zero as $x \to 0$ for all values of the parameter $\alpha \in \mathbb{R}$.

Answer. From

$$\cos x - 1 = -\frac{1}{2}x^2 + \frac{1}{4!}x^4 + o(x^5) = -\frac{1}{2}x^2 + o(x^3)$$

$$\log(1 + y) = y - \frac{1}{2}y^2 + \frac{1}{3}y^3 - \frac{1}{4}y^4 + o(y^4)$$

it follows in particular that $\cos x - 1 = o(x)$. By substitution,

$$\log(\cos x) = \log(1 + (\cos x - 1))$$

$$= (\cos x - 1) - \frac{1}{2}(\cos x - 1)^2 + \frac{1}{3}(\cos x - 1)^3 - \frac{1}{4}(\cos x - 1)^4 + o((\cos x - 1)^4)$$

$$= \left(-\frac{1}{2}x^2 + \frac{1}{4!}x^4 + o(x^5)\right) - \frac{1}{2}\left(-\frac{1}{2}x^2 + o(x^3)\right)^2$$

$$+ \frac{1}{3}\left(-\frac{1}{2}x^2 + o(x^3)\right)^3 - \frac{1}{4}\left(-\frac{1}{2}x^2 + o(x^3)\right)^4 + o(x^8)$$

$$= -\frac{1}{2}x^2 - \frac{1}{12}x^4 + o(x^4).$$

Hence

$$g(x) = \left(\alpha - \frac{1}{2}\right) x^2 - \frac{1}{12} x^4 + o(x^4).$$

Therefore, for $\alpha \neq 1/2$, g vanishes of order 2, while for $\alpha = 1/2$, it vanishes of order 4.

This is very similar to the previous one. It also illustrates how to compute the Taylor expansion of a non trivial composition of functions, in this case $\log(\cos x)$.

8.6 Consider the function $f(x) = x^{\log x}$.

(a) Find the domain of f.
(b) Compute the second order Taylor polynomial of f at $x_0 = 1$.

Answer. (a) In order for $\log x$ to be defined, we must have $x > 0$. In this case, x^y is well defined for every real number y. In particular, $x^{\log x}$ makes sense if $x > 0$. Therefore, the domain of f is $(0, +\infty)$.
(b) Recall that whenever g is a strictly positive function on $I \subset \mathbb{R}$, the exponential formula

$$g(x)^{h(x)} = e^{h(x) \log(g(x))} \tag{8.2}$$

holds for every function $h : I \to \mathbb{R}$. Hence $f(x) = e^{\log^2 x}$ and therefore

$$f'(x) = e^{\log^2 x} \frac{2 \log x}{x}$$

$$f''(x) = e^{\log^2 x} \left(\frac{4 \log^2 x}{x^2} + 2 \frac{\frac{1}{x} x - \log x}{x^2}\right) = 2 \frac{e^{\log^2 x}}{x^2} \left(2 \log^2 x - \log x + 1\right).$$

It follows that $f(1) = 1$, $f'(1) = 0$ and $f''(1) = 2$, so that the desired polynomial is $p_2(x) = 1 + (x - 1)^2$.

The function of which one has to compute the Taylor polynomial is not easily written as a composition or sum of functions for which known expansions are available. Furthermore, since only two derivatives are needed, any cumbersome algebraic manipulation is not advisable, and a direct computation is safer.

8.7 Compute the sixth order McLaurin polynomial of the function $\sqrt[4]{1 - x^2}$.

Answer. The known expansion for $\sqrt{1 + x}$ implies

$$\sqrt{1 - x} = 1 - \frac{1}{2} x - \frac{1}{8} x^2 - \frac{1}{16} x^3 + o(x^3) \tag{8.3}$$

and hence

$$\sqrt{1 - x^2} = 1 - \frac{1}{2} x^2 - \frac{1}{8} x^4 - \frac{1}{16} x^6 + o(x^6) = 1 - \left(\frac{1}{2} x^2 + \frac{1}{8} x^4 + \frac{1}{16} x^6 + o(x^6)\right).$$

From (8.3), by composing, it follows

$$\sqrt[4]{1-x^2} = \sqrt{1 - \left(\frac{1}{2}x^2 + \frac{1}{8}x^4 + \frac{1}{16}x^6 + o(x^6)\right)}$$

$$= 1 - \frac{1}{2}\left(\frac{1}{2}x^2 + \frac{1}{8}x^4 + \frac{1}{16}x^6\right) - \frac{1}{8}\left(\frac{1}{2}x^2 + \frac{1}{8}x^4 + \frac{1}{16}x^6\right)^2$$

$$- \frac{1}{16}\left(\frac{1}{2}x^2 + \frac{1}{8}x^4 + \frac{1}{16}x^6\right)^3 + o(x^6)$$

$$= 1 - \frac{1}{4}x^2 - \frac{1}{16}x^4 - \frac{1}{32}x^6 - \frac{1}{8}\left(\frac{1}{4}x^4 + 2\cdot\frac{1}{2}x^2\cdot\frac{1}{8}x^4\right) - \frac{1}{16}\left(\frac{1}{8}x^6\right) + o(x^6)$$

$$= 1 - \frac{1}{4}x^2 - \frac{3}{32}x^4 - \frac{7}{128}x^6 + o(x^6).$$

The polynomial is therefore $p_6(x) = 1 - (1/4)x^2 - (3/32)x^4 - (7/128)x^6$.

This exercise is an application of the composition rule for Taylor polynomials. Since the target is the polynomial, and no estimate of the remainder is required, no particular form of the latter is needed. In this case, Landau's notation is perhaps the fastest.

8.8 (a) Write the fifth order McLaurin expansion of $f(x) = x - \arctan x$ and compute the order with which $\sqrt[5]{f(x)}$ vanishes at $x_0 = 0$.

(b) Compute $\displaystyle\lim_{x\to 0^+} \frac{\log \sqrt[5]{f(x)}}{x^2 + x}$.

(c) Compute $\displaystyle\lim_{x\to 0^+} \frac{x^\alpha \log \sqrt[5]{f(x)}}{x^2 + x}$ as α ranges in $\mathbb{R}$.

Answer. (a) From the expansion

$$\arctan x = x - \frac{1}{3}x^3 + \frac{1}{5}x^5 + o\left(x^6\right)$$

it follows immediately that

$$f(x) = x - \arctan x = \frac{1}{3}x^3 - \frac{1}{5}x^5 + o\left(x^6\right).$$

Now, f is positive to the right of $x_0 = 0$ (because $\arctan x \le x$ for $x > 0$) and vanishes of order 3. Therefore, its root of order 5 vanishes of order 3/5.

(b) The function $x \mapsto \log \sqrt[5]{f(x)} = (1/5)\log(f(x))$ tends to $-\infty$ as $x \to 0^+$ because $f(x) \to 0^+$, whereas $x \mapsto 1/(x^2+x)$ tends to $+\infty$. Therefore, their product tends to $-\infty$.

(c) Write

$$\frac{x^\alpha \log \sqrt[5]{f(x)}}{x^2 + x} = \frac{1}{5}\frac{1}{1+x}\left\{x^{\alpha-1}\log\left(\frac{1}{3}x^3 + o\left(x^4\right)\right)\right\}.$$

Since the first factor tends to $1/5$, only the term in curly brackets is relevant. From the limit $y^\varepsilon \log y \to 0$ for $y \to 0^+$ (valid for all $\varepsilon > 0$) it follows that it tends to zero whenever $\alpha - 1 > 0$. Hence the limit is zero if $\alpha > 1$. If $\alpha = 1$, then the term in curly brackets reduces to the logarithm alone and hence tends to $-\infty$ as $x \to 0^+$, while for $\alpha < 1$ it is the product of the logarithm (which goes to $-\infty$ as $x \to 0^+$) and of $x^{\alpha-1}$ (which goes to $+\infty$ as $x \to 0^+$). Hence, for $\alpha \le 1$ the limit is $-\infty$.

This is a standard exercise on the use of Taylor expansions in the calculation of limits.

8.9 For all values of the parameter $a > 0$, compute the limit

$$\lim_{x \to 0^+} \frac{x^x - \cos x - x \log x}{\tan x^a - \arctan x^a},$$

whenever it exits.

Answer. Both numerator and denominator tend to zero as $x \to 0^+$. The main goal here is to obtain precise informations on their orders. The exponential form (8.2) yields

$$g(x) = x^x - \cos x - x \log x = e^{x \log x} - x \log x - \cos x.$$

Observe that $x \log x \to 0$ as $x \to 0^+$, so that we can use the McLaurin expansion of order two of the exponential function and obtain

$$e^{x \log x} - x \log x = 1 + \frac{1}{2}x^2 \log^2 x + x^2 \log^2 x \cdot \omega(x \log x)$$

with $\omega(y) \to 0$ for $y \to 0$. Adding the expansion of $\cos x$ it then follows

$$g(x) = 1 + \frac{1}{2}x^2 \log^2 x + x^2 \log^2 x \cdot \omega_0(x \log x) - 1 + \frac{x^2}{2} - x^2 \cdot \omega_1(x)$$

$$= x^2 \log^2 x \left(\frac{1}{2} + \omega_2(x)\right)$$

where $\omega_i(y) \to 0$ for $y \to 0$. Here the fact that x^2 tends to zero faster than $x^2 \log^2 x$ was used. Consider next the denominator. By assumption $x > 0$, so that x^a is well defined for every $a > 0$. The following step is to compute the McLaurin expansion of the denominator, using those of the functions $\tan x$ and $\arctan x$. Thus

$$\tan x^a - \arctan x^a = x^a + \frac{1}{3}x^{3a} + x^{3a} \cdot \omega_3(x^a) - x^a + \frac{1}{3}x^{3a} - x^{3a} \cdot \omega_4(x^a)$$

$$= x^{3a}\left(\frac{2}{3} + \omega_5(x^a)\right),$$

where as usual $\omega_i(x) \to 0$ for $x \to 0$. Therefore

$$\frac{x^x - \cos x - x\log x}{\tan x^a - \arctan x^a} = x^{2-3a}\log^2 x \frac{1/2 + \omega_2(x)}{2/3 + \omega_5(x^a)}$$

and the conclusion is

$$\lim_{x\to 0^+}\frac{x^x - \cos x - x\log x}{\tan x^a - \arctan x^a} = \begin{cases} 0 & a < \frac{2}{3} \\ +\infty & a \geq \frac{2}{3}. \end{cases}$$

The key idea for this exercise is to use the McLaurin expansion of the exponential in order to get a simplified form for $e^{x\log x} - x\log x$. A plain use of McLaurin expansions for each summand of the numerator is impossible in this case. Indeed, neither x^x nor $x\log x$ admit such an expansion because they are not even defined at zero, nor in a left neighborhood of it. The reader is urged to check that a finer approximation of the exponential to order three, or higher, would have given no further useful information.

8.10 Suppose that f is defined and differentiable at least five times in $\mathbb{R}$ and assume that: $f(0) = 0$, $f^{(1)}(0) = f^{(2)}(0) = f^{(3)}(0) = 0$, $f^{(4)}(0) = 1$ and $|f^{(5)}(x)| \leq 2 + x^2$ for every $x \in \mathbb{R}$. Find a neighborhood of $x_0 = 0$ in which the fourth order McLaurin polynomial of f approximates f with an error less than or equal to 10^{-5}.

Answer. Lagrange's formula, for $x \in [-\delta, \delta]$ reads:

$$f(x) = f(0) + f^{(1)}(0)x + \frac{f^{(2)}(0)}{2}x^2 + \frac{f^{(3)}(0)}{6}x^3 + \frac{f^{(4)}(0)}{24}x^4 + \frac{f^{(5)}(c)}{5!}x^5$$

$$= 1 + \frac{x^4}{24} + \frac{f^{(5)}(c)}{5!}x^5$$

for some $c \in \mathbb{R}$ satisfying $0 \leq |c| \leq |x| \leq \delta$. Hence, when approximating $f(x)$ with its McLaurin polynomial $p_4(x) = 1 + x^4/24$ the error at x is:

$$E(x) = |f(x) - p_4(x)| = \left|\frac{f^{(5)}(c)}{120}x^5\right|. \tag{8.4}$$

Consequently, provided that $\delta = 10^{-1}$, the error in the interval $[-\delta, \delta]$ is

$$E = \sup{}_{x\in[-\delta,\delta]}E(x) = \sup{}_{x\in[-\delta,\delta]}\frac{2+c^2}{120}|x|^5 \leq \frac{2+\delta^2}{120}\delta^5 \leq \frac{1}{40}\delta^5.$$

Hence $p_4(x)$ approximates $f(x)$ in $[-1/10, 1/10]$ with an error smaller than 10^{-5}.

This exercise is based on the remainder formula of Lagrange, as many others in which a quantitative estimate on the approximation is required. The general formula (8.4) suggests that if some information on the (fifth) derivative is known, then this can be used to achieve the desired estimate. Indeed, the derivative depends on c, which satisfies $|c| < |x|$, and $|x|$ is bounded by a small positive δ that we are allowed to choose.

8.11 Let $f(x) = \sin(4x^2 - 1)/(2x^2 - x)$. Find a polynomial $P(x)$ of degree two and a neighborhood I of $x_0 = 1/2$ such that $|f(x) - P(x)| < 10^{-2}$ for every $x \in I \setminus \{1/2\}$.

Answer. The domain of f is

$$\mathrm{Dom}(f) = \{x \in \mathbb{R} : 2x^2 - x \neq 0\} = \mathbb{R} \setminus \{0, \tfrac{1}{2}\}.$$

For simplicity, it is enough to consider symmetric neighborhoods I around x_0, that is, of the form $I = (1/2 - \delta, 1/2 + \delta)$ with $0 < \delta < 1/2$. Put next $h(x) = \sin(4x^2 - 1)$ and $g(x) = 1/x$, so that

$$f(x) = \frac{h(x)g(x)}{2x - 1}.$$

Next, it is advisable to write the numerator using the second order Taylor expansions of h and g at x_0 and express the remainder in Lagrange's form. As far as h is concerned, the McLaurin expansion of the sine function can be used because its argument tends to zero as $x \to x_0$. Hence, for any fixed $t \in \mathbb{R}$ and $x > 0$, there exist $c_1, c_2 \in \mathbb{R}$ such that $0 < |c_1| < |t|, 0 < |c_2 - 1/2| < |x - 1/2|$ for which

$$\sin t = t - \frac{t^2}{2} \sin c_1$$

$$g(x) = g\left(\frac{1}{2}\right) + g'\left(\frac{1}{2}\right)\left(x - \frac{1}{2}\right) + \frac{g''(c_2)}{2}\left(x - \frac{1}{2}\right)^2.$$

Therefore

$$h(x) = 4x^2 - 1 - \frac{1}{2}\left(4x^2 - 1\right)^2 \sin c_1$$

$$g(x) = 2 - 4\left(x - \frac{1}{2}\right) + \frac{1}{c_2^3}\left(x - \frac{1}{2}\right)^2.$$

It follows that upon writing $h(x)/(2x - 1) = p_1(x) + R_1(x)$ and $g(x) = p_2(x) + R_2(x)$ with

$$p_1(x) = 2x + 1, \qquad p_2(x) = 2 - 4\left(x - \frac{1}{2}\right),$$

then $f(x) = [h(x)/(2x - 1)] \cdot g(x)$ takes the form

$$f(x) = p_1(x) \cdot p_2(x) + p_1(x)R_2(x) + p_2(x)R_1(x) + R_1(x)R_2(x).$$

Here $p(x) = p_1(x) \cdot p_2(x)$ is the required second order polynomial. The error in approximating f with p is evidently

$$|f - p| = |p_1 R_2 + p_2 R_1 + R_1 R_2|.$$

In order to estimate this error as $x \in I$, it is useful to impose an *a priori* bound on δ. For example, choose $\delta < 1/4$. In this case $I \subset (1/4, 3/4)$, so that if $x \in I$, then

$$|p_1(x)| = |2x + 1| \leq \frac{5}{2}$$

$$|p_2(x)| = |2 - 4(x - \frac{1}{2})| \leq 2 + 4\delta \leq 3$$

$$|R_1(x)| = |-\frac{1}{2}(2x + 1)(4x^2 - 1)\sin c_1| \leq |2x + 1|^2 |x - \frac{1}{2}| \leq \frac{25}{4} \cdot \delta$$

$$|R_2(x) \leq |\frac{1}{c_2^3}(x - \frac{1}{2})^2| \leq 4^3\delta^2.$$

Hence, if $x \in I$, then

$$|f(x) - p(x)| \leq \frac{5}{2}4^3\delta^2 + 3\frac{25}{4}\delta + 4^2 \cdot 25\delta^3 \leq 580\delta$$

because $\delta^3 < \delta^2 < \delta$. Finally $580\delta < 10^{-2}$ if $\delta < 10^{-2}/580$. In conclusion, the desired estimate $|f(x) - p(x)| < 10^{-2}$ for every $x \in I \setminus \{1/2\}$ holds provided that

$$\delta < \min\left\{\frac{1}{4}, \frac{1}{58}10^{-3}\right\} = \frac{1}{58}10^{-3},$$

where $p(x) = p_1(x)p_2(x) = -8x^2 + 4x + 4$.

This exercise has two main difficulties. The first is that it is required to approximate a function in a punctured neighborhood of a point that is not in the domain of the function, namely $x_0 = 1/2$. At x_0, however, the numerator vanishes with a higher order than the denominator, as the Taylor expansion reveals. For this reason the ratio $h(x)/(2x - 1)$ plays a prominent role. The second is of more technical nature, and consists in finding a way to determine δ small enough so that the required accuracy is met. The strategy is to estimate all the powers of $(x - 1/2)$ by δ because $\delta < 1$, and hence $\delta^n < \delta$. In the end, one gets a multiple of δ, and it is therefore obvious how to select δ in such a way that this multiple is of the desired magnitude.

8.3 Problems on Taylor Expansions

8.12 Compute the second order McLaurin polynomial of $f(x) = \log(1 + \tan x)$.

8.13 Compute the third order Taylor polynomial of $f(x) = (x - \log x)^2$ at the point $x_0 = 1$.

8.14 Compute the fourth order McLaurin polynomial of $f(x) = \log(1 + \sin x) - (11/12)\tan^4 x$.

8.15 Compute the McLaurin polynomial of the indicated order of the functions:

$$f(x) = \arctan\left(e^x - \cos x\right) \tag{4}$$

$$g(x) = \sqrt{1 + \tan(2x)} \tag{3}$$

$$h(x) = \cos\left(\sqrt{1+x} - 1\right) \tag{4}$$

$$k(x) = \cos(\sin x) - e^{\sin^2 x} \tag{6}$$

$$\ell(x) = e^x/(1 - x^2) \tag{3}$$

$$m(x) = \sqrt{\cos x} - 1. \tag{4}$$

8.16 Compute the order with which each of the functions

$$f(x) = \tan(\lambda x) - \log(1 + \sin x)$$

$$g(x) = (\sin x)\log(1 + x^2) - \lambda x^3$$

$$h(x) = \log(\cos x) + \lambda x^2$$

$$k(x) = (\sin(\lambda x))^2 - \sin(x^2)$$

$$\ell(x) = \log(\cos(\lambda x)) + x^2$$

$$m(x) = 1 - (\cos 2x)^2 - \lambda(\sin x)^4$$

$$n(x) = \log(1 + x) - \sin x + e^{x^2/2} - 1 - \lambda x^3$$

vanishes at $x_0 = 0$, for all values of $\lambda \in \mathbb{R}$.

8.17 Let $f(x) = e^{1-\cos x} - (1 + (1/8)x^2)$.

(a) Compute the order with which f vanishes at $x_0 = 0$.
(b) Compute $\lim\limits_{x \to 0} f(x)/(1 - \cos x)$.

8.18 Let $f(x) = \log(1 + x^2) + \cos(\lambda x) - 1$, with $\lambda \in \mathbb{R}$.

(a) Compute the order with which f vanishes at $x_0 = 0$, for all values of $\lambda \in \mathbb{R}$.
(b) Compute $\lim\limits_{x \to 0} f(x)/x^2$ for $\lambda = 1$.

8.19 Let $f(x) = \sin x - x \cos x + \lambda x^3$, with $\lambda \in \mathbb{R}$.

(a) Compute the order with which f vanishes at $x_0 = 0$, for all values of $\lambda \in \mathbb{R}$.
(b) Compute $\lim\limits_{x \to 0} f(x)/x^3$ for $\lambda = 2$.

8.20 Compute the following limits as the parameter λ varies in $\mathbb{R}$

$$\lim_{x \to 0} \frac{x - \tan x + e^{\lambda x^2/2} - \cos \lambda x}{x^3 - \lambda x^2}$$

$$\lim_{x \to 0^+} \frac{e^x - \cos\left(\lambda\sqrt{x}\right)}{\sqrt{1 + \sin(\lambda x)} - 1}$$

$$\lim_{x \to 0} \frac{e^{1-\cos(\lambda x)} - 1}{x^2}$$

$$\lim_{x \to 0^+} \frac{\log(1 + x^2) - \cos x + e^{x^2} + \lambda x^2}{x^4}.$$

8.21 Compute, if it exists, the limit

$$\lim_{x \to 0^+} \left(\frac{\sin x}{x}\right)^{1/(\cos x - 1)}.$$

8.22 Let $f(x) = |e^x - \cos x|^\alpha / x$. Determine for which values of the parameter $\alpha \in \mathbb{R}$ the function f admits a finite limit ℓ at $x_0 = 0$ and, in the affirmative cases, decide if the continuous function, obtained by extending f with the value ℓ at $x_0 = 0$, is also differentiable.

8.23 Let $f(x) = \sqrt{1 - x^2}$. Find a polynomial $P(x)$ and a neighborhood I of $x_0 = 0$ in such a way that $P(x)$ approximates f in I with an error less than or equal to 10^{-3}.

8.24 Let $f(x) = x^3 \arctan(1/x)$. Find a polynomial $P(x)$ such that $|f(x) - P(x)| < 10^{-2}$ for every $x \geq 10$.

8.25 Let

$$f(x) = \begin{cases} \dfrac{|e^x - 1 - x - (x^2/2)|^\alpha}{x} & x \neq 0 \\ \beta & x = 0. \end{cases}$$

Determine for which values of the parameters $\alpha, \beta \in \mathbb{R}$, the function f is continuous at $x_0 = 0$ and for which values it is also differentiable at x_0.

8.26 Decide if the function $f(x) = \sqrt[3]{\cos(x^2)} - e^{x^3}$ is differentiable at $x_0 = 0$ and, in the affirmative case, write its first order McLaurin polynomial.

8.27 Is there a continuous extension of the function $f(x) = (\log(x - \sin x))^{-1}$ at $x_0 = 0$ which is also differentiable at that point?

8.28 Find a rational number that approximates $\log(\pi/2)$ with an error less than 10^{-2}.

8.29 Find a rational number that approximates $\sqrt{11}$ with an error less than 10^{-2}.

8.30 Consider the function

$$f(x) = \begin{cases} \cos x & x > 0 \\ 1 - \dfrac{x^2}{2} + x^4 & x \leq 0. \end{cases}$$

Find a polynomial P of degree four and an interval $I = (-\delta, \delta)$, with $\delta > 10^{-1}$, such that $|f(x) - P(x)| < 10^{-4}$ for every $x \in I$.

8.31 Find a polynomial P such that $|P(x) - \cos(x - \sin x)| < 4 \cdot 10^{-5}$ for every $x \in [-1/2, 1/2]$.

8.32 Determine where the function

$$f(x) = \begin{cases} \cos \sqrt{x} - \dfrac{\sin \sqrt{x}}{\sqrt{x}} & x > 0 \\ 0 & x \le 0 \end{cases}$$

is differentiable.

8.33 Let $f(x) = \sin(\log x) - \log x$. Find $a, b, c, d \in \mathbb{R}$ such that

$$\lim_{x \to 1} \frac{f(x) + ax^3 + bx^2 + cx + d}{(x-1)^3} = 0.$$

8.34 Consider the function

$$f(x) = \begin{cases} \dfrac{e^{x^2} - e^x}{x - 1} & x \ne 1 \\ k & x = 1, \end{cases}$$

where k is a real parameter. Write the second order Taylor polynomial centered at $x_0 = 1$ of the function f for those values of $k \in \mathbb{R}$ for which it exists.

Chapter 9
The Geometry of Functions

This Chapter is devoted to the study of geometric properties of functions, namely properties that reflect, and are reflected in, the shape of the graph of a given function $f : I \to \mathbb{R}$, such as the presence of *asymptotes*, the *convexity* or *concavity* of f, the presence of local or global extreme points or of *inflection points*.

9.1 Asymptotes

Roughly speaking, an asymptote is a straight line to which the graph of f becomes arbitrary close. Asymptotes, or *asymptotic lines*, are classically divided in two categories, the *vertical asymptotes* and the *oblique asymptotes*, the latter including the *horizontal asymptotes*. The definitions are as follows:

Definition 9.1 (*Asymptotes*) Consider the function $f : I \to \mathbb{R}$.

(i) The vertical line $x = x_0$ is called a veritical asymptote for f if $x_0 \in \mathbb{R}$ is a limit point of I and if:

- either $\lim\limits_{x \to x_0^-} f(x) = \pm\infty$
- or $\lim\limits_{x \to x_0^+} f(x) = \pm\infty$.

(ii) The straight line with equation $y = mx + n$ is an oblique asymptote for f at $+\infty$ if $+\infty$ is a limit point of I and if

$$\lim_{x \to +\infty} [f(x) - (mx + n)] = 0,$$

and similarly at $-\infty$. If the straight line with equation $y = n$ is an oblique asymptote, that is, if $m = 0$, then the asymptote is called a horizontal asymptote.

© Springer International Publishing Switzerland 2016
M. Baronti et al., *Calculus Problems*, UNITEXT - La Matematica per il 3+2 101,
DOI 10.1007/978-3-319-15428-2_9

A trivial example exhibiting both vertical and horizontal asymptotes is the function
$f(x) = 1/x$. Indeed, $f(x) \to \pm\infty$ for $x \to 0^{\pm}$, so that $x = 0$ is a vertical asymptote,
and $f(x) \to 0$ as $x \to \pm\infty$, so that the line $y = 0$ is a horizontal asymptote for f,
both at $+\infty$ and at $-\infty$. A function with infinitely many vertical asymptotes is
$f(x) = \tan x$ and a function with different horizontal asymptotes at $+\infty$ and $-\infty$
is $f(x) = \arctan x$.

While the search for vertical or horizontal asymptotes is a rather straightforward
issue, the determination of (properly) oblique asymptotes requires a little work, and
can be performed with the help of the result that follows.

Proposition 9.1 *Consider the function* $f : I \to \mathbb{R}$.

 (i) *Suppose that the straight line with equation* $y = mx + n$ *is an oblique asymptote
    for f at $+\infty$. Then*

$$\lim_{x \to +\infty} \frac{f(x)}{x} = m,$$

 and

$$\lim_{x \to +\infty} (f(x) - mx) = n,$$

 and similarly if the asymptote is at $-\infty$.
 (ii) *Conversely, if there exists $m \in \mathbb{R}$ such that*

$$\lim_{x \to +\infty} \frac{f(x)}{x} = m,$$

 and if there exists $n \in \mathbb{R}$ such that

$$\lim_{x \to +\infty} (f(x) - mx) = n,$$

 *then the straight line with equation $y = mx + n$ is an oblique asymptote for f
 at $+\infty$, and a similar statement holds at $-\infty$.*

Observe that both conditions listed in (ii) of the above proposition must be satisfied
in order for $y = mx + n$ to be an oblique asymptote for f. For example, $f(x) = \log x$
clearly satisfies the first at $+\infty$, but not the second, for $(\log x - 0) \to +\infty$ as
$x \to +\infty$.

The first limiting property, that is $f(x)/x \to m$, says that if f diverges as $x \to$
$\pm\infty$ then it does so with order exactly 1, so this is another way of spelling a necessary
condition for the existence of an oblique asymptote.

A class of functions that often exhibit asymptotes is the rational functions. Indeed,
suppose that

$$f(x) = \frac{P(x)}{Q(x)} = \frac{a_0 x^p + a_1 x^{p-1} + \cdots + a_{p-1} x + a_p}{b_0 x^q + b_1 x^{q-1} + \cdots + b_{q-1} x + b_q}, \qquad a_0 b_0 \neq 0,$$

Table 9.1 Asymptotes at $\pm\infty$ for rational functions

Degree comparison	Type of asymptote	Equation
$p < q$	Horizontal	$y = 0$
$p = q$	Horizontal	$y = \dfrac{a_0}{b_0}$
$p = q + 1$	Oblique	$y = \frac{a_0}{b_0}x + n$ (n is the rest in long division)
$p > q + 1$	No asymptote	Not applicable

where evidently P and Q are the polynomials at the numerator and denominator, respectively. The presence or not of asymptotic lines at $\pm\infty$ is summarized in Table 9.1. In the case when $p = q + 1$, then upon performing the long division, $P(x) = (mx + n)Q(x) + R(x)$ for some polynomial $R(x)$ with degree less than q, and where $m = a_0/b_0$. Hence the rational function $f = P/Q$ is actually of the form $(mx + n) + R(x)Q(x)^{-1}$, and $R(x)Q(x)^{-1} \to 0$ as $x \to \pm\infty$. Evidently, $y = mx + n$ is the asymptote.

9.2 Convexity and Concavity, Inflection Points

A region of the plane, or of the three-dimensional space is convex if every pair of points in it can be joined by a segment completely contained in the region. This notion is applicable to functions $f : I \to \mathbb{R}$ defined on an interval I by looking at their *epigraph*, namely the set of point above the graph, that is

$$\mathrm{epi}(f) = \{(x, y) \in \mathbb{R}^2 : x \in I, \ y \geq f(x)\}.$$

It is easy to see that asking for $\mathrm{epi}(f)$ to be a convex subset of $\mathbb{R}^2$ is equivalent to the definition that follows.

Definition 9.2 (*Convex function*) A function f defined on the interval I is called *convex* in I if for every $x_0, x_1 \in I$

$$f\left((1 - t)x_0 + tx_1\right) \leq (1 - t)f(x_0) + tf(x_1), \qquad t \in [0, 1].$$

The function is called *strictly convex* if the above inequality holds strictly for every $t \in (0, 1)$. The function is called *concave* if the opposite inequality holds, namely if

$$f\left((1 - t)x_0 + tx_1\right) \geq (1 - t)f(x_0) + tf(x_1), \qquad t \in [0, 1]$$

and is *strictly concave* if the above inequality holds strictly for every $t \in (0, 1)$.

Observe that f is convex if and only if $-f$ is concave, and the analogous statement hold for strict convexity. Thus, every statement about convex functions has a counterpart for concave functions. Hence, below only convex functions are treated.

Proposition 9.2 *The function f defined on the interval I is convex if and only if the difference quotient*

$$R_{x_0}(x) = \frac{f(x) - f(x_0)}{x - x_0}, \qquad x \in I \setminus \{x_0\}$$

is increasing for every $x_0 \in I$, and it is strictly convex if and only if all of the above functions are strictly increasing.

Corollary 9.1 *If the function f defined on the interval I is convex, then in every interior point x_0 of I both the left derivative $f'_-(x_0)$ and the right derivative $f'_+(x_0)$ exist and $f'_-(x_0) \le f'_+(x_0)$, and, furthermore, f is continuous at x_0.*

A function can be convex in $[a, b]$ and yet be discontinuous at a or b. For example

$$f(x) = \begin{cases} x^2 & |x| < 1 \\ 2 & x = \pm 1 \end{cases}$$

is defined in $[-1, 1]$, where it is convex; however, f is discontinuous at ± 1.

Theorem 9.1 *Suppose that $f : [a, b] \to \mathbb{R}$ is differentiable in $I = [a, b]$. Then the following are equivalent*

 (i) *f is convex in I;*
 (ii) *f' is increasing in I;*
 (iii) *$f(x) \ge f(y) + f'(y)(x - y)$ for every $x, y \in I$.*

Geometrically speaking, the above theorem states that the tangent line to the graph of a differentiable and convex function lies below the graph of the function. An example of this inequality, for $y = 0$ and $f(x) = e^x$, is

$$e^x \ge 1 + x$$

and is listed in the Appendix. Analogous results hold for strictly convex functions.

Theorem 9.2 *Suppose that $f : [a, b] \to \mathbb{R}$ is differentiable in $I = [a, b]$. Then the following are equivalent*

 (i) *f is strictly convex in I;*
 (ii) *f' is strictly increasing in I;*
 (iii) *$f(x) > f(y) + f'(y)(x - y)$ for every $x, y \in I$ with $x \ne y$.*

More can be said if the function is actually twice differentiable in I.

Theorem 9.3 *Suppose that $f : [a, b] \to \mathbb{R}$ is twice differentiable in $I = [a, b]$. Then*

(i) *f is convex in I if and only if $f''(x) \geq 0$ for every $x \in I$;*
(ii) *f is strictly convex in I if and only if $f''(x) \geq 0$ for every $x \in I$ and the set of zeroes of f'' has no interior points.*

Definition 9.3 (*Inflection point*) A point x_0 that is an interior point of the domain of the function f is called an inflection point for f if there exists $\delta > 0$ such that f is convex in $[x_0 - \delta, x_0]$ and concave in $[x_0, x_0 + \delta]$, or *viceversa*. More precisely, the inflection point is

(i) *ascending* if f is concave in $[x_0 - \delta, x_0]$ and convex in $[x_0, x_0 + \delta]$;
(ii) *descending* if f is convex in $[x_0 - \delta, x_0]$ and concave in $[x_0, x_0 + \delta]$.

Notice that if f is twice differentiable at the inflection point x_0, then $f''(x_0) = 0$. Indeed, by Taylor's expansion

$$f(x) - \left[f(x_0) + f'(x_0)(x - x_0) \right] = \frac{1}{2} f''(x_0)(x - x_0)^2 + o\left((x - x_0)^2 \right)$$

so that if $f''(x_0) \neq 0$, then the left-hand side would have the same sign as that of $f''(x_0)$ in a neighborhood of x_0, contrary to the assumption that x_0 is an inflection point. Conversely, if f is differentiable n times at x_0 and if

$$f''(x_0) = \cdots = f^{(n-1)}(x_0) = 0, \qquad f^{(n)}(x_0) \neq 0,$$

then, by Taylor's theorem

$$f(x) - \left[f(x_0) + f'(x_0)(x - x_0) \right] = \frac{1}{n!} f^{(n)}(x_0)(x - x_0)^n + o\left((x - x_0)^n \right). \quad (9.1)$$

Now, for x sufficiently close to x_0 the right-hand side has the same sign as that of $f^{(n)}(x_0)(x - x_0)^n$. It follows that if n is even, then the sign of the left-hand side is the same for $x < x_0$ and for $x > x_0$, whereas if n is odd, then the sign of the left-hand side for $x < x_0$ and for $x > x_0$ is different. In the latter case x_0 is an inflection point by Theorem 9.3, in the former x_0 is not an inflection point, because f is either convex or concave in a neighborhood of x_0. The situation is summarized in Table 9.2.

Table 9.2 Points for which $f''(x_0) = 0$: $n \geq 3$ is the minimum for which $f^{(n)}(x_0) \neq 0$

Parity of n in (9.1)	Sign of $f^{(n)}(x_0)$	Type of inflection point
Odd ($n \geq 3$)	$f^{(n)}(x_0) > 0$	Ascending inflection point
Odd ($n \geq 3$)	$f^{(n)}(x_0) < 0$	Descending inflection point
Even ($n \geq 4$)	$f^{(n)}(x_0) > 0$	No inflection point: f is convex near x_0
Even ($n \geq 4$)	$f^{(n)}(x_0) < 0$	No inflection point: f is concave near x_0

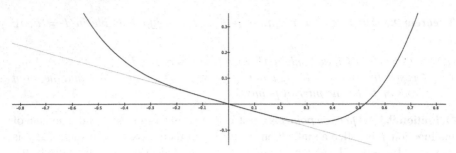

Fig. 9.1 Graph of $f(x) = 2(x^4 - x/7)$ near the origin

The function $f(x) = 2(x^4 - x/7)$, for example, has $f^{(2)}(0) = f^{(3)}(0) = 0$, but $f^{(4)}(0) = 48 > 0$. Thus the origin is not an inflection point, nor is it a local extreme point, because $f'(0) = -2/7$. In Fig. 9.1 the convexity near the origin is visible.

9.3 The Nature of Critical Points

Taylor's expansions can efficiently reveal the nature of a critical point. Suppose that f is differentiable n times at the point x_0 and assume that x_0 is an interior point of its domain. Suppose further that x_0 is a critical point for f, and that in fact

$$f'(x_0) = \cdots = f^{(n-1)}(x_0) = 0, \qquad f^{(n)}(x_0) \neq 0.$$

Then, by Taylor's theorem

$$f(x) - f(x_0) = \frac{1}{n!} f^{(n)}(x_0)(x - x_0)^n + o\left((x - x_0)^n\right) \qquad (9.2)$$

and for x sufficiently close to x_0 the right-hand side has the same sign as that of $f^{(n)}(x_0)(x - x_0)^n$. It follows that if n is even, then the sign of the left-hand side is the same for $x < x_0$ and for $x > x_0$, whereas if n is odd, then the sign of the left-hand side for $x < x_0$ is the opposite of the sign that it has for $x > x_0$. In the former case, x_0 is a local extreme point, in the latter it is not. However, the minimum odd value of n is 3 and hence in particular $f''(x_0) = 0$. Therefore the results of the previous section apply (see Table 9.2) and we may conclude that x_0 is an inflection point. For critical points, the situation is thus described in Table 9.3.

Table 9.3 Critical points: $n \geq 2$ is the order of the first non-zero derivative at x_0

Parity of n in (9.2)	Sign of $f^{(n)}(x_0)$	Type of critical point
Even ($n \geq 2$)	$f^{(n)}(x_0) > 0$	Local maximum
Even ($n \geq 2$)	$f^{(n)}(x_0) < 0$	Local minimum
Odd ($n \geq 3$)	$f^{(n)}(x_0) > 0$	Ascending inflection point
Odd ($n \geq 3$)	$f^{(n)}(x_0) < 0$	Descending inflection point

9.4 Guided Exercises on the Geometry of Functions

9.1 Consider the function $f(x) = x^2 + k \arctan(x + 1)$.

(a) For which $k \in \mathbb{R}$ is $x_0 = 0$ a local maximum or minimum?
(b) For which $k \in \mathbb{R}$ is f convex in a neighborhood of 0?
(c) For which $k \in \mathbb{R}$ is 0 an inflection point?

Answer. Clearly $f \in C^\infty(\mathbb{R})$ for every $k \in \mathbb{R}$ and

$$f'(x) = 2x + \frac{k}{1 + (x+1)^2}, \qquad f''(x) = 2 - \frac{2k(x+1)}{(1 + (x+1)^2)^2},$$

$$f'''(x) = -\frac{2k}{(1 + (x+1)^2)^2} + \frac{8k(x+1)^2}{(1 + (x+1)^2)^3}.$$

It follows that

$$f(0) = k\frac{\pi}{4}, \qquad f'(0) = \frac{k}{2}, \qquad f''(0) = \frac{4 - k}{2}, \qquad f'''(0) = \frac{k}{2}.$$

(a) In order for the origin to be a local extreme point it must be $f'(0) = 0$ and hence $k = 0$. In this case $f(x) = x^2$, so that $x_0 = 0$ is a global minimum.
(b) If $k < 4$, then f is convex in a neighborhood of 0. Indeed, in this case $f''(0) = (4 - k)/2 > 0$ and by permanence of sign $f''(x) > 0$ in a neighborhood of $x_0 = 0$. Similarly, if $k > 4$, then f is concave in a neighborhood of $x_0 = 0$.
(c) If $k = 4$, then $x_0 = 0$ is an ascending inflection point for f because $f''(0) = 0$ and $f'''(0) = 2 > 0$, so that the results in Table 9.2 apply.
This is a basic exercise on the analysis of the local geometry of a function. Since $f \in C^\infty(\mathbb{R})$, one may take as many derivatives as needed, three in the case at hand. For (a), the question is to see when $f'(0) = 0$, and this has a very simple answer. For (b), one has to look at $f''(0)$, which has a very clear dependency on k. Finally, to answer question (c) only the case $k = 4$ is possible, for this is the only case in which $f''(0) = 0$. But then $f'''(0) > 0$, and Table 9.2 gives the answer.

9.2 Determine whether the function $f(x) = x/\sqrt{x+1}$ has local or global extreme points and find the solutions of the equation

$$\frac{x}{\sqrt{x+1}} = \frac{3}{2} + \frac{5}{16}(x - 3).$$

Answer. The domain of f is $(-1, +\infty)$, and $f \in C^\infty((-1, +\infty))$ because it is a composition and ratio of functions in $C^\infty((-1, +\infty))$. Now,

$$f'(x) = \frac{x + 2}{2(x+1)^{3/2}}, \qquad f''(x) = -\frac{1}{4}\frac{x + 4}{(x+1)^{5/2}}.$$

Since $x > -1$, it follows $f'(x) > 0$ and $f''(x) < 0$ for every $x \in (-1, +\infty)$, so that f is strictly increasing and concave in $I = (-1, +\infty)$. As the function is differentiable in the open interval I and has no critical points, there are no extreme points, neither local nor global. Since $f(3) = 3/2$ and $f'(3) = 5/16$, the tangent line to the graph of f in the point $(3, 3/2)$ has equation $y = 3/2 + 5/16(x - 3)$. Since f is strictly concave,

$$f(x) < \frac{3}{2} + \frac{5}{16}(x - 3)$$

for $x \in I \setminus \{3\}$. Hence the only solution is $x = 3$.

Question (a) has an immediate answer because the first derivative is easy to compute and to analyze. As for (b), one has to understand how the line with equation $y = 3/2 + 5/16(x - 3)$ relates to the graph of f. As f' has already been computed, one realizes at once that the line is just the tangent to the graph at the point $(3, 3/2)$, hence of course what matters is whether f is convex or concave. The second derivative of f is clearly everywhere strictly negative, so the answer is that there is a unique intersection, which occurs at the point $(3, 3/2)$.

9.3 Consider the function

$$f(x) = \begin{cases} e^x + \beta & x < 0 \\ \alpha & x = 0 \\ \arctan x & x > 0. \end{cases}$$

(a) Find, if existing, $\alpha, \beta \in \mathbb{R}$ such that f is differentiable at $x_0 = 0$.
(b) For the values of $\alpha, \beta \in \mathbb{R}$ found in (a), say if f has at $x_0 = 0$ a local extreme.
(c) For the values of $\alpha, \beta \in \mathbb{R}$ found in (a), say if f has at $x_0 = 0$ an inflection point.

Answer. (a) Evidently, f is continuous at 0 provided that

$$\lim_{x \to 0^-} f(x) = \lim_{x \to 0^+} f(x) = f(0) \qquad \Longleftrightarrow \qquad \alpha = 0, \ \beta = -1.$$

Further, f is differentiable at least in $\mathbb{R} \setminus \{0\}$: for $x < 0$ it is $f'(x) = e^x$ and for $x > 0$ it is $f'(x) = 1/(1 + x^2)$. Since

$$\lim_{x \to 0^-} f'(x) = \lim_{x \to 0^+} f'(x) = 1$$

the function is differentiable also at $x_0 = 0$, hence in $\mathbb{R}$.
(b) For $\alpha = 0$ and $\beta = -1$ the function is differentiable, but $f'(0) = 1 \neq 0$. Hence for no value of the parameters f has at $x_0 = 0$ a local extreme.
(c) For $\alpha = 0$ and $\beta = -1$ the function is twice differentiable at least in $\mathbb{R} \setminus \{0\}$ and

$$f''(x) = \begin{cases} e^x & x < 0 \\ -\dfrac{2x}{(1 + x^2)^2} & x > 0. \end{cases}$$

Thus, f is convex in $(-\infty, 0)$ and concave in $(0, +\infty)$, hence $x_0 = 0$ is an inflection point. Observe also that the tangent line to the graph of f at $(0, 0)$ has equation $y = x$, and that $f(x) > x$ for $x \in (-\infty, 0)$ and $f(x) < x$ for $x \in (0, +\infty)$. Observe also that by the theorem on the limit of the derivative, f is not twice differentiable at the origin.

This exercise is about tuning the parameters $\alpha, \beta \in \mathbb{R}$ in such a way that f is continuous at $x_0 = 0$. There is only one such choice, which yields a differentiable function at $x_0 = 0$. This, however, is not a local minimum because the "glued" function is differentiable but $f'(0) = 1$. The convexity of $x \mapsto e^x$ together with the concavity of $x \mapsto \arctan x$ for positive x imply that the origin is an inflection point. No appeal to the second derivative at the origin is either necessary or possible, because the geometry is clear and because the second derivative at $x_0 = 0$ does not exist.

9.4 Take a function $f : \mathbb{R} \to \mathbb{R}$ which is strictly increasing and convex, and suppose further that $f \in C^2(\mathbb{R})$. Prove that $k(x) = (f \circ f)(x) + x^2$ is strictly convex in $\mathbb{R}$.

Answer. Using the regularity of f and the chain rule, if $h(x) = (f \circ f)(x)$, then

$$h'(x) = f'(f(x)) f'(x), \qquad h''(x) = f''(f(x))(f'(x))^2 + f'(f(x)) f''(x).$$

The assumptions on f give that $f'(x) \geq 0$, $f''(f(x)) \geq 0$ and $f''(x) \geq 0$ for every $x \in \mathbb{R}$. Therefore $h''(x) \geq 0$ for every $x \in \mathbb{R}$. Thus, $k''(x) = h''(x) + 2 \geq 2$ and this shows that k is strictly convex in $\mathbb{R}$.

It is asked to compute the second derivative of $f \circ f$ and to observe that if both f' and f'' are non-negative, such is also $(f \circ f)''$. Adding a strictly convex function such $x \mapsto x^2$ yields a strictly convex function.

9.5 Draw the graph of the function $f(x) = x - 2x \log\left(1 + \dfrac{1}{x}\right)$.

Answer. The domain of f is $(-\infty, -1) \cup (0, +\infty)$ and $f \in C^\infty((-\infty, -1) \cup (0, +\infty))$ being obtained by summing, multiplying and composing functions that are C^∞ in the domain. The behaviour at the boundary is:

$$\lim_{x \to -\infty} f(x) = \lim_{x \to -\infty} \left(x - 2\frac{\log(1 + 1/x)}{1/x}\right) = -\infty$$

$$\lim_{x \to -1^-} f(x) = -\infty$$

$$\lim_{x \to 0^+} f(x) = 0$$

$$\lim_{x \to +\infty} f(x) = \lim_{x \to +\infty} \left(x - 2\frac{\log(1 + 1/x)}{1/x}\right) = +\infty.$$

As for asymptotes, the line $x = -1$ is a vertical asymptote, and since $f(x) \to \pm\infty$ as $x \to \pm\infty$, there might be oblique asymptotes. Now, from

$$\lim_{x \to \pm\infty} \frac{f(x)}{x} = \lim_{x \to \pm\infty} (1 - 2\log(1 + 1/x)) = 1$$

$$\lim_{x \to \pm\infty} (f(x) - x) = \lim_{x \to \pm\infty} -2\frac{\log(1 + 1/x)}{1/x} = -2$$

it follows that the line $y = x - 2$ is an oblique asymptote for $x \to \pm\infty$.
Consider next the derivative, namely

$$f'(x) = 1 - 2\log\left(1 + \frac{1}{x}\right) + \frac{2}{x+1}.$$

Its sign can be determined studying its derivative, that is

$$f''(x) = \frac{2}{x(x+1)} - \frac{2}{(x+1)^2} = \frac{2}{x(x+1)^2}.$$

Now, $f''(x) > 0$ for $x > 0$ and $f''(x) < 0$ for $x < -1$, which entails that f is strictly convex in $(0, +\infty)$ and strictly concave in $(-\infty, -1)$. Furthermore, f' is strictly decreasing in $(-\infty, -1)$ and strictly increasing in $(0, +\infty)$. Since f' is continuous in $(-\infty, -1)$ and

$$\lim_{x \to -\infty} f'(x) = 1, \qquad \lim_{x \to -1^-} f'(x) = -\infty,$$

there exists $x_1 \in (-\infty, -1)$ such that $f'(x) > 0$ for $x \in (-\infty, x_1)$ and $f'(x) < 0$ for $x \in (x_1, -1)$. Analogously, since f' is continuous in $(0, +\infty)$ and

$$\lim_{x \to 0^+} f'(x) = -\infty, \qquad \lim_{x \to +\infty} f'(x) = 1,$$

there exists $x_2 \in (0, +\infty)$ such that $f'(x) < 0$ for $x \in (0, x_2)$ and $f'(x) > 0$ for $x \in (x_2, +\infty)$. The sign of f' has thus been established, and it is possible to infer that f is strictly increasing in $(-\infty, x_1)$ and in $(x_2, +\infty)$, and it and strictly decreasing in $(x_1, -1)$ and in $(0, x_2)$.

Evidently, x_1 is a local maximum and x_2 is a local minimum. Since f is unbounded above and below, it has no global extreme points. Finally, observe that $f(x) < 0$ for $x < -1$ and for $x \in (0, 1/(\sqrt{e} - 1))$. The graph of f is in Fig. 9.2.

9.6 Consider the function $f(x) = k(x+1)\log(x+1) - x^2$.

(a) For which values of the parameter $k > 0$ the point $x_0 = e - 1$ is a local extreme point?
(b) Put $k = 1$. Determine the maximal intervals in which f is concave and those in which it is convex, and find the inflection points.
(c) Prove that when $n \geq 2$ is even, then $f^{(n)}$ is convex in its domain, and when $n \geq 2$ is odd, then $f^{(n)}$ is concave in its domain.

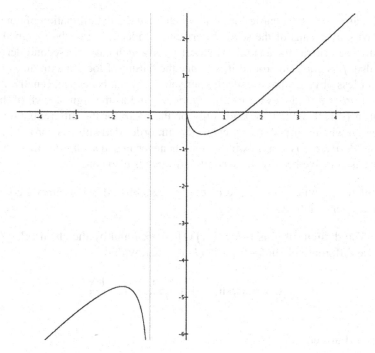

Fig. 9.2 The function of Exercise 9.5

Answer. (a) First of all, $f \in C^\infty((-1, +\infty))$ for every value of $k > 0$. Furthermore,

$$f'(x) = k(\log(x+1) + 1) - 2x, \qquad f''(x) = \frac{k}{x+1} - 2.$$

Now, $x_0 = e - 1$ is a local extreme point only if $f'(e-1) = 2k - 2e + 2 = 0$, that is $k = e - 1$. In this case $f''(e-1) = (e-1)/e - 2 < 0$, and by permanence of sign the same holds in a neighborhood of $x_0 = e - 1$. Thus f is is concave around $x_0 = e - 1$, which is therefore a local maximum.

(b) Put $k = 1$. Now, $f''(x) > 0$ implies $x < -1/2$, hence f is convex in $(-1, -1/2)$ and concave in $(-1/2, +\infty)$, so that $x_0 = -1/2$ is an inflection point.

(c) Put $k = 1$. Observe that

$$f^{(3)}(x) = -(x+1)^{-2}, \qquad f^{(4)}(x) = 2(x+1)^{-3}, \qquad f^{(5)}(x) = -6(x+1)^{-4}.$$

and, by induction,

$$f^{(n)}(x) = (-1)^n (n-2)!(x+1)^{1-n}$$

for $n \geq 3$. Therefore $f^{(n)}(x) > 0$ for $n \geq 3$ even, while $f^{(n)}(x) < 0$ for $n \geq 3$ odd and $x > -1$. Therefore the derivatives of odd order are concave and those of even order are convex in their domains.

This is an exercise the basic focus of which is on the determination of convexity or concavity by means of the second derivative. Indeed, in (a), the assigned point is found to be critical for a single value of k, in which case the second derivative reveals that f is concave around it, whence the nature of local maximum. In (b) it is more or less obvious that the second derivative is what is needed. Finally, for (c), the higher order derivatives are easily computed, and as the sign of each of them is constant in $(-1, +\infty)$ and depends only on the parity of the differentiation order n. Precisely, whenever positive (n even) the nth order derivative is convex because its second derivative is also positive, and whenever negative (n odd) the nth order derivative is concave because its second derivative is also concave.

9.7 Draw the graph of $f(x) = x \arcsin(1/x)$. Establish if f has either a convex or a concave extension to $\mathbb{R}$.

Answer. The domain of f is $(-\infty, -1] \cup [1, +\infty)$ and by the chain rule, f is at least twice differentiable in $(-\infty, -1) \cup (1, +\infty)$, with

$$f'(x) = \arcsin \frac{1}{x} + x \frac{1}{\sqrt{1 - \frac{1}{x^2}}}\left(-\frac{1}{x^2}\right)$$

for $|x| > 1$. Further,

$$f''(x) = \begin{cases} \dfrac{1}{x(x^2 - 1)\sqrt{x^2 - 1}} & x > 1 \\[3mm] -\dfrac{1}{x(x^2 - 1)\sqrt{x^2 - 1}} & x < -1. \end{cases}$$

Hence, f is convex in $(-\infty, -1]$ and in $[1, +\infty)$. The extreme points ± 1 are indeed included because the function is continuous at ± 1. The sign of f'' says that f' is strictly increasing in $[1, +\infty)$ and in $[-\infty, -1)$, and from the limits

$$\lim_{x \to -\infty} f'(x) = \lim_{x \to +\infty} f'(x) = 0$$

it follows that $f'(x) > 0$ in $(-\infty, -1)$ and $f'(x) < 0$ in $(1, +\infty)$. Therefore f is strictly increasing in $(-\infty, -1)$ and strictly decreasing in $(1, +\infty)$. Also,

$$\lim_{x \to -\infty} f(x) = \lim_{x \to -\infty} \frac{\arcsin \frac{1}{x}}{\frac{1}{x}} = \lim_{y \to 0^-} \frac{\arcsin y}{y} = 1$$

and similarly $\lim_{x \to +\infty} f(x) = 1$. Thus, the line with equation $y = 1$ is a horizontal asymptote. At the points $x = \pm 1$ the function attains its global maximum, namely $\pi/2$, whereas it has neither global no local minimum points. The graph of f is in Fig. 9.3. Consider now $g : \mathbb{R} \to \mathbb{R}$ convex and such that $g(x) = f(x)$ for every

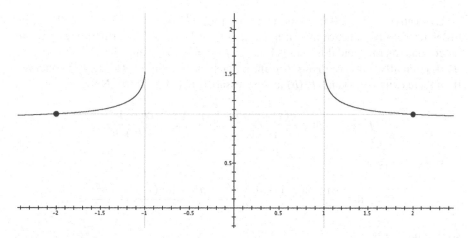

Fig. 9.3 The function of Exercise 9.7, together with the secant joining two symmetric points

$x \in (-\infty, -1] \cup [1, +\infty)$. Then g satisfies the inequality $g(tx + (1 - t)y) \le tg(x)$ $+ (1 - t)g(y)$ for every $x, y \in \mathbb{R}$ and every $t \in [0, 1]$. Take now two points x and y that are symmetric relative to the origin, for example $x = -2$ and $y = 2$ and take t such that $tx + (1 - t)y = 1$, that is, $t = 1/4$. For these values of x, y and t, the inequality $g(tx + (1 - t)y) \le tg(x) + (1 - t)g(y)$ yields $\pi/2 \le \pi/3$, which is false. Hence f has no convex extension in $\mathbb{R}$. Obviously, f has no concave extension either because f itself has convex restrictions, and this is not possible for a globally concave function.

In this exercise the problem of finding a convex extension is considered. The given function is convex on two separate intervals. The graph of f strongly suggests that f has no convex extensions because there are many secant horizontal lines (those joining symmetric point with respect to the origin) with plenty of points on the graph that lie above the points on the secant.

9.8 Consider the function

$$f(x) = \begin{cases} \log(1 + x + x^2) & x \le b \\ ax + c & x > b. \end{cases}$$

Find the values of the parameters $a, b, c \in \mathbb{R}$ such that f is concave in $\mathbb{R}$. Put $a = -1$, $b = 1$ and $c = \log 3 + 1$, and draw the graph of f.

Answer. Define first $g(x) = \log(1 + x + x^2)$. The domain of g is $\mathbb{R}$ because $1 + x + x^2 > 0$ for every $x \in \mathbb{R}$, and $g \in C^2(\mathbb{R})$. Look next at the first and second derivatives:

$$g'(x) = \frac{2x + 1}{1 + x + x^2}, \qquad g''(x) = -\frac{2x^2 + 2x - 1}{(1 + x + x^2)^2}.$$

Evidently, $g''(x) < 0$ if either $x < -(1 + \sqrt{3})/2$ or $x > (\sqrt{3} - 1)/2$. Thus, in order for f to be concave in $\mathbb{R}$ it must be $b \leq -(1 + \sqrt{3})/2$. Furthermore, f must be continuous in $\mathbb{R}$ and this entails $\lim_{x \to b} f(x) = f(b)$, that is, $\log(1 + b + b^2) = ab + c$. Finally, observe that f is differentiable at least in $\mathbb{R} \setminus \{b\}$. If f is concave, then there exist $f'_-(b)$ and $f'_+(b)$ and necessarily $f'_-(b) \geq f'_+(b)$. Now,

$$f'_-(b) = \lim_{x \to b^-} \frac{g(x) - g(b)}{x - b} = g'(b) = \frac{2b + 1}{1 + b + b^2}.$$

Furthermore

$$f'_+(b) = \lim_{x \to b^+} \frac{ax + \log(1 + b + b^2) - ab - \log(1 + b + b^2)}{x - b} = a,$$

so that it must be

$$\frac{2b + 1}{1 + b + b^2} \geq a.$$

Therefore, f is concave in $\mathbb{R}$ provided that

$$\begin{cases} b \leq -\dfrac{1 + \sqrt{3}}{2} \\[2mm] a \leq \dfrac{2b + 1}{1 + b + b^2} \\[2mm] c = \log(1 + b + b^2) - ab. \end{cases}$$

Put now $a = -1, b = 1$ and $c = \log 3 + 1$, so that

$$f(x) = \begin{cases} \log(1 + x + x^2) & x \leq 1 \\ -x + \log 3 + 1 & x > 1. \end{cases}$$

As discussed above, the domain is $\mathbb{R}$, where f is continuous. Furthermore, it is differentiable in $(-\infty, 1)$ and in $(1, +\infty)$, but is not differentiable at $x_0 = 1$. Indeed, it has already been proved that $f'_-(1) = 1$ and that $f'_+(1) = -1$.

Look next at the monotonicity issue. For $x \in (-\infty, 1)$

$$f'(x) = g'(x) = \frac{2x + 1}{1 + x + x^2}$$

and so $f'(x) > 0$ if $x \in (-1/2, 1)$. Hence f is strictly increasing in $[-1/2, 1]$ and strictly decreasing in $(-\infty, -1/2)$. It is also manifestly decreasing in $[1, +\infty)$.

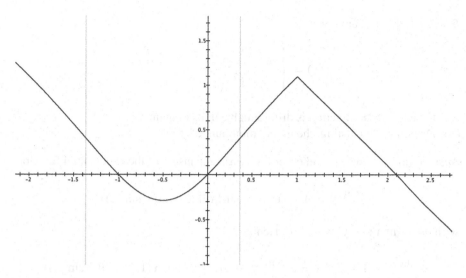

Fig. 9.4 The function of Exercise 9.8; the *vertical lines* have abscissae x_1 and x_2, respectively

As for the behaviour at $\pm\infty$,

$$\lim_{x \to -\infty} f(x) = +\infty, \qquad \lim_{x \to +\infty} f(x) = -\infty.$$

Since $f(x)/x \to 0$ as $x \to -\infty$, there are no oblique asymptotes at $-\infty$, whereas f is linear in $(1, +\infty)$, hence it is its own asymptote at $+\infty$.

It has already been proved that f is concave in the intervals $(-\infty, -(1+\sqrt{3})/2)$, $(-(1-\sqrt{3})/2, 1)$ and $(1, +\infty)$ and convex in $(-(1+\sqrt{3})/2, -(1-\sqrt{3})/2)$. Since $f'_+(1) \leq f'_-(1)$, it is possible to infer that f is concave in $(-(1-\sqrt{3})/2, +\infty)$. The inflection points are $x_1 = -(1+\sqrt{3})/2$ and $x_2 = -(1-\sqrt{3})/2$. Finally, considering the monotonicity it is clear that $x_3 = -1/2$ is a local minimum and that $x_4 = 1$ is a local maximum. Being unbounded both above and below, f has no global extreme points. The graph of f is in Fig. 9.4.

The first part of this exercise asks to understand that a piecewise defined function is globally concave provided that both branches of the function are concave and provided that at the "gluing" point b the function behaves nicely, that is: f must be continuous and the left and right derivatives must exist and $f'_-(b) \geq f'_+(b)$. Together, these are several conditions, one of which (concavity of the right branch) is obvious. In the second part, the issue is the shape of the graph of f for fixed values of the parameter, values for which, by the way, f is not concave but continuous. Asymptotes, convexity, local extreme points and inflection points are all easy to determine because first and second derivatives have already been computed (where they exist) and are actually easy to analyze. There is a corner at $x_0 = b = 1$.

9.9 Consider the function

$$f(x) = \begin{cases} \dfrac{e^{-\sin^2 x} - \cos^2 x}{x^3} & x > 0 \\ 0 & x = 0. \end{cases}$$

(a) Establish if the function is differentiable in its domain.
(b) Determine, if existing, the global minimum of f.

Answer. (a) From the second order McLaurin expansion of the exponential function

$$e^{-\sin^2 x} = 1 - \sin^2 x + \sin^4 x (1/2 + \omega(-\sin^2 x))$$

with $\omega(-\sin^2 x) \to 0$ as $x \to 0$. Hence

$$e^{-\sin^2 x} - 1 + \sin^2 x = e^{-\sin^2 x} - \cos^2 x = \sin^4 x (1/2 + \omega(-\sin^2 x))$$

and consequently

$$\lim_{x \to 0^+} f(x) = \lim_{x \to 0^+} \frac{\sin^4 x (1/2 + \omega(-\sin^2 x))}{x^3} = 0.$$

This proves that f is continuous at $x_0 = 0$ and

$$\lim_{x \to 0^+} \frac{f(x) - f(0)}{x} = \lim_{x \to 0^+} \frac{f(x)}{x} = \lim_{x \to 0^+} \frac{\sin^4 x (1/2 + \omega(-\sin^2 x))}{x^4} = 1/2.$$

shows that at $x_0 = 0$ it is also differentiable. For $x > 0$, the function is differentiable because it is the ratio of differentiable functions with non vanishing denominator. It follows that f is differentiable in its domain.

(b) The inequality $e^x \geq 1 + x$ valid for every $x \in \mathbb{R}$ (see the Appendix) entails

$$e^{-\sin^2 x} \geq 1 - \sin^2 x,$$

which implies that

$$e^{-\sin^2 x} - \cos^2 x \geq 0 = f(0).$$

This proves that 0 is the minimum value of the function. Observe that f actually attains this value infinitely many times: at $x_n = n\pi$ for every $n \in \mathbb{N}$ with $n > 1$.

Part (a) of this problem is the computation of a limit using, for example, a McLaurin expansion. Part (b) may be solved with a little ingenuity, via the inequality $e^x \geq 1 + x$, without appealing to critical point analysis. Being the extreme point of the interval $[0, +\infty)$, the point $x_0 = 0$ may well be a (global) minimum of f without being a critical point for f. The other minima are interior points and in fact

$f'(x_n) = 0$ for every $n \in \mathbb{N}$ with $n > 1$, but $f'_+(0) = 1/2 \neq 0$. There is obviously no contradiction with Fermat's Theorem 7.2.

9.10 Draw the graph of the function $f(x) = -x + \log x + e^x$ and find $a > 1$ such that $f(x) \leq 3(x - 1) + e - 1$ for every $x \in [1, a]$.

Answer. First of all, $f \in C^\infty((0, +\infty))$, and in particular is continuous. Furthermore, $f(x) \to -\infty$ as $x \to 0^+$ and $f(x) \to +\infty$ as $x \to +\infty$. Hence f is unbounded below and above, and therefore there are no global extreme points. Also,

$$f'(x) = -1 + \frac{1}{x} + e^x, \qquad f''(x) = -\frac{1}{x^2} + e^x = \frac{-1 + x^2 e^x}{x^2} = \frac{\varphi(x)}{x^2},$$

where $\varphi(x) = -1 + x^2 e^x$. Now, $e^x \geq x + 1 > e - 1/x$. Indeed, the first inequality is in in the Appendix, while for $x > 0$ it is always $x + 1 > e - 1/x$. Therefore $f'(x) > 0$ for $x > 0$, and it follows that, being strictly increasing, f has no local extreme point either.

Now, since $x^2 > 0$, in order to study $f''(x)$ it is enough to look at $\varphi(x)$. Evidently, $\varphi'(x) = 2xe^x + x^2 e^x > 0$ for every $x > 0$. Hence φ is strictly increasing in $[0, +\infty)$. Since $\varphi \in C(\mathbb{R})$, $\varphi(0) = -1$ and $\varphi(x) \to +\infty$ as $x \to +\infty$, by the intermediate zero theorem there is a unique $\overline{x} \in (0, +\infty)$ such that $\varphi(x) < 0$ for $x \in (0, \overline{x})$ and $\varphi(x) > 0$ for $x \in (\overline{x}, +\infty)$. It follows that f is concave in $(0, \overline{x}]$ and convex in $[\overline{x}, +\infty)$ and that $\overline{x}$ is the unique inflection point. Observe that $\overline{x} < 1$, because $\varphi(1) = e - 1 > 0$.

The behaviour as $x \to 0^+$ shows that the y-axis is a vertical asymptote. There are no asymptote as $x \to +\infty$ because f diverges to $+\infty$ as $x \to +\infty$ with order greater than 1. A qualitative graph of f is in Fig. 9.5.

The line with equation $y = 3(x - 1) + e - 1$ meets the graph of f at $(1, e - 1)$ but its slope is 3, which is larger than the slope of the tangent line at $(1, e - 1)$, which is e. The line $y = 3(x - 1) + e - 1$ will eventually meet again the graph of f at some point $(b, f(b))$ with $b > 1$ and all the points on it that have abscissa $x \in (\overline{x}, b)$ are above the graph of f because f is convex in the interval $[\overline{x}, +\infty)$ and thus satisfy the inequality $f(x) \leq 3(x - 1) + e - 1$. By the intermediate value theorem, for every $x > 1$ there exists a point $c \in (1, x)$ such that

$$f(x) - f(1) = f(x) - (e - 1) = f'(c)(x - 1) < f'(x)(x - 1) < f'(a)(x - 1)$$

whenever $x < a$ because f' is strictly increasing in $(1, +\infty)$. Any point a such that $f'(a) < 3$ satisfies the requirement. A possible value of a is $9/8$.

In the first picture in Fig. 9.5, the graph of f together with the vertical line $x = \overline{x}$ and the point $(1, f(1))$ are drawn; in the second, which is at a finer scale, again the point $(1, f(1))$, the line $y = 3(x - 1) + e - 1$, which visibly crosses the graph of f a second time to the right of 1 (at the point $(b, f(b))$), and the vertical line $x = 9/8$.

This is a slightly demanding exercise. The first part is just a thorough analysis of the function, and contains no difficulties. The sign of the second derivative is understood by studying the function φ. The geometry involved in the final question

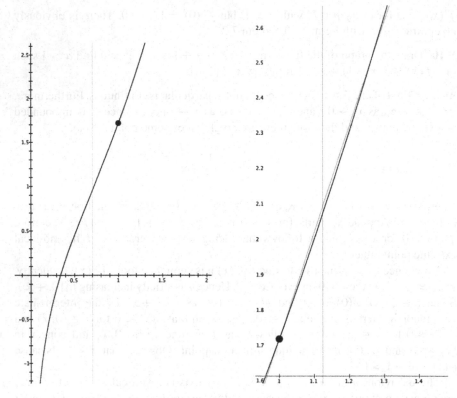

Fig. 9.5 The graph of the function f in Exercise 9.10 (*left*) and the line $y = 3(x-1)+\mathrm{e}-1$ (*right*)

Fig. 9.6 The final argument
in Exercise 9.10

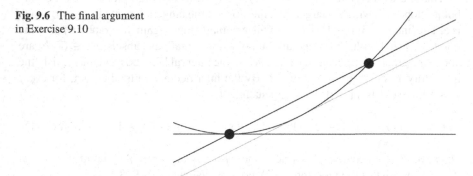

is subsumed in the observation that the line $y = 3(x-1)+\mathrm{e}-1$ describes a secant to the graph, hence the points on it are above the graph because f is convex in the region considered. The geometry is illustrated abstractly in Fig. 9.6. The two emphasized dots correspond to the points $(1, f(1))$ and $(b, f(b))$ in the exercise; the segment joining them corresponds to the given line $y = 3(x - 1) + e - 1$, and the parallel to it meets the graph at $(c, f(c))$.

9.5   Problems on the Geometry of Functions

9.11 Study the convexity and draw the graph of $f(x) = \arcsin\left(\dfrac{x^2 - 1}{x^2 + 1}\right)$.

9.12 Draw the graph of $f(x) = e^x \arctan x$.

9.13 Consider the function: $f(x) = x + x \log x + kx^2$, where $k \in \mathbb{R}$ is a parameter.

(a) Study where f is convex, as k varies in $\mathbb{R}$.
(b) For which $k \in \mathbb{R}$ is f monotonic in its domain?
(c) Put $k = -1$. Draw the graph of f.

9.14 Draw the graph of $f(x) = \sqrt{x + x \log x}$, with particular attention to convexity.

9.15 Draw the graph of $f(x) = \dfrac{2x^2 - x^3}{x^2 - 3}$, with particular attention to convexity.

9.16 Take two intervals I and J, a convex function $f : I \to \mathbb{R}$ and a function $g : J \to \mathbb{R}$ convex and increasing in $f(I) \subset J$. Prove that $g \circ f$ is convex in I.

9.17 Study convexity and inflection points of the function $f(x) = (1 + x - x^2)e^{-x}$.

Chapter 10
Indefinite and Definite Integrals

10.1 Primitive Functions

Definition 10.1 (*Primitive function*) Let I be an interval and let $f : I \to \mathbb{R}$ be a function defined on I. A function $F : I \to \mathbb{R}$ is called a *primitive function* of f if it is differentiable in I and if $F'(x) = f(x)$ for every $x \in I$. The set of all the primitive functions of f is denoted by

$$\int f(x)\,dx$$

and is called the *indefinite integral* of f.

For rather obvious reasons, a primitive function is also called an *antiderivative* of f. The reason behind the symbol used will be illustrated below in Sect. 10.4, where the fundamental theorem of calculus is discussed.

The indefinite integral of a function f has the structure described in the next theorem, which is a direct consequence of the intermediate value theorem.

Theorem 10.1 (Structure of the indefinite integral) *Let I be an interval and let $f : I \to \mathbb{R}$ be a function defined on I. If F is a primitive function of f on I, then*

$$\int f(x)\,dx = \{F + c : c \in \mathbb{R}\}.$$

The notion of primitive function can be generalized to the case where $\mathrm{Dom}(f)$ is not necessarily an interval but a disjoint union on intervals. For our purposes, the case of a finite (disjoint) union of intervals

$$\mathrm{Dom}(f) = I_1 \cup I_2 \cup \cdots \cup I_n \tag{10.1}$$

is enough. Then, a function $F : \mathrm{Dom}(f) \to \mathbb{R}$ is called a primitive function of f if $F'(x) = f(x)$ for every $x \in \mathrm{Dom}(f)$. In this situation, the structure theorem must be handled with care, because it no longer holds in the form that any primitive is

© Springer International Publishing Switzerland 2016
M. Baronti et al., *Calculus Problems*, UNITEXT - La Matematica per il 3+2 101,
DOI 10.1007/978-3-319-15428-2_10

obtained by adding to a single primitive F an arbitrary constant. The point is that the structure theorem has to be applied in each interval I_j separately, adding an arbitrary constant for each interval. If $\text{Dom}(f)$ is as in (10.1) and F is a primitive function of f, it is customary to write

$$\int f(x)\,dx = F(x) + c$$

with the understanding that c is constant in each interval, but not necessarily in $\text{Dom}(f)$, that is

$$c = c(x) = \begin{cases} c_1 & x \in I_1 \\ \vdots \\ c_n & x \in I_n, \end{cases} \tag{10.2}$$

where $c_1, \ldots, c_n$ are arbitrary real numbers.

An important fact is that not every function admits a primitive function. Indeed, consider for example $f : (-2, 2) \to \mathbb{R}$ given by

$$f(x) = \begin{cases} -1 & -2 < x < 0 \\ 1 & 0 \le x < 2. \end{cases}$$

If $F : (-2, 2) \to \mathbb{R}$ were an antiderivative of f, then since $F'(-1) = f(-1) = -1$ and $F'(1) = f(1) = 1$, by Darboux' theorem, namely Corollary 7.5 of Chap. 7, F' would necessarily attain every value in the interval $(-1, 1)$. However, f does not attain any of those values. A careful analysis reveals that whenever f has a point of discountinuity $x_0 \in I$ for which either the limit as $x \to x_0^+$ or the limit as $x \to x_0^-$ exist, then f does not have primitive functions in I. In particular, as in the previous example, functions with jump discountinuities do not admit primitive functions.

As we shall see below (see Corollary 10.1 in Sect. 10.4), continuous functions do have antiderivatives. This does not mean that they can be computed explicitly in terms of elementary functions. For example, the functions e^{x^2} and $\text{sinc}(x)$ are continuous on $\mathbb{R}$ and hence admit primitive functions, but no explicit expression can be found involving compositions, sums or products of elementary functions. A precise formulation of this statement defies the purpose of this book, but the reader should be aware that it is not always possible to obtain concrete expressions for primitive functions.

Proposition 10.1 *Let I be an interval and suppose that $f, g : I \to \mathbb{R}$ admit primitive functions in I. Then for every $\alpha, \beta \in \mathbb{R}$, not both zero, the function $\alpha f + \beta g$ admits a primitive function in I. Furthermore*

$$\int (\alpha f + \beta g)(x)\,dx = \alpha \int f(x)\,dx + \beta \int g(x)\,dx. \tag{10.3}$$

The set on the right hand side of (10.3) must be interpreted as the set consisting of the linear combinations $\alpha F + \beta G$ where F is a primitive function of f and G is a primitive function of g. If either $\alpha = 0$ or $\beta = 0$ the formula reduces to the simple fact that $(\alpha f)' = \alpha f'$.

10.2 Computing Indefinite Integrals

10.2.1 General Techniques

The actual computation of primitive functions can be a rather difficult task. Generally speaking, the process of finding one, hence in fact all, primitive functions of a given f is called *integration* process, or integration technique. Many of them have been developed over the years and nowadays rather exhaustive tables are available, so that most primitive functions of reasonable functions are known. The most common techniques, however, are easy to master. Among these, of foremost importance are those illustrated in the two theorems that follow, known as *integration by substitution* and *integration by parts*.

Theorem 10.2 (Integration by substitution, I) *Suppose that I and J are two intervals and suppose that $f : I \to \mathbb{R}$ and $g : J \to I$ are such that f admits a primitive function F on I and that g is differentiable on J. Then $F \circ g$ is a primitive function of $x \to f(g(x))g'(x)$ in I and*

$$\int f(g(x))g'(x)\,dx = \{F \circ g + c : c \in \mathbb{R}\}. \qquad (10.4)$$

Behind this theorem is simply the chain rule, because evidently $\frac{d}{dx}F \circ g(x) = F'(g(x))g'(x) = f(g(x))g'(x)$. Equality (10.4) can also be written

$$\int f(g(x))g'(x)\,dx = \left[\int f(y)\,dy\right]_{y=g(x)} \qquad (10.5)$$

and both (10.4) and (10.5) must be interpreted as follows: if the function to be integrated is of the form $f(g(x))g'(x)$, then one can formally substitute $g(x) = y$ and $g'(x)\,dx = dy$ so as to get $\int f(y)\,dy$. Then one computes a primitive function $F(y)$ of f, which is supposed to be easier and known, and finally get a primitive function of $f(g(x))g'(x)$ by substituting again $g(x) = y$, that is, writing $F(g(x))$.

Theorem 10.3 (Integration by parts, I) *Suppose that I is an interval and suppose that $f, g : I \to \mathbb{R}$ are differentiable on I and that fg' has a primitive function on I. Then so does $f'g$ and*

$$\int f'(x)g(x)\,dx = fg - \int f(x)g'(x)\,dx. \qquad (10.6)$$

Behind formula (10.6) is Leibnitz' rule, namely $\frac{d}{dx}(fg) = f'g + fg'$.

10.2.2   Rational Functions

A particularly important class of functions that are continuous on their domains, and
hence have primitive functions, is the class of *rational functions*, namely

$$f(x) = \frac{P(x)}{Q(x)}$$

where P and Q are polynomials, of degrees $\deg(P) = m$ and $\deg(Q) = n$, respectively. If $n \geq m$, then by long division it is possible to find polynomials D and R
such that $P(x) = Q(x)D(x) + R(x)$, where $\deg(D) = m - n$ and $\deg(R) < n$, so
that

$$f(x) = D(x) + \frac{R(x)}{Q(x)}.$$

Thus, f is the sum of a polynomial and of a *proper rational function*, namely one
where the degree of the numerator is smaller than the degree of the denominator.
Now, polynomials are easy to integrate because for every non negative integer d

$$\int x^d \, dx = \frac{x^{d+1}}{d+1} + c$$

so that the indefinite integral of a polynomial is computed applying Proposition 10.1.
Therefore, the computation of the indefinite integral of a rational function reduces
to the case of proper rational functions. We collect here the most elementary cases.

Proposition 10.2 (Simple factors) *If $a \in \mathbb{R}$, and $n > 1$ is an integer, then*

$$\int \frac{1}{x - a} \, dx = \log|x - a| + c$$

$$\int \frac{1}{(x - a)^n} \, dx = \frac{1}{1 - n} \frac{1}{(x - a)^{n-1}} + c$$

where c denotes independent constants to the left and to the right of a, as in (10.2).

Proposition 10.3 (Irreducible factors of degree two) *Suppose that $p, q \in \mathbb{R}$ are such
that $\Delta = p^2 - q < 0$, then for every $a, b \in \mathbb{R}$*

$$\int \frac{1}{x^2 + 2px + q} \, dx = \frac{1}{\sqrt{-\Delta}} \arctan\left(\frac{x + p}{\sqrt{-\Delta}}\right) + c$$

$$\int \frac{ax + b}{x^2 + 2px + q} \, dx = \frac{a}{2} \log(x^2 + 2px + q) + \frac{b - ap}{\sqrt{-\Delta}} \arctan\left(\frac{x + p}{\sqrt{-\Delta}}\right) + c$$

as c ranges in $\mathbb{R}$.

It is actually not difficult, but more laborious, to compute

$$\int \frac{ax+b}{(x^2+2px+q)^n}\,dx$$

for every positive integer n. The first step is to notice that it can be written as

$$\frac{a}{2}\int \frac{2x+2p}{(x^2+2px+q)^n}\,dx + (b-ap)\int \frac{1}{(x^2+2px+q)^n}\,dx.$$

Up to additive constants, the first summand is $a(x^2+2px+q)^{1-n}/2(1-n)$. The second can be reduced to the computation of an integral of the form

$$I_n = \int \frac{1}{(x^2+1)^n}\,dx$$

after a suitable substitution. The latter fulfills the recursion relation

$$I_n = \frac{x(x^2+1)^{1-n}}{2(n-1)} + \frac{2n-3}{2n-2}I_{n-1}$$

and can therefore be calculated at least for small values of n. We shall not delve into these complicated cases.

The integration of an arbitrary proper rational function is carried out combining the elementary integrals shown above with the following basic result, which will be applied to the denominator $Q(x)$ of the given rational function.

Theorem 10.4 (Decomposition into irreducible factors) *Every polynomial Q of degree m with real coefficients can be written as a product of irreducible factors of the form*

$$Q(x) = d(x-a_1)^{r_1}\cdots(x-a_h)^{r_h}(x^2+2p_1x+q_1)^{s_1}\cdots(x^2+2p_kx+q_k)^{s_k},$$

$$(10.7)$$

where $r_1+\cdots+r_h+2(s_1+\cdots+s_k)=m$ *and* $p_j^2-q_j<0$ *for every* $j=1,\ldots,k$.

Using this decomposition, one infers that the given proper rational function $f = P/Q$ can be written in the form

$$\frac{P(x)}{Q(x)} = \frac{1}{d}\big[F_1(x)+\cdots+F_h(x)+G_1(x)+\cdots+G_k(x)\big].$$

Each $F_i(x)$ corresponds to one of the h distinct real roots of Q and is of the form

$$F(x) = \frac{A_1}{(x-a)} + \frac{A_2}{(x-a)^2} + \cdots + \frac{A_r}{(x-a)^r},$$

where a is the root and r its multiplicity. Each $G_j(x)$ corresponds to one of the k irreducible factors of degree two, none of which has real roots and each of which has the form

$$G(x) = \frac{B_1 x + C_1}{(x^2 + 2px + q)} + \frac{B_2 x + C_2}{(x^2 + 2px + q)^2} + \cdots + \frac{B_s x + C_s}{(x^2 + 2px + q)^s}.$$

Evidently, s is the multiplicity of the irreducible factor $x^2 + 2px + q$ in Q. The coefficients A_i, B_j and C_j that appear in these formulae are found by solving linear systems. The decomposition thus achieved is called the *decomposition into partial fractions* of f.

We illustrate the partial fraction decomposition, and the subsequent integration, with a very simple example. Consider the rational function $f(x) = (2x + 3)/(x^3 - 1)$. Since $Q(x) = x^3 - 1 = (x - 1)(x^2 + x + 1)$, we look for A, B and C such that

$$\frac{2x + 3}{x^3 - 1} = \frac{A}{x - 1} + \frac{Bx + C}{x^2 + x + 1} = \frac{(A + B)x^2 + (A - B + C)x + (A - C)}{x^3 - 1}.$$

This holds if and only if

$$\begin{cases} A + B = 0 \\ A - B + C = 2 \\ A - C = 3 \end{cases} \iff A = \frac{5}{3}, \quad B = -\frac{5}{3}, \quad C = -\frac{4}{3}.$$

Therefore

$$\int f(x)\,dx = \frac{5}{3} \int \frac{1}{x - 1}\,dx - \frac{1}{3} \int \frac{5x + 4}{x^2 + x + 1}\,dx$$

$$= \frac{5}{3} \log|x - 1| - \frac{5}{6} \log(x^2 + x + 1) - \frac{\sqrt{3}}{3} \arctan\left(\frac{2x + 1}{\sqrt{3}}\right) + c.$$

Notice that $\mathrm{Dom}(f) = (-\infty, 1) \cup (1, +\infty)$, the disjoint union of two intervals. Hence the constant c appearing above must be interpreted as in (10.2) with $n = 2$.

10.3 Riemann Integrals

The Riemann integral of a bounded function f on an interval $[a, b]$, if existing, is meant to capture the area lying below its graph, if f is non negative. To the portions of the graph that lie below the x-axis will be assigned a negative value. This notion is defined via an approximation process, by considering partitions of $[a, b]$ into smaller subintervals on each of which the function is replaced by its minimum, thereby approximating it from below by a piecewise constant function, and similarly from above. A natural notion of area is associated to each such approximation, that of the skyscraper-like profile thus obtained: the sum of the areas of the rectangles whose bases are the subintervals and whose heights are the infima (or suprema) of f in the intervals. By refining the subdivisions the approximations get better and under favourable circumstances the process converges. More precisely, the approximations from below increase as the partition gets finer, and those from above decrease, and a sound notion of integral is met when the two limiting processes reach a common value.

Definition 10.2 (*Partitions*) A *partition* of the interval $[a, b]$ is a finite collection of points $P = \{x_0, x_1, \ldots, x_n\}$ satisfying

$$a = x_0 < x_1 < x_2 < \cdots < x_n = b.$$

The collection of all partitions of $[a, b]$ will be denoted $\mathscr{P}$.

Definition 10.3 (*Riemann sums*) Let $f : [a, b] \to \mathbb{R}$ be a bounded function, and let $P \in \mathscr{P}$ be a partition of $[a, b]$, say $P = \{x_0, x_1, \ldots, x_n\}$, and put

$$m_i = \inf\{f(x) : x \in [x_{i-1}, x_i]\}, \qquad M_i = \sup\{f(x) : x \in [x_{i-1}, x_i]\}.$$

The real number

$$s(P, f) = \sum_{i=1}^{n} m_i(x_i - x_{i-1})$$

is called the *lower Riemann sum* of f associated to P, and the real number

$$S(P, f) = \sum_{i=1}^{n} M_i(x_i - x_{i-1})$$

is called the *upper Riemann sum* of f associated to P.

Definition 10.4 (*Riemann integral*) Let $f : [a, b] \to \mathbb{R}$ be a bounded function. The real number

$$s(f) = \sup\{s(P, f) : P \in \mathscr{P}\}$$

is called the *lower Riemann integral* of f on $[a, b]$, and the real number

$$S(f) = \inf\{S(P, f) : P \in \mathscr{P}\}$$

is called the *upper Riemann integral* of f on $[a, b]$. The function f is said to be *Riemann integrable* (or simply *integrable*) on $[a, b]$ provided that $s(f) = S(f)$. In this case, the real number $s(f) = S(f)$ is denoted

$$\int_a^b f(x)\,dx$$

and is called the *Riemann integral* of f on $[a, b]$, or the *definite integral* of f on $[a, b]$, or simply the *integral* of f on $[a, b]$. The collection of all Riemann integrable functions on $[a, b]$ will be denoted $\mathscr{R}([a, b])$.

Notice the difference between the notions of indefinite and definite integral. Their very close relation is the content of the fundamental theorem of calculus discussed below in Sect. 10.4. If $f \in \mathscr{R}([a, b])$, then, by definition

$$\int_b^a f(x)\,\mathrm{d}x = -\int_a^b f(x)\,\mathrm{d}x, \qquad \int_c^c f(x)\,\mathrm{d}x = 0$$

for every $c \in [a, b]$. If $f(x) \geq 0$ for every $x \in [a, b]$, the subset of the plane

$$T = \left\{(x, y) \in \mathbb{R}^2 : 0 \leq y \leq f(x)\right\}$$

is called the *trapezoid* under the graph of f and, by definition

$$\mathrm{Area}(T) = \int_a^b f(x)\,\mathrm{d}x.$$

The following theorem exhibits a rich class of examples of integrable functions.

Theorem 10.5 *(i) If $f : [a, b] \to \mathbb{R}$ is continuous in $[a, b]$, then it is integrable on $[a, b]$;*

(ii) if $f : [a, b] \to \mathbb{R}$ is monotonic and bounded on $[a, b]$, then it is integrable on $[a, b]$;

(iii) if $f : [a, b] \to \mathbb{R}$ is piecewise continuous and bounded in $[a, b]$, then it is integrable on $[a, b]$.

A classical example of a function that is not integrable on an interval is the *Dirichlet function*, namely

$$D(x) = \begin{cases} 1 & x \in [0, 1] \cap \mathbb{Q} \\ 0 & x \in [0, 1] \setminus \mathbb{Q}. \end{cases}$$

It is immediate to check that $s(D) = 0$ and $S(D) = 1$.

Theorem 10.6 *Suppose that $f, g \in \mathscr{R}([a, b])$. Then*

(i) $f + g \in \mathscr{R}([a, b])$ and $\int_a^b (f(x) + g(x))\,\mathrm{d}x = \int_a^b f(x)\,\mathrm{d}x + \int_a^b g(x)\,\mathrm{d}x$;

(ii) $\lambda f \in \mathscr{R}([a, b])$ for every $\lambda \in \mathbb{R}$ and $\int_a^b \lambda f(x)\,\mathrm{d}x = \lambda \int_a^b f(x)\,\mathrm{d}x$;

(iii) if $f(x) \leq g(x)$ for every $x \in [a, b]$, then $\int_a^b f(x)\,\mathrm{d}x \leq \int_a^b g(x)\,\mathrm{d}x$;

(iv) $|f| \in \mathscr{R}([a, b])$ and $\left| \int_a^b f(x)\,\mathrm{d}x \right| \leq \int_a^b |f(x)|\,\mathrm{d}x$;

(v) if $c \in (a, b)$, then $f \in \mathscr{R}([a, c]) \cap \mathscr{R}([c, b])$ and $\int_a^b f(x)\,\mathrm{d}x = \int_a^c f(x)\,\mathrm{d}x + \int_c^b f(x)\,\mathrm{d}x$.

Notice that item (c) and (d) of the above theorem imply in particular that

$$\left| \int_a^b f(x)\,\mathrm{d}x \right| \le (b-a) \sup_{x\in[a,b]} |f(x)|.$$

Item (e) expresses the additivity of the integral with respect to the domain of integration.

Theorem 10.7 (Mean value theorem) *Suppose that* $f, g \in \mathscr{R}([a, b])$. *Then*

(i) if $M = \sup\limits_{x\in[a,b]} f(x)$ *and* $m = \inf\limits_{x\in[a,b]} f(x)$, *then*

$$m \le \frac{1}{b-a} \int_a^b f(x)\,\mathrm{d}x \le M;$$

(ii) if f *is continuous in* $[a, b]$, *then there exists* $c \in [a, b]$ *such that*

$$\int_a^b f(x)\,\mathrm{d}x = f(c)(b-a).$$

10.4 The Fundamental Theorem of Calculus

The following theorem clarifies, together with its corollaries, the relation between indefinite and definite (Riemann) integrals.

Theorem 10.8 (Fundamental theorem of calculus) *Suppose that* $f : [a, b] \to \mathbb{R}$ *is continuous and take* $c \in [a, b]$. *Then the function*

$$F(x) = \int_c^x f(t)\,\mathrm{d}t \tag{10.8}$$

is differentiable in $[a, b]$ *and* $F'(x) = f(x)$ *for every* $x \in [a, b]$.

Formula (10.8) permits at once to see that continuous functions are derivatives, a fact that was anticipated in Sect. 10.1. Here is the formal statement.

Corollary 10.1 *Let* I *be an interval and let* $f : I \to \mathbb{R}$ *be continuous on* I. *Then* f *admits primitive functions in* I.

The next corollary is most useful in the actual computation of integrals: just find a primitive function and evaluate the difference at the endpoints.

Corollary 10.2 *Suppose that* $f : [a, b] \to \mathbb{R}$ *is continuous and let* F *be a primitive function of* f *in* $[a, b]$. *Then*

$$\int_a^b f(x)\,\mathrm{d}x = F(b) - F(a).$$

Because of the above result, integration by substitution and by parts have "definite" counterparts.

Theorem 10.9 (Integration by substitution, II) *Suppose that $f : [a, b] \to \mathbb{R}$ is continuous and that and $g : [\alpha, \beta] \to [a, b]$ is differentiable on $[\alpha, \beta]$ with continuous derivative and satisfies $g(\alpha) = a$, $g(\beta) = b$. Then*

$$\int_a^b f(x)\,\mathrm{d}x = \int_\alpha^\beta f(g(x))g'(x)\,\mathrm{d}x. \tag{10.9}$$

Theorem 10.10 (Integration by parts, II) *Suppose that $f, g : [a, b] \to \mathbb{R}$ are differentiable on $[a, b]$ with continuous derivatives. Then*

$$\int_a^b f'(x)g(x)\,\mathrm{d}x = f(b)g(b) - f(a)g(a) - \int_a^b f(x)g'(x)\,\mathrm{d}x. \tag{10.10}$$

10.5   Guided Exercises on Integration

10.1 Consider the function

$$f(x) = \begin{cases} \dfrac{x+1}{\sqrt{x-1}} & x > 1 \\ 0 & x = 1 \\ e^x & x < 1. \end{cases}$$

(a) Determine the intervals $[a, b]$ in which f has a primitive function.
(b) Find the explicit expression of the primitive F of f in $(1, +\infty)$ such that $F(2) = 0$.

Answer. (a) Clearly, f is defined in $\mathbb{R}$ and continuous in $\mathbb{R} \setminus \{1\}$ because $f(x) \to +\infty$ as $x \to 1^+$. Therefore, f has a primitive function in every interval $[a, b]$ for which either $a < b < 1$ or $1 < a < b$. Furthermore, f has a left jump discontinuity, that is $f(x) \to e$ as $x \to 1^-$, while $f(1) = 0$. Thus, the previous intervals are the only ones in which a primitive can be found.
(b) Since f is continuous in $(1, +\infty)$, it does have primitive functions in this interval. From the substitution $x(t) = t^2 + 1$ with $t > 0$ it follows $t(x) = \sqrt{x - 1}$ and hence

$$F(x) = \int \frac{x+1}{\sqrt{x-1}}\,\mathrm{d}x$$

$$= \int \frac{x(t)+1}{\sqrt{x(t)-1}}\,x'(t)\,\mathrm{d}t$$

$$= 2 \int (2 + t^2) \, dt$$

$$= 2 \left(2\sqrt{x-1} + \frac{(x-1)}{3} \sqrt{x-1} + c \right)$$

with $c \in \mathbb{R}$. Now, $F(2) = \frac{7}{3} + c = 0$ implies $c = -7/3$ and therefore

$$F(x) = 2 \left(2\sqrt{x-1} + \frac{(x-1)}{3} \sqrt{x-1} - \frac{7}{3} \right).$$

This is a basic exercise on indefinite integration. In part (a) the task is to recall that a function that is continuous in an interval admits primitive functions in it, whereas a function diverging at point (say from the right) cannot have a primitive function in a neighborhood of that point. In part (b) one first computes an indefinite integral and then applies the structure theorem on primitive functions on an interval, finding the constant c that meets the requirement $F(2) = 0$. As for the computation, the substitution is guided by the observation that

$$\frac{d}{dx} \sqrt{x-1} = \frac{1}{2\sqrt{x-1}}$$

which leads to the natural choice $t(x) = \sqrt{x-1}$.

10.2 Draw the graph of the primitive F of $f(x) = \sqrt{\dfrac{x - x^3 + 6}{x}}$ that satisfies $F(1) = 0$.

Answer. Since

$$\sqrt{\frac{x - x^3 + 6}{x}} = \sqrt{\frac{(2-x)(x^2 + 2x + 3)}{x}}$$

and the polynomial $x^2 + 2x + 3$ has negative discriminant, it is easy to see that $\text{Dom}(f) = (0, 2]$. Furthermore, $f \in C^0((0, 2])$ because it is a composition of continuous functions. Therefore f has antiderivatives in $(0, 2]$. Denote by F the antiderivative of f such that $F(1) = 0$. It follows in particular that

$$F'(x) = f(x) = \sqrt{\frac{x - x^3 + 6}{x}},$$

so that F is strictly increasing in $(0, 2]$. Further, it is twice differentiable in $(0, 2)$ and

$$F''(x) = f'(x) = \frac{-x^3 - 3}{x^2} \sqrt{\frac{x}{x - x^3 + 6}} < 0.$$

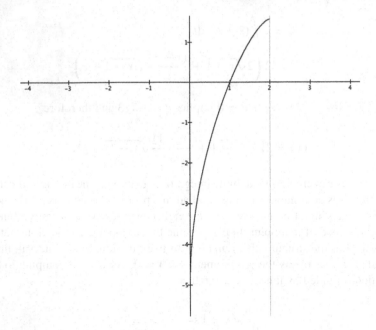

Fig. 10.1 The graph of F, Exercise 10.2

Therefore F is a strictly increasing and concave function in $(0, 2]$. Its graph is thus as in Fig. 10.1. In this standard exercise the idea is that the information concerning the derivative of a function which is defined as antiderivative is readily available, while its second derivative is computed with a single differentiation. In the case at hand, F is increasing because f is positive and concave because f' is negative.

10.3 Consider the function $f(x) = \dfrac{e^x}{e^{2x} + e^x + k}$.

(a) For which values of $k \in \mathbb{R}$ does f admit primitive functions defined in $\mathbb{R}$?
(b) Put $k = 5/4$. Find all the primitive functions of f, specifying whether they are bounded on their domains.

Answer. (a) Clearly $\text{Dom}(f) = \{x \in \mathbb{R} : e^{2x} + e^x + k \neq 0\}$. Now, if $k \geq 0$, then $\text{Dom}(f) = \mathbb{R}$ and f has primitive functions defined on $\mathbb{R}$ because it is everywhere continuous. If $k < 0$, since $f(x) \to k$ as $x \to -\infty$ and $f(x) \to +\infty$ as $x \to +\infty$, by the intermediate value theorem there exists at least a point where the denominator vanishes, at which f is not defined. Consequently it does not admit globally defined primitive functions.
(b) Put $k = 5/4$. The primitive functions of f are found by computing the indefinite integral of f by means of the substitution $e^x = t$, which yields

$$\int \frac{e^x}{e^{2x} + e^x + k} \, dx = \left[\int \frac{4}{(2t+1)^2 + 4} \, dt \right]_{t=e^x}.$$

Substituting again $u = t + 1/2$ it follows that

$$\int \frac{4}{(2t+1)^2 + 4} \, dt = \left[\int \frac{4}{4u^2 + 4} \, du \right]_{u=t+1/2} = \arctan\left(t + \frac{1}{2}\right) + c.$$

Therefore

$$\int f(x) \, dx = \arctan\left(e^x + \frac{1}{2}\right) + c$$

with $c \in \mathbb{R}$. Observe that these are all bounded in $\mathbb{R}$, because

$$|\arctan\left(e^x + \frac{1}{2}\right)| < \frac{\pi}{2}$$

for every $x \in \mathbb{R}$.

This exercise asks first to determine when the denominator of the given function never vanishes, which is at a glance achieved for non negative k. After a quick computation, $k \geq 0$ is found to be necessary as well. In this case, f is continuous on $\mathbb{R}$ and therefore has globally defined antiderivatives. The second request is the computation of an indefinite integral that clearly amounts to integrating a rational function. One should keep in mind that the prototypical bounded antiderivatives of rational functions arise as compositions with the arctangent function, that is the antiderivative of $(1 + u^2)^{-1}$. In turn, this is the lowest degree prototypical rational function that is globally defined, that is, without zeroes at the denominator.

10.4 Consider the function $f(x) = \dfrac{x^2 - 4}{|x + 2| + k}$.

(a) For which values of $k \in \mathbb{R}$ does f admit primitive functions defined on $\mathbb{R}$?
(b) Put $k = 1$. Find, if existing, the primitive functions F of f for which $F(0) = 0$.

Answer. (a) Evidently, $\text{Dom}(f) = \{x \in \mathbb{R} : |x + 2| + k \neq 0\}$. Hence, if $k > 0$ then $\text{Dom}(f) = \mathbb{R}$ and $f \in C^0(\mathbb{R})$. If $k \leq 0$ then $\text{Dom}(f)$ is not $\mathbb{R}$ and therefore f does not have globally defined primitive functions
(b) If $k = 1$, then

$$f(x) = \begin{cases} \dfrac{x^2 - 4}{x + 3} & x \geq -2 \\[3mm] \dfrac{4 - x^2}{x + 1} & x < -2. \end{cases}$$

By (a), f has globally defined primitive functions. For $x \geq -2$

$$\int \frac{x^2 - 4}{x + 3} \, dx = \int \left(x - 3 + \frac{5}{x + 3}\right) dx = \frac{x^2}{2} - 3x + 5 \log(x + 3) + c_1$$

with $c_1 \in \mathbb{R}$, whereas for $x < -2$

$$\int \frac{-x^2 + 4}{x + 1}\,dx = \int (-x + 1 + \frac{3}{x + 1})\,dx = -\frac{x^2}{2} + x + 3\log|x + 1| + c_2$$

with $c_2 \in \mathbb{R}$. Put

$$F(x) = \begin{cases} \frac{x^2}{2} - 3x + 5\log(x + 3) + c_1 & x > -2 \\ c_3 & x = -2 \\ -\frac{x^2}{2} + x + 3\log|x + 1| + c_2 & x < -2 \end{cases}$$

with $c_1, c_2, c_3 \in \mathbb{R}$. Now, F is a primitive of f in $\mathbb{R}$ if $F'(x) = f(x)$ for every $x \in \mathbb{R}$. As shown above, $F'(x) = f(x)$ for every $x \in \mathbb{R} \setminus \{-2\}$. Furthermore, F is continuous at -2 provided that

$$\lim_{x \to -2^-} F(x) = \lim_{x \to -2^+} F(x) = F(-2),$$

which implies $c_2 - 4 = c_3 = c_1 + 8$. By the limit of the derivative criterion, namely Corollary 7.6 in Chap. 7, F is differentiable at -2 with $F'(-2) = f(-2)$. It follows that the globally defined primitive functions of f are

$$F(x) = \begin{cases} \frac{x^2}{2} - 3x + 5\log(x + 3) + c_1 & x > -2 \\ 8 + c_1 & x = -2 \\ -\frac{x^2}{2} + x + 3\log|x + 1| + 12 + c_1 & x < -2 \end{cases}$$

with $c_1 \in \mathbb{R}$. Finally, $F(0) = 0$ if and only if $5\log 3 + c_1 = 0$, which implies $c_1 = -5\log 3$.

This is a typical exercise on gluing antiderivatives. The point is that if one explicitly finds all the primitive functions on $(-\infty, a)$, which are parametrized by a real constant, and all the primitive functions on $(a, +\infty)$, which are parametrized by another real constant, then tuning the constants in such a way that the two branches of F meet at a, and furthermore fixing $F(a)$ in such a way that F is actually continuous at a, then the limit of the derivative criterion, namely Corollary 7.6 in Chap. 7, guarantees that the so obtained continuous function F is automatically a differentiable function which is therefore a global primitive function of f. This is because the existence and coincidence of the left and right limits of F at a entails the existence and coincidence of the left and right limits of f' at a; since f is continuous at a to begin with, the limit of the derivative criterion applies and yields the differentiability of f at a.

10.5 Find all the primitive functions of $f(x) = ||e^x - 1| - 1|$.

Answer. Being the composition of continuous functions on $\mathbb{R}$, f is continuous on $\mathbb{R}$ and can be written as:

$$f(x) = \begin{cases} |e^x - 2| & x \geq 0 \\ e^x & x < 0 \end{cases} = \begin{cases} e^x - 2 & x \geq \log 2 \\ 2 - e^x & 0 \leq x < \log 2 \\ e^x & x < 0. \end{cases}$$

Since

$$\int e^x \, dx = e^x + c_1, \qquad \int (2 - e^x) \, dx = 2x - e^x + c_2, \qquad \int (e^x - 2) \, dx = e^x - 2x + c_3,$$

we are led to define

$$F(x) = \begin{cases} e^x - 2x + c_3 & x > \log 2 \\ c_5 & x = \log 2 \\ 2x - e^x + c_2 & 0 < x < \log 2 \\ c_4 & x = 0 \\ e^x + c_1 & x < 0 \end{cases}$$

with $c_i \in \mathbb{R}$ for $i = 1, \ldots, 5$. Recall that F is a primitive of f in $\mathbb{R}$ provided that $F'(x) = f(x)$ for every $x \in \mathbb{R}$, and by the above computations this is certainly true for $x \in \mathbb{R} \setminus \{0, \log 2\}$. In order for F to be continuous at 0 and $\log 2$ it must hold

$$\lim_{x \to 0^-} F(x) = \lim_{x \to 0^+} F(x) = F(0), \qquad \lim_{x \to \log 2^-} F(x) = \lim_{x \to \log 2^+} F(x) = F(\log 2).$$

This implies

$$\begin{cases} 1 + c_1 = c_2 - 1 = c_4 \\ 2 \log 2 - 2 + c_2 = 2 - 2 \log 2 + c_3 = c_5 \end{cases}$$

and hence $c_2 = 2 + c_1$, $c_3 = 4 \log 2 - 2 + c_1$, $c_4 = c_1 + 1$ and $c_5 = 2 \log 2 + c_1$. By the limit of the derivative criterion (Corollary 7.6 in Chap. 7), for these values of c_2, c_3, c_4, c_5 the function F is differentiable in 0 and $\log 2$ as well. As c_1 ranges in $\mathbb{R}$, these are all the globally defined primitive functions of f.

This exercise is just a variation on the theme played in Exercise 10.4.

10.6 Compute $\displaystyle\int_4^9 \frac{\log(\sqrt{x} - 1)}{(x + 1)^2} \, dx$.

Answer. Denote by f the integrand. Since $f \in C^0([4, 9])$, its definite integral on $[4, 9]$ exists. It is advisable to use integration by parts setting $g(x) = -1/(1 + x)$ and $h(x) = \log(\sqrt{x} - 1)$, so that $f(x) = g'(x) \cdot h(x)$ and

$$\int_4^9 f(x) \, dx = g(9)h(9) - g(4)h(4) + \int_4^9 \frac{1}{x + 1} \frac{1}{\sqrt{x} - 1} \frac{1}{2\sqrt{x}} \, dx.$$

The last integral calls for the natural substitution $z(x) = \sqrt{x}$, that leads to

$$\int_4^9 \frac{1}{x+1} \frac{1}{\sqrt{x}-1} \frac{dx}{2\sqrt{x}} = \int_4^9 \frac{1}{z^2(x)+1} \frac{1}{z(x)-1} z'(x)\,dx = \int_2^3 \frac{1}{z^2+1} \frac{1}{z-1}\,dz.$$

Finally, decomposing into partial fractions

$$\frac{1}{z^2+1} \cdot \frac{1}{z-1} = \frac{Az+B}{z^2+1} + \frac{C}{z-1}$$
$$= \frac{Az^2+Bz-Az-B+Cz^2+C}{(z^2+1)(z-1)}$$
$$= \frac{(A+C)z^2+(B-A)z+C-B}{(z^2+1)(z-1)}.$$

These rational functions are equal if and only if

$$\begin{cases} A+C=0 \\ B-A=0 \\ C-B=1. \end{cases}$$

Hence $A = -\frac{1}{2} = B$ and $C = \frac{1}{2}$, so that

$$\int_2^3 \frac{1}{z^2+1} \cdot \frac{1}{z-1}\,dz = -\frac{1}{2}\int_2^3 \frac{z+1}{z^2+1} + \frac{1}{2}\int_2^3 \frac{1}{z-1}\,dz$$
$$= -\frac{1}{2}\left[\frac{1}{2}\int_2^3 \frac{2z}{z^2+1} + \int_2^3 \frac{1}{z^2+1}\,dz\right] + \frac{1}{2}\log 2$$
$$= -\frac{1}{2}\left[\frac{1}{2}(\log 10 - \log 5) + \arctan 3 - \arctan 2\right] + \frac{1}{2}\log 2$$
$$= -\frac{1}{4}\log 2 - \frac{1}{2}(\arctan 3 - \arctan 2) + \frac{1}{2}\log 2.$$

In conclusion

$$\int_4^9 f(x)\,dx = -\frac{\log 2}{10} + \frac{1}{4}\log 2 - \frac{1}{2}(\arctan 3 - \arctan 2).$$

This exercise displays the use of three integration techniques: first an integration by parts (which is dictated by the fact that $1/(x+1)^2$ is a derivative and by the fact that logarithms "simplify" under differentiation), then a substitution (which is dictated by the many square roots and in particular by the multiplicative factor $1/2\sqrt{x}$ that nicely translates the change of variable) and finally a partial fraction decomposition that handles the rational function to which the previous operations have led.

10.7 Consider the function

$$f(x) = \begin{cases} (3x^2 - 1) \arccos x & x > 0 \\ \dfrac{x - r}{x^2 - 1} & x \le 0 \end{cases}$$

where $r \in \mathbb{R}$.

(a) Determine for which values of r, if any, f has primitive functions in $[-\frac{1}{2}, \frac{1}{2}]$.

(b) Determine for which values of r, if any, f is integrable in $[-\frac{1}{2}, \frac{1}{2}]$.

(c) Put $r = 0$. Compute $\int_{-1/2}^{1/2} f(x)\,dx$.

Answer. (a) For any $r \in \mathbb{R}$, f is continuous at least in $[-1/2, 1/2] \setminus \{0\}$. Furthermore, $f(x) \to -\pi/2$ as $x \to 0^+$ and $f(x) \to r$ as $x \to 0^-$. Therefore, if $r = -\pi/2$ then f is continuous in $[-1/2, 1/2]$ and thus has primitive functions. If $r \ne -\pi/2$, then f has a jump discontinuity at 0 and therefore it does not admit any primitive in any neighborhood of 0.

(b) Certainly f is continuous in $[-1/2, 1/2]$ except for at most a point, where it admits both left and right (finite) limits. In particular f is bounded. Therefore f is integrable in $[-1/2, 1/2]$ for every r.

(c) Put $r = 0$. Then

$$\int_{-\frac{1}{2}}^{\frac{1}{2}} f(x)\,dx = \int_{-\frac{1}{2}}^{0} \frac{x}{x^2 - 1}\,dx + \int_{0}^{\frac{1}{2}} (3x^2 - 1) \arccos x\,dx$$

$$= \frac{1}{2} \log |x^2 - 1| \Big|_{-\frac{1}{2}}^{0} + \int_{0}^{\frac{1}{2}} (3x^2 - 1) \arccos x\,dx.$$

The last integral can be computed by parts, putting $g(x) = x^3 - x$ and $h(x) = \arccos x$, so that $(3x^2 - 1) \arccos x = g'(x)h(x)$ and

$$\int_{0}^{\frac{1}{2}} (3x^2 - 1) \arccos x\,dx = g\left(\frac{1}{2}\right) h\left(\frac{1}{2}\right) - g(0)h(0) + \int_{0}^{\frac{1}{2}} (x^3 - x)\frac{1}{\sqrt{1 - x^2}}\,dx$$

$$= -\frac{\pi}{8} - \int_{0}^{\frac{1}{2}} x\sqrt{1 - x^2}\,dx$$

$$= -\frac{\pi}{8} + \frac{1}{2}(1 - x^2)^{\frac{3}{2}} \frac{2}{3} \Big|_{0}^{\frac{1}{2}}$$

$$= -\frac{\pi}{8} + \frac{1}{3}\left(\frac{3}{4}\frac{\sqrt{3}}{2} - 1\right).$$

It follows that

$$\int_{-\frac{1}{2}}^{\frac{1}{2}} f(x)\,dx = -\frac{1}{2} \log \frac{3}{4} - \frac{\pi}{8} + \frac{\sqrt{3}}{8} - \frac{1}{3}.$$

The first two questions are designed to stress the difference between indefinite and definite integrability: in the former case a jump discontinuity is an obstacle, in the latter no. The third question is about integration by parts.

10.8 Compute, if existing, $\displaystyle\int_0^{\frac{\pi}{2}} \frac{\sin 2x}{\cos^2 x + 2\sin x - 3}\,dx$.

Answer. The integrand $f(x)$ is globally defined and furthermore $f \in C^0(\mathbb{R})$. Therefore f is integrable in $[0, \pi/2]$. Computing,

$$\int_0^{\frac{\pi}{2}} f(x)\,dx = \int_0^{\frac{\pi}{2}} \frac{2\sin x \cos x}{1 - \sin^2 x + 2\sin x - 3}\,dx$$

$$= \int_0^{\frac{\pi}{2}} \frac{2\sin x}{-\sin^2 x + 2\sin x - 2}\cos x\,dx.$$

Under the substitution $z(x) = \sin x$ this becomes

$$\int_0^{\frac{\pi}{2}} \frac{2\sin x}{-\sin^2 x + 2\sin x - 2}\cos x\,dx = \int_0^{\frac{\pi}{2}} \frac{2z(x)}{-z^2(x) + 2z(x) - 2}z'(x)\,dz$$

$$= \int_0^1 \frac{2z}{-z^2 + 2z - 2}\,dz$$

$$= -\int_0^1 \frac{2z - 2 + 2}{(z-1)^2 + 1}\,dz$$

$$= -\int_0^1 \frac{2(z-1)}{(z-1)^2 + 1}\,dz - 2\int_0^1 \frac{1}{(z-1)^2 + 1}\,dz$$

$$= -\log[(z-1)^2 + 1]\Big|_0^1 - 2\arctan(z-1)\Big|_0^1$$

$$= \log 2 + 2\frac{\pi}{4}.$$

This exercise shows an example on how to convert the integral of a function that is rational as a function of sines and cosines into an integral of an ordinary rational function. A key simplifying fact is the multiplicative factor $\cos x$, that nicely helps in the subsitution.

10.9 Compute, if existing, $\displaystyle\int_0^{\frac{1}{2}\log 3} e^x \arctan e^x\,dx$.

Answer. Evidently, the integrand f is in $C^0(\mathbb{R})$, hence it is integrable in $[0, \frac{1}{2}\log 3]$. Upon setting $z(x) = e^x$ it follows that

$$\int_0^{\frac{1}{2}\log 3} e^x \arctan e^x\,dx = \int_0^{\frac{1}{2}\log 3} \arctan(z(x))z'(x)\,dx = \int_1^{\sqrt{3}} \arctan z\,dz.$$

The latter integral can be integrated by parts. If $h(z) = z$ and $g(z) = \arctan z$, then

$$\int_1^{\sqrt{3}} \arctan z \, dz = h(\sqrt{3})g(\sqrt{3}) - g(1)h(1) - \int_1^{\sqrt{3}} h(z)g'(z) \, dz$$

$$= \sqrt{3}\frac{\pi}{3} - \frac{\pi}{4} - \int_1^{\sqrt{3}} \frac{z}{1+z^2} \, dz$$

$$= \sqrt{3}\frac{\pi}{3} - \frac{\pi}{4} - \frac{1}{2}\log(z^2+1)|_1^{\sqrt{3}}$$

$$= \sqrt{3}\frac{\pi}{3} - \frac{\pi}{4} - \frac{1}{2}\log 4 + \frac{1}{2}\log 2.$$

This is a very simple exercise on basic integration techniques: substitution, parts.

10.10 Compute, if existing, $\displaystyle\int_0^1 x \arctan \sqrt{x} \, dx$.

Answer. The function $f(x) = x \arctan \sqrt{x}$ is in $C^0([0, +\infty))$, hence it is integrable in $[0, 1]$. If $h(x) = \frac{x^2}{2}$ and $g(x) = \arctan\sqrt{x}$ then $h'(x)g(x) = f(x)$ and integrating by parts

$$\int_0^1 f(x) \, dx = h(1)g(1) - h(0)g(0) - \int_0^1 h(x)g'(x) \, dx$$

$$= \frac{\pi}{8} - \int_0^1 \frac{x^2}{2} \cdot \frac{1}{1+x} \cdot \frac{1}{2\sqrt{x}} \, dx.$$

Under the substitution $z(x) = \sqrt{x}$ the integral appearing above becomes

$$\int_0^1 \frac{x^2}{2(1+x)} \cdot \frac{1}{2\sqrt{x}} \, dx = \int_0^1 \frac{z^4(x)}{2(1+z^2(x))} z'(x) \, dx$$

$$= \int_0^1 \frac{z^4}{2(1+z^2)} \, dz$$

$$= \frac{1}{2} \int_0^1 (z^2 - 1 + \frac{1}{z^2+1}) \, dz$$

$$= \frac{1}{2}\left[\frac{z^3}{3} - z + \arctan z\right]_0^1.$$

Hence

$$\int_0^1 f(x) \, dx = \frac{1}{3}.$$

This is again an exercise on basic integration techniques, this time in reverse order: parts, substitution.

10.11 Consider the function

$$f(x) = \begin{cases} x \sin^2(x) + \alpha & x \le 0 \\[2mm] \dfrac{\beta}{\sin x + \cos x} & x > 0. \end{cases}$$

(a) Find the values of $\alpha, \beta \in \mathbb{R}$, if any, for which f has a primitive function in $[-\pi/2, \pi/3]$.
(b) Find the values of $\alpha, \beta \in \mathbb{R}$, if any, for which f is integrable in $[-\pi/2, \pi/3]$.
(c) Compute, if existing, $\int_{-\pi/2}^{\pi/3} f(x)\, dx$ with $\alpha = 0$ and $\beta = 1$.

Answer. (a) The function f is continuous at least in $[-\pi/2, \pi/3] \setminus \{0\}$. Furthermore, $f(x) \to \beta$ as $x \to 0^+$ and $f(x) \to \alpha$ as $x \to 0^-$. Thus, if $\alpha = \beta$, then f is continuous in $[-\pi/2, \pi/3]$ where it therefore admits primitive functions. If $\alpha \ne \beta$ f, then f has a jump discontinuity at 0 and therefore it does not admit primitive functions in $[-\pi/2, \pi/3]$.

(b) As shown in (a), f is continuous in $[-\pi/2, \pi/3] \setminus \{0\}$ and bounded, hence integrable in $[-\pi/2, \pi/3]$ for every α and β.

(c) For $\alpha = 0$ and $\beta = 1$ the indicated integral is equal to

$$\int_{-\frac{\pi}{2}}^{\frac{\pi}{3}} f(x)\, dx = \int_{-\frac{\pi}{2}}^{0} x \sin^2 x\, dx + \int_{0}^{\frac{\pi}{3}} \frac{1}{\sin x + \cos x}\, dx.$$

As far as the first integral is concerned, if $h(x) = \sin^2(x)$, then a primitive function of h is $(x - \sin x \cos x)/2$, so that a simple integration by parts gives

$$\int_{-\frac{\pi}{2}}^{0} x \sin^2 x\, dx = -\frac{\pi}{2}\frac{\pi}{4} - \int_{-\frac{\pi}{2}}^{0} \left(\frac{x}{2} - \frac{\sin x \cos x}{2} \right) dx$$

$$= -\frac{\pi^2}{8} - \frac{x^2}{4}\Big|_{-\frac{\pi}{2}}^{0} + \frac{1}{2}\frac{\sin^2 x}{2}\Big|_{-\frac{\pi}{2}}^{0}$$

$$= -\frac{\pi^2}{16} - \frac{1}{4}.$$

For the second integral, the substitution $x(t) = 2 \arctan t$ yields

$$\int_{0}^{\frac{\pi}{3}} \frac{1}{\sin x + \cos x}\, dx = \int_{0}^{\frac{\sqrt{3}}{3}} \frac{1}{\sin x(t) + \cos x(t)} x'(t)\, dt$$

$$= \int_{0}^{\frac{\sqrt{3}}{3}} \frac{1}{\left(\frac{2t}{1+t^2} \right) + \left(\frac{1-t^2}{1+t^2} \right)} \frac{2}{1+t^2}\, dt$$

$$= \int_{0}^{\frac{\sqrt{3}}{3}} \frac{2}{-t^2 + 2t + 1}\, dt.$$

The next step is to decompose the rational function to be integrated into partial fractions. The denominator has the real roots $1 \pm \sqrt{2}$, hence

$$\frac{-2}{t^2 - 2t - 1} = \frac{A}{t - (1 - \sqrt{2})} + \frac{B}{t - (1 + \sqrt{2})} = \frac{At - A(1 + \sqrt{2}) + Bt - B(1 + \sqrt{2})}{(t - (1 - \sqrt{2}))(t - (1 + \sqrt{2}))}.$$

The indicated identity holds identically if and only if

$$\begin{cases} A + B = 0 \\ -A(1 + \sqrt{2}) - B(1 - \sqrt{2}) = -2 \end{cases}$$

that is, if $A = \sqrt{2}/2$ and $B = -\sqrt{2}/2$. Therefore

$$\int_0^{\frac{\sqrt{3}}{3}} \frac{-2}{t^2 - 2t - 1}\, dt = \frac{\sqrt{2}}{2} \int_0^{\frac{\sqrt{3}}{3}} \frac{dt}{t - (1 - \sqrt{2})} - \frac{\sqrt{2}}{2} \int_0^{\frac{\sqrt{3}}{3}} \frac{dt}{t - (1 + \sqrt{2})}$$

$$= \frac{\sqrt{2}}{2} \left\{ \log\left|\frac{\sqrt{3}}{3} - (1 - \sqrt{2})\right| - \log\left|1 - \sqrt{2}\right| - \log\left|\frac{\sqrt{3}}{3} - (1 + \sqrt{2})\right| + \log\left|1 + \sqrt{2}\right| \right\}$$

$$= \frac{\sqrt{2}}{2} \log\left[\frac{3 + \sqrt{3} + \sqrt{6}}{3 + \sqrt{3} - \sqrt{6}} \right].$$

Finally

$$\int_{-\frac{\pi}{2}}^{\frac{\pi}{3}} f(x)\, dx = -\frac{\pi^2}{16} - \frac{1}{4} + \frac{\sqrt{2}}{2} \log\left[\frac{3 + \sqrt{3} + \sqrt{6}}{3 + \sqrt{3} - \sqrt{6}} \right].$$

This exercise wraps up a few basic theoretical and practical issues on indefinte and definite integration: continuity versus piecewise continuity in item (a) and a combination of integration by parts, substitution and partial fractions in item (b). The substitution is sometimes expressed as $t = \tan(x/2)$, which is typical when the integrand is a ratio of linear combinations of sines and cosines.

10.6 Problems on Integration

10.12 Find all the primitive functions of $f(x) = ||x| + x^2 - 1|$.

10.13 In which intervals $[a, b]$ does the function $f(x) = 1/(x^2 + 2x - 3)$ admit primitive functions?

10.14 Find all the primitive functions of $f(x) = 1/|x^2 - 1|$.

10.15 Compute the following integrals:

(a) $\displaystyle\int_1^2 \log(4x^2 - 3x)\, dx$

(b) $\displaystyle\int_2^{e^\pi} \frac{(x^3 - 1)\sin^3(\log x) - 1}{x^4 - x}\, dx$

(c) $\displaystyle\int_{-\frac{2\pi}{3}}^{\frac{2\pi}{3}} \frac{1}{1 - \sqrt{2}\cos x}\, dx$

(d) $\displaystyle\int_1^3 \frac{1}{\sqrt{x}} \arctan \frac{1}{\sqrt{x}}\, dx$

(e) $\displaystyle\int_{\frac{\pi}{4}}^{\frac{\pi}{2}} \frac{1}{\sin x}\, dx$

(f) $\displaystyle\int_{\frac{\pi^2}{16}}^{\frac{\pi^2}{9}} \frac{1}{\sqrt{x}\sin\sqrt{x}}\, dx.$

(g) $\displaystyle\int_0^{\frac{\pi}{4}} \sin 2x \log(\cos x)\, dx$

(h) $\displaystyle\int_0^1 \arcsin\sqrt{x}\, dx$

(i) $\displaystyle\int_{-1}^0 \frac{x}{x^2 - 3x + 2}\, dx$

(l) $\displaystyle\int_1^2 x \arcsin\frac{1}{x}\, dx.$

10.16 In which intervals $[a, b]$ does

$$f(x) = \begin{cases} 2 & 1 < x \leq 2 \\ (x - 1)^2 & 0 \leq x \leq 1 \\ e^x - 1 & x < 0 \end{cases}$$

admit primitive functions?

10.17 Consider the function

$$f(x) = \begin{cases} x \sin x & x \geq 0 \\ \dfrac{1 - \cos x}{x^2} + k & x < 0. \end{cases}$$

(a) For which $k \in \mathbb{R}$ is f bounded in $[-1, 1]$?
(b) For which $k \in \mathbb{R}$ does f admit primitive functions in $[-1, 1]$?

10.18 Consider the function

$$f(x) = \begin{cases} \alpha \log(1 + x) & -1 < x \leq 0 \\ \beta x^{\frac{3}{2}} \sin(1/x) & x > 0. \end{cases}$$

(a) For which $\alpha, \beta \in \mathbb{R}$ does f admit primitive functions in $[-1/2, 1]$?
(b) Put $\alpha = \beta = 1$. Draw, in a neighborhood of 0, the graph of the primitive function F of f for which $F(0) = 1$.

10.19 Find all the primitive functions of $f(x) = e^{-x} \arctan(1 - e^x)$ and draw the graph of the primitive F such that $F(0) = 0$.

10.20 Find the domain of $f(x) = \sqrt{2x^2 - \sin^3 x}$ and write the second order McLaurin polynomial of the primitive function of f that vanishes at 0.

10.21 Find the primitive function of $f(x) = \sqrt[4]{x} \log x$ that vanishes at 1 and draw its graph.

10.22 Find the primitive function of $f(x) = \log(x - \frac{1}{x}) - x^2$ that vanishes at e and determine where it is concave.

10.23 Draw the graph of the primitive F of $f(x) = \log(e^x + x) - \log(e^x - xe)$ that vanishes at 0.

10.24 Write the secon order Taylor polynomial centered at $e - 1$ of the primitive function of $f(x) = \log(\log(x + 1))$ that vanishes at $e - 1$.

10.25 Compute, if existing, the integral $\displaystyle\int_{\frac{3}{4}\pi}^{\frac{5}{4}\pi} \frac{\tan x}{1 - \cos x}\, dx.$

10.26 Compute, if existing, the integral $\displaystyle\int_0^{\frac{1}{2}} \frac{\arcsin x}{\sqrt{1 - x^2}}\, dx.$

10.27 Compute, if existing, the integral $\displaystyle\int_{\frac{\pi}{6}}^{\frac{\pi}{3}} \frac{1}{\log(\sin x)\tan x}\, dx.$

10.28 Consider the function

$$f(x) = \begin{cases} \log|\sin x| \cdot \sin x & x \neq 0 \\ r & x = 0 \end{cases}$$

with $r \in \mathbb{R}$.

(a) For which $r \in \mathbb{R}$ does f admit primitive functions in $[-\pi/2, \pi/2]$?
(b) For which $r \in \mathbb{R}$ is f integrable in $[-\pi/2, \pi/2]$?
(c) Put $r = 1$. Compute, if existing, the integral $\int_{-\pi/4}^{\pi/2} f(x)\, dx.$

10.29 Consider the function

$$f(x) = \begin{cases} \dfrac{\sin x}{x} & x \neq 0 \\ 1 & x = 0. \end{cases}$$

Compute, if existing, the integral $\displaystyle\int_{-\frac{\pi}{2}}^{\pi} f(x) f'(x)\, dx.$

Chapter 11
Improper Integrals and Integral Functions

11.1 Improper Integrals

The notion of improper integral extends the notion of Riemann integral when either f is not bounded on some bounded interval, or the domain of the function is itself not bounded. Again, the notion that lies behind is that of the area below the graph of, say, a non negative function. If the domain is a single interval, the idea is to consider smaller intervals, obtained by shrinking the original interval at one side, and to assume that the given function is Riemann integrable on each such smaller interval. Then one takes the limit as the endpoint of the shrunken interval reaches the boundary. More general situations are handled by using the main additivity property of the integral with respect to domains, in the sense of item (v) of Theorem 10.6 in Chap. 10.

One of the most relevant applications of the notion of improper integral is in the study of *integral functions*, namely functions of the form

$$F(x) = \int_a^x f(t)\, dt. \tag{11.1}$$

Indeed, depending on the nature of $\mathrm{Dom}(f)$, it is possible to ask what happens as x approaches either a point at which f diverges or as x approaches $\pm\infty$, thereby exploring the natural domain of F. Also, the properties of F that depend on its derivative are reduced to those of f, by the fundamental theorem of calculus.

Definition 11.1 (*Unbounded functions*)

(i) Suppose that $f : (a, b] \to \mathbb{R}$ is bounded and Riemann integrable on $[x, b]$ for every $x \in (a, b)$. If the limit

$$\lim_{x \to a^+} \int_x^b f(t)\, dt \tag{11.2}$$

© Springer International Publishing Switzerland 2016
M. Baronti et al., *Calculus Problems*, UNITEXT - La Matematica per il 3+2 101,
DOI 10.1007/978-3-319-15428-2_11

exists then this limit is called the *improper integral* of f on $[a, b]$ and one writes

$$\int_a^b f(t)\,dt = \lim_{x \to a^+} \int_x^b f(t)\,dt.$$

If the limit is a real number, then it is said that f has *convergent improper integral* on $[a, b]$. If the limit (11.2) diverges, then f has *divergent improper integral* on $[a, b]$.

(ii) Similarly, if $f : [a, b) \to \mathbb{R}$ is bounded and Riemann integrable on $[a, x]$ for every $x \in (a, b)$ and if the limit

$$\lim_{x \to b^-} \int_a^x f(t)\,dt \qquad\qquad (11.3)$$

exists, then this limit is called improper integral of f on $[a, b]$ and one writes

$$\int_a^b f(t)\,dt = \lim_{x \to b^-} \int_a^x f(t)\,dt.$$

If the limit is a real number, then it is said that f has convergent improper integral on $[a, b]$, and if the limit (11.3) diverges, then f has divergent improper integral on $[a, b]$.

With the above notions, it is possible to look at the improper integral of a function that is defined on more general bounded sets. First, consider the case where the interval is (a, b). Take $f : (a, b) \to \mathbb{R}$, and assume that f is bounded and integrable on each subinterval $[x, y] \subset (a, b)$. In this case, the observation that the equality

$$\int_x^y f(t)\,dt = \int_x^c f(t)\,dt + \int_c^y f(t)\,dt$$

holds true for any $c \in [x, y]$ leads to defining

$$\int_a^b f(t)\,dt = \lim_{x \to a^+} \int_x^c f(t)\,dt + \lim_{y \to b^-} \int_c^y f(t)\,dt,$$

provided that both limits exist. If both are finite, then the improper integral is convergent, and if only one of them diverges, then the improper integral is divergent. The integral is also divergent when both limits diverge to $+\infty$ or both diverge to $-\infty$. If they diverge with opposite sign, then an indeterminacy occurs. It is possible to show that when the integral converges or diverges, the choice of the point c is irrelevant, in the sense that the above sum is the same for any choice of c.

Secondly, it is possible to further extend the notion of improper integral to the case where the domain of f is a union of bounded intervals on each of which one of the above situations occur. Then the improper integral of f on its domain converges if, by definition, each improper integral on each subinterval converges, and the improper integral is the sum of the individual limits.

Definition 11.2 (*Unbounded intervals*)

(i) Suppose that $f : [a, +\infty) \to \mathbb{R}$ is bounded and Riemann integrable on $[a, x]$ for every $x \in (a, +\infty)$. If the limit

$$\lim_{x \to +\infty} \int_a^x f(t) \, dt \tag{11.4}$$

exists and is a real number, then this limit is called the *improper integral* of f on $[a, +\infty)$ and one writes

$$\int_a^{+\infty} f(t) \, dt = \lim_{x \to +\infty} \int_a^x f(t) \, dt.$$

In this case, it is also said that f has *convergent improper integral* on $[a, +\infty)$. If the limit (11.4) diverges, then f has *divergent improper integral* on $[a, +\infty)$.

(ii) Similarly, if $f : (-\infty, b] \to \mathbb{R}$ is bounded and Riemann integrable on $[x, b]$ for every $x \in (-\infty, b]$ and if the limit

$$\lim_{x \to -\infty} \int_x^b f(t) \, dt \tag{11.5}$$

exists, then this limit is called improper integral of f on $(-\infty, b]$ and one writes

$$\int_{-\infty}^b f(t) \, dt = \lim_{x \to -\infty} \int_x^b f(t) \, dt.$$

If the limit is a real number, then it is said that f has convergent improper integral on $(-\infty, b]$, and if the limit (11.5) diverges, then f has divergent improper integral on $(-\infty, b]$.

As in the case of unbounded functions, one uses the one-sided notions to define two-sided limiting processes. Suppose that $f : \mathbb{R} \to \mathbb{R}$ is bounded and integrable on each interval $[x, y]$ and suppose further that for some (hence for every) $a \in \mathbb{R}$ both the improper integrals of f on $(-\infty, a]$ and on $[a, +\infty)$ converge. Then, by definition

$$\int_{-\infty}^{+\infty} f(t) \, dt = \int_{-\infty}^a f(t) \, dt + \int_a^{+\infty} f(t) \, dt.$$

The notion of divergent integral in $(-\infty, +\infty)$ is handled similarly as before. Summarizing, if $\mathrm{Dom}(f)$ is a finite disjoint union of adjacent intervals, then it is meaningful to speak about the possible convergence of the improper integral of f on

Dom(f): it is required that the improper integral of f converges on each (possibly unbounded) interval and, in this case, the improper integral of f will then be the (finite) sum of all the improper integrals over the various intervals.

11.2   Convergence Criteria

In this section it is assumed for simplicity that I is a possibly unbounded interval. If a function f has convergent improper integral on I, it is denoted

$$\int_I f(t)\,dt$$

to mean either of the various improper integrals introduced thus far. Several convergence (or divergence) criteria are discussed.

Theorem 11.1 *Suppose that $f, g : I \to \mathbb{R}$ are bounded and integrable on each interval $[x, y] \subset I$. Then*

(i) if $|f|$ has convergent improper integral on I, so does f and

$$\left| \int_I f(t)\,dt \right| \leq \int_I |f(t)|\,dt;$$

(ii) if $|f(x)| \leq g(x)$ for every $x \in I$ and g has convergent improper integral on I, then also $|f|$ has convergent improper integral on I, and

$$\int_I |f(t)|\,dt \leq \int_I g(t)\,dt.$$

It is to be noticed that, unlike the case of the (proper) Riemann integral, the integrability of f does not imply that of $|f|$. A very famous example of that of the sinc function, which is integrable in $[0, +\infty)$ but whose absolute value is not.

Below the case in which f is a non-negative is treated. The case $f(x) \leq 0$ follows from the previous one by a simple change of sign.

Theorem 11.2 (Order criteria, bounded intervals) *Suppose that $f : (a, b] \to [0, +\infty)$ is bounded and Riemann integrable on $[x, b]$ for every $x \in (a, b)$ and assume further that either $\lim_{x \to a^+} f(x) = +\infty$ or that $\lim_{x \to a^+} f(x) = -\infty$.*

(i) If there exists $\alpha \in (0, 1)$ such that $\lim_{x \to a^+} f(x)(x - a)^\alpha = 0$, then f has convergent improper integral in $[a, b]$;

(ii) if there exists $\beta \geq 1$ such that $\lim_{x \to a^+} f(x)(x - a)^\beta$ exists, finite or infinite, and is not equal to 0, then f has divergent improper integral in $[a, b]$.

Of course, the above theorem admits a left-sided version, for intervals $[a, b]$. In the current jargon, Theorem 11.2 is formulated by saying that if f diverges at a point of order $\leq \alpha$, for some $\alpha < 1$, then it is integrable; if it diverges with order $\geq \alpha$, for some $\alpha \geq 1$, then it is not integrable.

The following is a necessary, but not sufficient, condition for the convergence of improper integrals over unbounded intervals.

Theorem 11.3 *Suppose that* $f : [a, +\infty) \to [0, +\infty)$ *is bounded and Riemann integrable on* $[a, b]$ *for every* $b \in (a, +\infty)$. *If* f *has convergent improper integral in* $[a, +\infty)$ *and if the limit* $\lim_{x \to +\infty} f(x) = \ell$ *exists, then* $\ell = 0$.

Theorem 11.4 (Order criteria, unbounded intervals) *Suppose that the function* $f : [a, +\infty) \to [0, +\infty)$ *is bounded and Riemann integrable on* $[a, b]$ *for every* $b \in (a, +\infty)$ *and assume further that* $\lim_{x \to +\infty} f(x) = 0$.

(i) *If there exists* $\alpha > 1$ *such that* $\lim_{x \to +\infty} x^\alpha f(x) = 0$, *then* f *has convergent improper integral in* $[a, +\infty)$;

(ii) *if there exists* $\beta \in (0, 1]$ *such that* $\lim_{x \to +\infty} x^\beta f(x)$ *exists, finite or infinite, and is not equal to 0, then* f *has divergent improper integral in* $[a, +\infty)$.

Also the above theorem is formulated by saying that if f converges to 0 at (plus or minus) infinity of order $\geq \alpha$, for some $\alpha > 1$, then it is integrable; if it diverges with order $\leq \alpha$, for some $\alpha \leq 1$, then it is not integrable.

11.3 Integral Functions

By integral function it is actually meant a function of a slightly more general form than that of (11.1), namely a function of the form

$$G(x) = \int_{\varphi(x)}^{\psi(x)} f(t) \, dt. \tag{11.6}$$

A detailed discussion of all that must, or can, be said in order for (11.6) to make sense is beyond the scope of this book. It is clear, though, that the domain of G will be contained in $A = \mathrm{Dom}(\varphi) \cap \mathrm{Dom}(\psi)$. Further, A must be compared to the domain of f, and whether f is integrable in which subsets of A, possibly in the sense of converging improper integrals, is of course another sensible issue. In the guided exercises 11.14 and 11.15 below it is shown how to treat the case in which either φ or ψ is neither a constant nor simply the function x, as in (11.1). A basic guideline is to observe that if for example $\varphi(x) = a$, so that the integral function is

$$G(x) = \int_a^{\psi(x)} f(t) \, dt,$$

then G is actually of the form $G = F \circ \psi$, where F is precisely as in (11.1). Similar considerations hold for the general case (11.6). Thus the case (11.1) stands out as particularly important.

11.4 Guided Exercises on Improper Integrals and Integral Functions

11.4.1 Improper Integrals

11.1 Compute, if existing, $\displaystyle\int_0^{+\infty} x e^{-x}\, dx$.

Answer. The function $g(x) = x e^{-x}$ is defined and continuous on $\mathbb{R}$. Furthermore, $\lim_{x \to +\infty} g(x) = 0$ with order greater than any real power $\alpha > 0$, because

$$\lim_{x \to +\infty} \frac{x e^{-x}}{(1/x)^\alpha} = 0$$

for every $\alpha > 0$. Hence the integral converges. Integrating by parts,

$$\int g(x)\, dx = -x e^{-x} - e^{-x} + c$$

with $c \in \mathbb{R}$. Hence a primitive function of $g(x)$ is given by $e^{-x}(-x-1)$ and it follows that

$$\int_0^{+\infty} x e^{-x}\, dx = \lim_{y \to +\infty} \int_0^y x e^{-x}\, dx = \lim_{y \to +\infty} e^{-y}(-y-1) + 1 = 1.$$

This is a computation of the value of an improper integral, calculating the limit as $y \to +\infty$ of $\int_0^y g(t)\, dt$, using its explicit expression which is known for every $y > 0$.

11.2 Show that the improper integral $\displaystyle\int_1^{+\infty} \frac{\arctan x}{x^4 + 1}\, dx$ converges, and find an upper bound for it.

Answer. The function $g(x) = \arctan x /(x^4 + 1)$ is continuous in $\mathbb{R}$ and $g(x) \to 0$ as $x \to +\infty$ with order 4, because

$$\lim_{x \to +\infty} \frac{g(x)}{\frac{1}{x^4}} = \frac{\pi}{2} \neq 0.$$

Therefore the improper integral converges. Now, if $x \geq 1$, then

$$0 < g(x) < \frac{\pi}{2}\frac{1}{x^4+1} \leq \frac{\pi}{2}\frac{1}{x^2+1}.$$

It follows that

$$\int_1^{+\infty} g(x)\, dx < \int_1^{+\infty} \frac{\pi}{2}\frac{1}{x^2+1}\, dx$$
$$= \lim_{y\to+\infty}\int_1^y \frac{\pi}{2}\frac{1}{x^2+1}\, dx$$
$$= \lim_{y\to+\infty}\frac{\pi}{2}(\arctan y - \frac{\pi}{4})$$
$$= \frac{\pi}{2}\cdot\frac{\pi}{4} = \frac{\pi^2}{8}.$$

This is an easy exercise where the main point is that the arctangent function is bounded and hence the behaviour of the improper integral is mainly dictated by the rapidly decaying rational function $1/(x^4+1)$. This, in turn, is bounded above by $1/(x^2+1)$, whose integral is easily computed because its primitive functions are known.

11.3 Compute, if existing, $\int_2^{+\infty} \frac{t}{t^3+t-2}\, dt.$

Answer. The function $g(t) = t/(t^3+t-2)$ is defined and continuous on $\mathbb{R}\setminus\{1\}$. Furthermore, $g(t) \to 0$ as $t \to +\infty$ with order 2, hence the improper integral converges. Now, since $t^3+t-2 = (t-1)(t^2+t+2)$, the function g is equal to

$$\frac{t}{(t-1)(t^2+t+2)} = \frac{1}{4}\left(\frac{1}{t-1} - \frac{t-2}{t^2+t+2}\right)$$
$$= \frac{1}{4}\left(\frac{1}{t-1} - \frac{1}{2}\left(\frac{2t+1}{t^2+t+2} - \frac{5}{t^2+t+2}\right)\right).$$
$$= \frac{1}{4}\frac{1}{t-1} - \frac{1}{8}\frac{2t+1}{t^2+t+2} + \frac{5}{8}\frac{1}{t^2+t+2}$$

In the open interval $(1,+\infty)$, the first two summands have easily computed indefinite integrals, because

$$\int \frac{1}{t-1}\, dt = \log(t-1) + c_1, \qquad \int \frac{2t+1}{t^2+t+2}\, dt = \log(t^2+t+2) + c_2$$

with $c_1, c_2 \in \mathbb{R}$. The third is computed by observing that

$$\int \frac{1}{t^2+t+2}\, dt = \int \frac{1}{(t+\frac{1}{2})^2+\frac{7}{4}}\, dt.$$

The substitution $t = \frac{\sqrt{7}}{2}u - \frac{1}{2}$ yields

$$\frac{2\sqrt{7}}{7} \int \frac{1}{u^2 + 1}\, du = \frac{2\sqrt{7}}{7} \arctan \frac{2t + 1}{\sqrt{7}} + c_3$$

with $c_3 \in \mathbb{R}$. It follows that

$$\int \frac{t}{t^3 + t - 2}\, dt = \frac{1}{8} \log \frac{(t - 1)^2}{t^2 + t + 2} + \frac{5\sqrt{7}}{28} \arctan \frac{2t + 1}{\sqrt{7}} + c$$

with $c \in \mathbb{R}$, so that

$$\int_2^{+\infty} \frac{t}{t^3 + t - 2}\, dt = \lim_{x \to +\infty} \int_2^x \frac{t}{t^3 + t - 2}\, dt$$

$$= \lim_{x \to +\infty} \left[\frac{1}{8} \log \frac{(t - 1)^2}{t^2 + t + 2} + \frac{5\sqrt{7}}{28} \arctan \frac{2t + 1}{\sqrt{7}} \right]_2^x$$

$$= \frac{5\sqrt{7}}{28} \frac{\pi}{2} - \frac{1}{8} \log 8 - \frac{5\sqrt{7}}{28} \arctan \frac{5}{\sqrt{7}}.$$

This is an explicit computation of a converging improper integral of a rational function. First one obtains the indefinite integral by partial fraction decomposition in the sensible interval, then one computes the limit of the appropriate definite integral.

11.4 Show that if $a, b \in \mathbb{R}^+$ satisfy $a < b - 1$, then $\displaystyle \int_1^{+\infty} x^a \cos x^b\, dx$ converges.

Answer. Put $f(x) = x^a \cos x^b$. Clearly, $f \in C^0([1, +\infty))$ and by definition

$$\int_1^{+\infty} f(x)\, dx = \lim_{y \to +\infty} \int_1^y f(x)\, dx,$$

provided that the limit exists. It is useful to write

$$f(x) = \frac{1}{b} x^{a+1-b} b x^{b-1} \cos x^b = \frac{1}{b} x^{a+1-b} \frac{d}{dx}(\sin x^b)$$

so that, integrating by parts

$$\int_1^y f(x)\, dx = \int_1^y \frac{1}{b} x^{a+1-b}(b x^{b-1} \cos x^b)\, dx$$

$$= \left[\frac{1}{b} x^{a+1-b} \sin x^b \right]_1^y - \int_1^y \frac{1}{b}(a + 1 - b) x^{a-b} \sin x^b\, dx$$

$$= \frac{1}{b}\frac{1}{y^{b-a-1}}\sin y^b - \frac{1}{b}\sin 1 - \int_1^y \frac{a+1-b}{b} x^{a-b}\sin x^b\,dx$$

$$= \frac{1}{b}\frac{1}{y^{b-a-1}}\sin y^b - \frac{1}{b}\sin 1 + \frac{b-a-1}{b}\int_1^y \frac{\sin x^b}{x^{b-a}}\,dx.$$

If $b-1-a > 0$, then

$$\lim_{y\to+\infty} \frac{1}{y^{b-a-1}}\sin y^b = 0.$$

Furthermore, as $x \to +\infty$,

$$\left| \frac{\sin x^b}{x^{b-a}} \right| \le \frac{1}{x^{b-a}} \to 0$$

with order $b-a > 1$. This implies that $\int_1^{+\infty} x^{a-b}\,dx$ converges, and hence the continuous function $x^{a-b}\sin x^b$, together with its absolute value, has convergent improper integral in $[1, +\infty)$. Therefore

$$\int_1^{+\infty} f(x)\,dx = \lim_{y\to+\infty}\int_1^y f(y)\,dy = -\frac{1}{b}\sin 1 + \frac{b-a-1}{b}\int_1^{+\infty} x^{a-b}\sin x^b\,dx$$

converges.

Here the main idea is to use integration by parts, multiplying and dividing by bx^{b-1} so as to obtain the derivative of $\sin x^b$.

11.5 Determine if the improper integral

$$\int_1^{+\infty} \frac{\sqrt{x}\sqrt{e^{\frac{1}{x}}-1}\sqrt[3]{\log^2(1+\frac{1}{x})}}{\sqrt{x-1}}\,dx$$

converges.

Answer. Put

$$f(x) = \frac{\sqrt{x}\sqrt{e^{\frac{1}{x}}-1}\sqrt[3]{\log^2(1+\frac{1}{x})}}{\sqrt{x-1}}.$$

Evidently, $\mathrm{Dom}(f) = (1, +\infty)$ and $f \in C^0((1, +\infty))$ because it is a ratio of continuous functions. The integral converges if for some (or any) fixed $\bar{x} \in (1, +\infty)$ both integrals $\int_1^{\bar{x}} f(x)\,dx$ and $\int_{\bar{x}}^{+\infty} f(x)\,dx$ are such. For the first, the behaviour as $x \to 1^+$ must be determined. Since $\lim_{x\to 1^+} f(x) = +\infty$ with order $1/2 < 1$, the first integral converges. As for the second, the behaviour as $x \to +\infty$ must be determined. Using the first order McLaurin expansions of the exponential and of $t \mapsto \log(1+t)$ it follows that

$$e^{\frac{1}{x}} = 1 + \frac{1}{x} + \frac{1}{x}\omega\left(\frac{1}{x}\right)$$

$$\log\left(1 + \frac{1}{x}\right) = \frac{1}{x} + \frac{1}{x}\omega_1\left(\frac{1}{x}\right)$$

for $x > 0$, where $\lim_{t \to 0} \omega(t) = \lim_{t \to 0} \omega_1(t) = 0$. Therefore the function

$$x \mapsto \sqrt{e^{\frac{1}{x}} - 1}\sqrt[3]{\log^2\left(1 + \frac{1}{x}\right)}$$

tends to 0 as $x \to +\infty$ with order $1/2 + 2/3 = 7/6 > 1$. It follows that $f(x) \to 0$ as $x \to +\infty$ with order $7/6 > 1$, and hence the second integral converges as well.

This is a standard exercise on integrals that are improper on both sides: the indicated integral is improper both because f is not bounded as $x \to 1^+$ and because the integration interval is unbounded. Therefore the first step is to consider the two problems separately. Each of them is then handled using order criteria, because exact orders can be established.

11.6 Determine if the improper integral $\displaystyle\int_0^{\log 2} \frac{x^{\sqrt{x}} - 1}{\log(2 - e^x)}\, dx$ converges.

Answer. Put $g(x) = x^{\sqrt{x}} - 1/\log(2 - e^x)$. Evidently,

$$\mathrm{Dom}(g) = \left\{x \in \mathbb{R} : x > 0, 2 - e^x > 0, 2 - e^x \neq 1\right\} = (0, \log 2)$$

and $g \in C^0((0, \log 2))$ because g is a ratio of functions that are continuous in $(0, \log 2)$. The behaviour g near $x_0 = 0$ and $x_1 = \log 2$ must therefore be established. Now, $g(x) \to 0$ as $x \to \log 2^-$, so that g has a continuous extension at x_1. This implies that the integral under consideration converges if given any $\overline{x} \in (0, \log 2)$ the integral $\int_0^{\overline{x}} g(x)\, dx$ converges. Now, the denominator $h(x) = \log(2 - e^x)$ of g is differentiable at $x_0 = 0$ and

$$h'(x) = \frac{-e^x}{2 - e^x}$$

with $h'(0) = -1 \neq 0$. It follows that h vanishes as $x \to 0$ of order 1. Furthermore, the first order McLaurin expansion of the exponential function gives

$$x^{\sqrt{x}} - 1 = e^{\sqrt{x}\log x} - 1 = \sqrt{x}\log x + \sqrt{x}\log x \cdot \omega(\sqrt{x}\log x),$$

where $\lim_{t \to 0} \omega(t) = 0$. Therefore the numerator of g vanishes as $x \to 0$ of order less than $1/2$ and greater than any number less than $1/2$. It follows that g diverges, as $x \to 0^+$, of order greater than $1/2$ but less than of any number greater than $1/2$, in particular less than some $\alpha < 1$. Therefore g has convergent improper integral in $[0, \overline{x}]$, hence in $[0, \log 2]$.

Much like in Exercise 11.5 above, one has to split the integral in two. The leftmost of the two integrals poses actually no problem because the function is not unbounded near the origin. For the second, the order criteria must be used. The leading term in the numerator, however, is $\sqrt{x}\log x$, which does not have a precise order in the scale of powers as $x \to 0^+$. Thus, the convergence criteria must be used with care.

11.7 Determine if the improper integral $\displaystyle\int_0^{+\infty} \frac{1}{\sqrt[3]{x-1}(\sqrt[4]{x^3}+1)}\,dx$ converges.

Answer. Put $f(x) = 1/(\sqrt[3]{x-1}(\sqrt[4]{x^3}+1))$, a continuous function in $[0,1)\cup$ $(1,+\infty)$. The indicated improper integral converges if for any fixed $\overline{x} \in (1,+\infty)$

$$\int_0^1 f(x)\,dx, \qquad \int_1^{\overline{x}} f(x)\,dx, \qquad \int_{\overline{x}}^{+\infty} f(x)\,dx$$

are convergent improper integrals. Now, since

$$\lim_{x\to 1^-} \frac{1}{\sqrt[3]{x-1}(\sqrt[4]{x^3}+1)} = -\infty$$

with order $1/3 < 1$, the first integral converges. Similarly, the second integral converges, because

$$\lim_{x\to 1^+} \frac{1}{\sqrt[3]{x-1}(\sqrt[4]{x^3}+1)} = +\infty$$

with order $1/3 < 1$. As for the third integral, since

$$\lim_{x\to +\infty} \frac{1}{\sqrt[3]{x-1}(\sqrt[4]{x^3}+1)} = 0$$

with order $1/3 + 3/4 = 13/12 > 1$, the integral converges as well. In summary, the indicated improper integral converges.

   This is an improper integral over a union of two disjoint but adjacent intervals, the leftmost is bounded, but f isn't, and the rightmost is unbounded. On the latter, the integral must be split into two integrals. The exact order with which f diverges at the point 1, where the two intervals meet, is easily established from both sides, as well as the exact order with which f vanishes at $+\infty$, yielding a globally convergent improper integral.

11.4.2 Integral Functions

11.8 Consider the function

$$f(x) = \begin{cases} \dfrac{x}{x^2-4} & x < -3 \\ ax+b & -3 \le x \le 0 \\ b & x > 0. \end{cases}$$

(a) For which values of the real parameters a and b, if any, is f integrable in every closed and bounded interval?

(b) Put $a = 0$ and $b = 2$. Write an explicit analytic expression for $F(x) = \int_{-1}^{x} f(t)\, dt$.

Answer. (a) The function f is defined on $\mathbb{R}$ and, for every $a, b \in \mathbb{R}$, is integrable in any closed and bounded interval because it is a piecewise continuous and bounded function; actually continuous if $3a - b = 3/5$

(b) If $a = 0$ and $b = 2$, then

$$f(x) = \begin{cases} \dfrac{x}{x^2 - 4} & x < -3 \\ 2 & x \geq -3. \end{cases}$$

Therefore

$$F(x) = \begin{cases} \displaystyle\int_{-1}^{-3} 2\, dt + \int_{-3}^{x} \frac{t}{t^2 - 4}\, dt & x < -3 \\ \displaystyle\int_{-1}^{x} 2\, dt & x \geq -3 \end{cases}$$

$$= \begin{cases} -4 + \dfrac{1}{2} \log\left(\dfrac{x^2 - 4}{5}\right) & x < -3 \\ 2(x + 1) & x \geq -3. \end{cases}$$

This is a basic exercise on integral functions: just compute the integral of f and obtain the explicit expression of the integral function F.

11.9 Consider the function $f(x) = e^{kx} - \sin x + 2 \displaystyle\int_0^x e^{-t^2}\, dt + h$.

(a) For which values of $k, h \in \mathbb{R}$ does f tend to 0 as $x \to 0^+$?

(b) For those values, if any, establish the order with which f vanishes.

Answer. (a) Since $g(t) = e^{-t^2}$ is continuous in $\mathbb{R}$, it is integrable in every closed interval $[0, x]$ with $x \in \mathbb{R}$. Thus $\mathrm{Dom}(f) = \mathbb{R}$ and $\lim_{x \to 0^+} f(x) = f(0) = 1 + h$. It follows that f tends to zero as $x \to 0^+$ if $h = -1$, for every $k \in \mathbb{R}$.

(b) Fix now $h = -1$, so that

$$f(x) = e^{kx} - \sin x + 2 \int_0^x e^{-t^2}\, dt - 1.$$

By the fundamental theorem of calculus f is differentiable in $\mathbb{R}$ and

$$f'(x) = k e^{kx} - \cos x + 2e^{-x^2}.$$

Hence $f'(0) = k - 1 + 2 = k + 1$. Also, $f''(x) = k^2 e^{kx} + \sin x + 2e^{-x^2}(-2x)$, so that $f''(0) = k^2$. The second order McLaurin expansion is then

$$f(x) = (k+1)x + \frac{k^2 x^2}{2} + o(x^2).$$

Therefore, if $k \neq -1$ the order of f is 1, and if $k = -1$ the order is 2.

In this exercise the main issues are two: first, if the integral function is of the form $\varphi(x) = \int_0^x h(t)\, dt$, then $\varphi(0) = 0$ and, second, $\varphi'(0) = h(0)$, so that it is easy to write a Taylor expansion. Here f is the sum of such a function φ and other elementary functions.

11.10 Consider the function

$$f(x) = \int_{-1}^{x} \frac{\log(1-t)}{(t^2+1)\arctan(t\sqrt[3]{t})}\, dt.$$

Find the domain of f, determine where f is differentiable and draw its graph.

Answer. Put

$$g(t) = \frac{\log(1-t)}{(t^2+1)\arctan(t\sqrt[3]{t})}.$$

The domain of f consists of those $x \in \mathbb{R}$ for which g is either Riemann integrable or for which the improper integral on $[-1, x]$ converges. Now, g is continuous in $\mathrm{Dom}(g) = (-\infty, 0) \cup (0, 1)$. Since $-1 \in (-\infty, 0)$ and g is continuous, it is integrable in $[-1, x]$ for every $x < 0$. Hence $(-\infty, 0) \subset \mathrm{Dom}(f)$.

Consider next $x_0 = 0$. Evidently, $x_0 \in \mathrm{Dom}(f)$ provided that the improper integral of g in $[-1, 0]$ converges. Since

$$\lim_{t \to 0^-} \frac{\log(1-t)}{(t^2+1)\arctan(t\sqrt[3]{t})} = +\infty$$

with order $1/3 < 1$, the integral converges and $0 \in \mathrm{Dom}(f)$.

Similarly, $x \in (0, 1)$ is in $\mathrm{Dom}(f)$ if the improper integral of g converges in $[-1, x]$, namely both in $[-1, 0]$ and in $[0, x]$. As before,

$$\lim_{t \to 0^+} \frac{\log(1-t)}{(t^2+1)\arctan(t\sqrt[3]{t})} = -\infty$$

with order $1/3 < 1$. It follows that $x \in \mathrm{Dom}(f)$, so that $(-\infty, 1) \subset \mathrm{Dom}(f)$.

Finally, in order to see if $1 \in \mathrm{Dom}(f)$ it is necessary to establish whether the improper integral of g in $[\bar{x}, 1]$ converges, with $\bar{x} \in (0, 1)$. Since

$$\lim_{t \to 1^-} \frac{\log(1-t)}{(t^2+1)\arctan(t\sqrt[3]{t})} = -\infty$$

with order less than any number less than 1. Thus the improper integral of g in $[-1, 1]$ converges, and $1 \in \mathrm{Dom}(f)$. In conclusion $\mathrm{Dom}(f) = (-\infty, 1]$.

In order to establish the behaviour of f at $-\infty$, observe that

$$\lim_{x \to -\infty} f(x) = \int_{-1}^{-\infty} g(t)\,dt,$$

so that the existence and finiteness of the limit is, by definition, equivalent to the convergence of the improper integral of g in $(-\infty, -1]$. Now,

$$\lim_{x \to -\infty} \frac{\log(1 - t)}{(t^2 + 1)\arctan(t\sqrt[3]{t})} = 0$$

with order greater than any number in $(1, 2)$, so that the improper integral of g in $(-\infty, -1]$ converges. This shows that $\lim_{x \to -\infty} f(x)$ exists and is a real number.

Since $g \in C^0((-\infty, 0))$, by the fundamental theorem of calculus $f \in C^1((-\infty, 0))$ and $f'(x) = g(x)$ if $x < 0$. Fix next $\overline{x} \in (0, 1)$, so that

$$f(x) = \int_{-1}^{\overline{x}} g(t)\,dt + \int_{\overline{x}}^{x} g(t)\,dt.$$

Since $g \in C((0, 1))$, the fundamental theorem of calculus implies that the function $h(x) = \int_{\overline{x}}^{x} g(t)\,dt$ is in $C^1((0, 1))$ and $h'(x) = g(x)$ if $x \in (0, 1)$. Hence, it follows that $f \in C^1((-\infty, 0) \cup (0, 1))$.

Finally, consider the point $x_0 = 0$. First of all, f is continuous at 0 because by definition of improper integral

$$\lim_{x \to 0} f(x) = \lim_{x \to 0} \int_{-1}^{x} g(t)\,dt = \int_{-1}^{0} g(t)\,dt = f(0).$$

Therefore f is continuous in a neighbourhood of 0 and differentiable in a punctured neighbourhood of 0. Furthermore,

$$\lim_{x \to 0^+} f'(x) = \lim_{x \to 0^+} g(x) = -\infty$$

so that, by the theorem on the limit of the derivative

$$\lim_{x \to 0^+} \frac{f(x) - f(0)}{x} = -\infty,$$

so that f is not differentiable at 0. Similarly, f is not differentiable at 1. Finally,

$$f'(x) = \frac{\log(1 - x)}{(x^2 + 1)\arctan(x\sqrt[3]{x})}$$

for $x \in (-\infty, 0) \cup (0, 1)$, so that $f'(x) > 0$ if $x \in (-\infty, 0)$ and $f'(x) < 0$ if $x \in (0, 1)$. Therefore f is strictly increasing in $(-\infty, 0)$ and strictly decreasing $(0, 1)$. Evidently, $f(-1) = 0$. The graph of f is thus that in Fig. 11.1. This exercise requires to understand all the basic features of an integral function, namely its domain, the behaviour at the boundary of it, where it is differentiable and the sign of its derivative, whence its monotonicity properties. The integrand g is defined and continuous in $(-\infty, 0) \cup (0, 1)$ and the integral function is

$$f(x) = \int_{-1}^{x} g(t)\, \mathrm{d}t.$$

Thus, the first issue is to establish the integrability at the origin, both from the left and from the right, which is done by applying (exact) order criteria. The second issue is integrability at 1, which again follows by order criteria, though in this case the exact order with which g diverges cannot be established, but only safely bounded above, for example, by $1/2$. The domain of f is thus $(-\infty, 1]$, strictly larger than that of g, in fact including its boundary points in $\mathbb{R}$. The behaviour at $-\infty$ raises again an issue of convergence of the improper integral at $-\infty$, and this is handled with order criteria because the power 2 at the denominator of g beats the diverging logarithm at the numerator in an integrable fashion (the resulting order is greater than anything smaller than 2).

Next come differentiability issues. The fundamental theorem of calculus guarantees it in $(-\infty, 0) \cup (0, 1)$, whereas the diverging behaviour of g together with the theorem of the limit of the derivative imply that f is not differentiable at the origin. The sign of the derivative is that of g and is thus easily established.

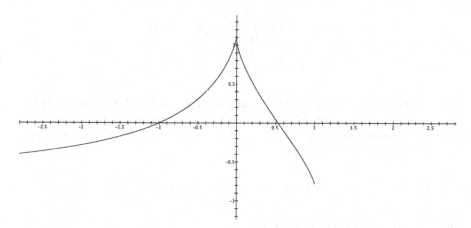

Fig. 11.1 Graph of f, Exercise 11.10

11.11   Consider the function $f(x) = \int_{x_0}^{x} \dfrac{te^t}{t^2 - 9}\, dt$.

(a) Determine the domain of f as x_0 ranges in $\mathbb{R}$.

(b) Let $x_0 = 0$. Compute the limits of f at the boundary of $\mathrm{Dom}(f)$.

(c) Let $x_0 = 0$. Compute $\lim_{x \to 0^+} (f(x) - kx)/x^2$ as k ranges in $\mathbb{R}$.

Answer. (a) Put $g(t) = te^t/(t^2 - 9)$. The domain of g is $\mathbb{R} \setminus \{\pm 3\}$, and g is continuous. Further, $\lim_{t \to \pm 3} |g(t)| = +\infty$ with order 1. Indeed,

$$\lim_{t \to 3^{\pm}} \frac{g(t)}{\frac{1}{t-3}} = \lim_{t \to 3^{\pm}} \frac{te^t}{(t-3)(t+3)}(t-3) = \frac{3e^3}{6} \in \mathbb{R} \setminus \{0\}.$$

Analogously

$$\lim_{t \to -3^{\pm}} \frac{g(t)}{\frac{1}{t-3}} = \frac{3e^{-3}}{6} \in \mathbb{R} \setminus \{0\}.$$

Hence g is not integrable in any interval containing either -3 or 3. It follows that

$$\mathrm{Dom}(f) = \begin{cases} (-\infty, -3) & x_0 < -3 \\ (-3, 3) & -3 < x_0 < 3 \\ (3, +\infty) & x_0 > 3 \\ \varnothing & x_0 = \pm 3. \end{cases}$$

(b) As shown in (a), if $x_0 = 0$ then $\mathrm{Dom}(f) = (-3, 3)$. Furthermore

$$\lim_{x \to 3^-} f(x) = \int_0^3 \frac{te^t}{t^2 - 9}\, dt = -\infty$$

$$\lim_{x \to -3^+} f(x) = \int_0^{-3} \frac{te^t}{t^2 - 9}\, dt = -\infty$$

because both improper integrals diverge, as g diverges of order 1 at ± 3, and $f'(x) = g(x) = xe^x/(x^2 - 9)$, which is non negative in $(-3, 0]$ and non positive in $[0, 3)$.

(c) It is easy to see that $f(0) = f'(0) = 0$ and $f''(0) = -1/9$. The second order McLaurin expansion of f is thus

$$f(x) = -\frac{1}{18}x^2 + x^2 \omega(x)$$

with $\lim_{x \to 0} \omega(x) = 0$, and it implies

$$\lim_{x \to 0^+} \frac{f(x) - kx}{x^2} = \lim_{x \to 0^+} \frac{-kx - \frac{1}{18}x^2 + x^2 \omega(x)}{x^2} = \begin{cases} +\infty & k < 0 \\ -\frac{1}{18} & k = 0 \\ -\infty & k > 0. \end{cases}$$

Here the integrand is everywhere defined except at ± 3, where it diverges of order 1. This means that no improper integral over an interval that contains either of these points can ever converge. Question (b) has already been almost answered in (a), because it has already been shown that the improper integrals diverge; it remains to be established whether f diverges positively or negatively, and this is done by inspecting its derivative, namely g. The final question is about the behaviour of $f - kx$ near the origin as compared to that of x^2, and calls for a McLaurin expansion of order two of f, which requires just one derivative of g at 0.

11.12 Consider the function

$$f(x) = \int_k^x \frac{\sqrt{t}}{e^t - \sqrt{t+1}}\, dt.$$

Find $\mathrm{Dom}(f)$ as k ranges in $\mathbb{R}$ and compute the limits at its boundary when $k = 1$.

Answer. Put

$$g(t) = \frac{\sqrt{t}}{e^t - \sqrt{t+1}} \qquad h(t) = e^t - \sqrt{t+1}.$$

Evidently, h is defined and continuous in $[-1, +\infty)$. Since $\mathrm{Dom}(g) \subset [0, +\infty)$, one has to look for the possible zeroes of h in $[0, +\infty)$. Since for all $t \geq 0$

$$e^t \geq t + 1 \geq \sqrt{t+1}$$

it follows that $h(t) \geq 0$, and $h(t) = 0$ if and only if $t = 0$. Therefore the domain of g is $(0, +\infty)$ and g is continuous. In order to establish the behaviour of g at the origin, it is useful to compute the first order McLaurin expansion of h. Since $h'(0) = 1/2$, the expansion is $h(t) = \frac{1}{2}t + o(t)$ and hence

$$\lim_{t \to 0^+} g(t) = \lim_{t \to 0^+} \frac{t^{\frac{1}{2}}}{\frac{1}{2}t + o(t)} = +\infty$$

with order $1/2 < 1$. The improper integral $\int_0^a g(t)\, dt$ is therefore convergent for every $a \geq 0$, so that $\mathrm{Dom}(f) = [0, +\infty)$ for $k \geq 0$ and $\mathrm{Dom}(f) = \emptyset$ for $k < 0$.

Put now $k = 1$. As shown above,

$$\lim_{x \to 0^+} f(x) = f(0) = \int_1^0 \frac{\sqrt{t}}{e^t - \sqrt{t+1}}\, dt < 0$$

and

$$\lim_{x \to +\infty} f(x) = \int_1^{+\infty} \frac{\sqrt{t}}{e^t - \sqrt{t+1}}\, dt.$$

The latter improper integral converges to a positive real number because $g(t) \to 0$ as $t \to +\infty$ with an order that is bigger than any positive number, and because g is positive on $(1, +\infty)$.

In this exercise, a little care must be paid in the preliminary step in which the domain of the integrand g is examined, and found to be $(0, +\infty)$. The analysis reveals that g diverges of exact order $1/2$ as $x \to 0^+$. Thus k can be taken to be either 0 or larger, but not negative because g is not defined for negative values. The convergence at $+\infty$ of the improper integral is a trivial matter, due to the presence of the exponential, hence f converges to a finite limit as $x \to +\infty$.

11.13 Consider the function $f(x) = \int_1^x \dfrac{t}{\sin t \sqrt{t+2}}\, dt$.

(a) Find the domain of f.
(b) Where is f continuous? Where is f differentiable?

Answer. (a) The function $g(t) = t/(\sqrt{t+2}\sin t)$ is continuous in

$$\mathrm{Dom}(g) = \{t \in \mathbb{R} : t > -2, \sin t \neq 0\} = \{t \in \mathbb{R} : t > -2, t \neq k\pi, k \in \mathbb{N}\}.$$

Notice that g has a continuous extension at $t = 0$ because $g(t) \to \sqrt{2}/2$ as $\to 0$. Furthermore

$$\lim_{t\to -2^+} g(t) = +\infty$$

with order $1/2$ and

$$\lim_{t\to \pi^-} g(t) = +\infty$$

with order 1. Therefore g is integrable, possibly with convergent improper integral, in every interval $[a, b] \subset [-2, \pi)$. It follows that $\mathrm{Dom}(f) = [-2, \pi)$.
(b) Applying the fundamental theorem of calculus as in Exercise 11.10,

$$f'(x) = \frac{x}{\sin x \sqrt{x+2}}$$

for $x \in (-2, \pi) \setminus \{0\}$. Hence f is continuous in $(-2, \pi) \setminus \{0\}$. Furthermore, f is continuous at $x = -2$ as well, because the improper integral $\int_{-1}^{-2} g(t)\, dt$ converges. Notice that f is differentiable at $x_0 = 0$ because g has a continuous extension at $t = 0$ and $f'(x) \to \sqrt{2}/2$ as $x \to 0$. However, f is not differentiable at $x = -2$ because $f'(x) \to +\infty$ as $x \to -2^+$ so that by the theorem on the limit of the derivative

$$\lim_{x\to -2^+} \frac{f(x) - f(-2)}{x+2} = +\infty.$$

In conclusion, f is continuous in $[-2, \pi)$ and differentiable in $(-2, \pi)$. Finally,

$$f'(x) = \begin{cases} \dfrac{x}{\sin x \sqrt{x+2}} & -2 < x < \pi, \ x \neq 0 \\[4mm] \dfrac{\sqrt{2}}{2} & x = 0. \end{cases}$$

The function g whose integral function $\int_1^x g(t)\, dt$ is studied in this exercise has a complicated domain, but the variable x, which must be greater than -2, cannot go past the first "serious" obstruction at π because the improper integral does not converge in $[\pi - \varepsilon, \pi]$. It can, however, be taken to be equal to -2 because the improper integral does converge. The issue at the origin is somehow fake, because g can be extended as a continuous function, as $x/\sin x$ tends to 1. The derivative exists in the full open interval, but it does not exist at the endpoints. This is actually a classical application of the theorem of the limit of the derivative, or, equivalently, of the de l'Hôpital's theorem.

11.14 Consider the function $f(x) = \displaystyle\int_x^{+\infty} \left(t \log \left(\dfrac{t+1}{t} \right) - e^{-1/2t} \right) dt.$

(a) Find the domain of f.
(b) Compute $\displaystyle\lim_{x \to +\infty} x^\alpha f(x)$ as α ranges in the positive real numbers.

Answer. (a) The function

$$g(t) = t \log(\frac{t+1}{t}) - e^{-\frac{1}{2t}}$$

is defined and continuous in $(-\infty, -1) \cup (0, +\infty)$. Since it is not defined in $(-1, 0)$ and since the integration interval is $(x, +\infty)$, the variable x cannot attain negative values. Put next $y = 1/t$ and $h(y) = (\log(y+1) - ye^{-\frac{y}{2}})/y$. Now,

$$\lim_{t \to +\infty} g(t) = \lim_{y \to 0^+} h(y) = 0$$

with order 2 because, using a third order McLaurin expansion at the numerator

$$\lim_{y \to 0^+} \frac{h(y)}{y^\alpha} = \lim_{y \to 0^+} \frac{y - \frac{y^2}{2} + \frac{y^3}{3} - y(1 - \frac{y}{2} + \frac{y^2}{8}) + o(y^3)}{y^{\alpha+1}}$$

$$= \lim_{y \to 0^+} \frac{5}{24} y^{2-\alpha} = \begin{cases} +\infty & \alpha > 2 \\ \frac{5}{24} & \alpha = 2 \\ 0 & \alpha < 2. \end{cases}$$

It follows that $\int_a^{+\infty} g(t)\,dt$ converges for any $a > 0$. Now,

$$\lim_{t\to 0^+} g(t) = \lim_{y\to +\infty} h(y) = 0$$

so that $g(t)$ admits a continuous extension at $t = 0$ and $\mathrm{Dom}(f) = [0, +\infty)$.

(b) For any fixed $\overline{x} > 0$

$$\lim_{x\to +\infty} f(x) = \lim_{x\to +\infty} \left(\int_x^{\overline{x}} g(t)\,dt + \int_{\overline{x}}^{+\infty} g(t)\,dt \right)$$

$$= \lim_{x\to +\infty} \left(-\int_{\overline{x}}^{x} g(t)\,dt + \int_{\overline{x}}^{+\infty} g(t)\,dt \right)$$

$$= -\int_{\overline{x}}^{+\infty} g(t)\,dt + \int_{\overline{x}}^{+\infty} g(t)\,dt = 0.$$

Therefore the limit

$$\lim_{x\to +\infty} x^\alpha f(x) = \lim_{x\to +\infty} \frac{f(x)}{x^{-\alpha}}$$

takes on an indeterminate form. The idea is to use the de l'Hôpital's theorem. First of all, f is differentiable in $(0, +\infty)$ because

$$f(x) = \int_{\overline{x}}^{+\infty} g(t)\,dt - \int_{\overline{x}}^{x} g(t)\,dt$$

and the function $x \mapsto \int_{\overline{x}}^{x} g(t)\,dt$ is of class $C^1((0, +\infty))$ by the fundamental theorem of calculus, since $g \in C^0((0, +\infty))$. Furthermore $f'(x) = -g(x)$ for every $x > 0$. Consider then the ratio

$$\frac{f'(x)}{-\alpha x^{-\alpha-1}} = \frac{g(x)}{\alpha x^{-\alpha-1}} = \frac{x^{-2}(\frac{5}{24} + \omega_2(x^{-1}))}{\alpha x^{-\alpha-1}}$$

where $\omega_2(x) \to 0$ as $x \to 0$, the limit of which is

$$\lim_{x\to +\infty} \frac{f'(x)}{-\alpha x^{-\alpha-1}} = \begin{cases} 0 & \alpha < 1 \\ \frac{5}{24} & \alpha = 1 \\ +\infty & \alpha > 1. \end{cases}$$

By the de l'Hôpital's theorem, the result is thus

$$\lim_{x\to +\infty} x^\alpha f(x) = \begin{cases} 0 & \alpha < 1 \\ \frac{5}{24} & \alpha = 1 \\ +\infty & \alpha > 1. \end{cases}$$

The first question requires first the observation that since g is defined in the union of two disjoint but not adjacent intervals, only one of them can actually be taken into consideration for the domain of f, namely $(0, +\infty)$. The convergence of the improper integral that defines f needs then a careful analysis on the order of g at $+\infty$. This can be carried out with a standard application of Taylor expansions, albeit after the sensible change of variable $y = 1/t$ that actually allows this technique (for Taylor expansions at $+\infty$ make no sense).

The second question is actually about the order with which f vanishes at $+\infty$, and is handled with a classical use of the de l'Hôpital's theorem, because the derivative of f is g, whose order at $+\infty$ has already been carefully investigated.

11.15 Consider the function $f(x) = \int_{-\frac{1}{2}}^{4x-3x^2} \dfrac{t+1}{\sqrt{|t|} - t^3}\, dt.$

(a) Find the domain of f.
(b) Compute the limits of f at the boundary points of its domain.
(c) Where is f monotonic?

Answer. (a) Denote by

$$g(t) = \frac{t+1}{\sqrt{|t|} - t^3}, \qquad h(y) = \int_{-\frac{1}{2}}^{y} g(t)\, dt, \qquad p(x) = 4x - 3x^2.$$

Then f is the composition $f = h \circ p$ and hence $\mathrm{Dom}(f) = \{x \in \mathbb{R} : p(x) \in \mathrm{Dom}(h)\}$. Thus, $\mathrm{Dom}(h)$ must be found, to which end the behaviour of g must be understood. First,

$$\mathrm{Dom}(g) = \left\{ t \in \mathbb{R} : \sqrt{|t|} - t^3 \neq 0 \right\} = \mathbb{R} \setminus \{0, 1\}$$

and $g \in C^0(\mathbb{R} \setminus \{0, 1\})$ as a ratio of continuous functions on $\mathbb{R} \setminus \{0, 1\}$. Therefore g is integrable in $[-1/2, y]$ for $y < 0$ and consequently $\mathrm{Dom}(h) \supset (-\infty, 0)$.

Next consider the improper integral of g on $[-1/2, 0]$. Since $g(t) \to +\infty$ as $t \to 0^-$ with order $1/2 < 1$, it follows that the improper integral of g on $[-1/2, 0]$ converges, so that $0 \in \mathrm{Dom}(h)$. Analogously, from the fact that $g(t) \to +\infty$ as $t \to 0^+$ of order $1/2 < 1$ it follows that $\mathrm{Dom}(h) \supset (-\infty, 1)$.

Finally, consider $[-1/2, 1]$. The function $q(t) = \sqrt{|t|} - t^3$ is differentiable at 1 and $q'(1) = -5/2 \neq 0$. Hence q vanishes with order 1 as $t \to 1$ and consequently $g(t) \to +\infty$ with order 1. Therefore the improper integral of g on $[-1/2, 1]$ diverges and $\mathrm{Dom}(h) = (-\infty, 1)$. From $\mathrm{Dom}(f) = \{x \in \mathbb{R} : p(x) \in \mathrm{Dom}(h)\}$ it follows that

$$\mathrm{Dom}(f) = \left\{ x \in \mathbb{R} : 4x - 3x^2 < 1 \right\} = (-\infty, \tfrac{1}{3}) \cup (1, +\infty).$$

(b) By the limit of composite functions

$$\lim_{x \to -\infty} f(x) = \lim_{y \to -\infty} h(y) = \int_{-\frac{1}{2}}^{-\infty} g(t)\,dt$$

whenever the latter exists. It is therefore necessary to establish whether the improper integral of g on $(-\infty, -1/2]$ converges. Now, $g(t) \to 0$ as $t \to -\infty$ with order $2 > 1$. Hence the improper integral of g on $(-\infty, -1/2]$ converges and will be denoted by I. Again by the limit of composite functions

$$\lim_{x \to +\infty} f(x) = \lim_{y \to -\infty} h(y) = I.$$

Similarly,

$$\lim_{x \to -\frac{1}{3}^-} f(x) = \lim_{y \to 1^-} h(y) = \int_{-\frac{1}{2}}^{1} g(t)\,dt = +\infty$$

and, for the reasons discussed in (a)

$$\lim_{x \to 1^+} f(x) = \lim_{y \to 1^-} h(y) = +\infty.$$

(c) In order to discuss the monotonicity of f, a preliminary study of the differentiability of h is needed. Since $g \in C^0(\mathbb{R} \setminus \{0, 1\})$, by the fundamental theorem of calculus $h \in C^1((-\infty, 0))$ and $h'(y) = g(y)$ for $y < 0$. Furthermore

$$h(y) = \int_{-\frac{1}{2}}^{\frac{1}{2}} g(t)\,dt + \int_{\frac{1}{2}}^{y} g(t)\,dt$$

and since $g \in C^0((0, 1))$ by the fundamental theorem of calculus $h \in C^1((0, 1))$ and $h'(y) = g(y)$ for $y \in (0, 1)$. Thus h is differentiable at least in $(-\infty, 0) \cup (0, 1)$. Observe that h is continuous at 0 by definition of improper integral, since

$$\lim_{y \to 0} h(y) = \lim_{y \to 0} \int_{-\frac{1}{2}}^{y} g(t)\,dt = \int_{-\frac{1}{2}}^{0} g(t)\,dt = h(0).$$

Using the theorems on continuity and differentiability of composite functions, f is continuous in $\mathrm{Dom}(f)$ and differentiable at least in

$$\left\{ x \in \mathbb{R} : 4x - 3x^2 < 1, 4x - 3x^2 \neq 0 \right\} = (\infty, 0) \cup (0, \frac{1}{3}) \cup (1, \frac{4}{3}) \cup (\frac{4}{3}, +\infty) := D.$$

The differentiability at 0 and 4/3 can be established using the theorem on the limit of the derivative. Evidently, for every $x \in D$

$$f'(x) = h'(p(x))p'(x) = g(p(x))p'(x) = \frac{4x - 3x^2 + 1}{\sqrt{|4x - 3x^2|} - (4x - 3x^2)^3}(4 - 6x),$$

so that

$$\lim_{x \to 0} f'(x) = +\infty = \lim_{x \to \frac{4}{3}} f'(x).$$

Hence f is not differentiable neither at 0 nor at $4/3$.

Since $f'(x) = g(p(x))p'(x)$, the sign of f' is found by studying separately the sign of $g(p(x))$ and that of $p'(x)$. Now, $g(t) > 0$ if and only if $0 < |t| < 1$ and $g(t) < 0$ if and only if $t < -1$. Therefore,

$$g(p(x)) > 0 \iff 0 < |4x - 3x^2| < 1$$

$$\iff x \in (\frac{2 - \sqrt{7}}{3}, 0) \cup (0, \frac{1}{3}) \cup (1, \frac{4}{3}) \cup (\frac{4}{3}, \frac{2 + \sqrt{7}}{3})$$

and

$$g(p(x)) < 0 \iff 4x - 3x^2 < -1 \iff x \in (-\infty, \frac{2 - \sqrt{7}}{3}) \cup (\frac{2 + \sqrt{7}}{3}, +\infty).$$

Finally,

$$p'(x) > 0 \iff x < \frac{2}{3}.$$

Combining these facts, it follows that f is strictly decreasing in $(-\infty, (2 - \sqrt{7})/3)$ and in $(1, (2 + \sqrt{7})/3)$, and that it is strictly increasing in $((2 - \sqrt{7})/3, 1/3)$ and in $((2 + \sqrt{7})/3, +\infty)$.

This exercise derives its complexity from the fact that the integral function to be studied is of the type

$$f(x) = \int_{x_0}^{p(x)} g(t) \, dx$$

and is therefore a composition $f = h \circ p$ where h is itself an integral function, of ordinary type. This calls for the following strategy: study h separately and then infer the properties of f by systematically applying the theorems on composite functions.

11.5 Problems on Improper Integrals and Integral Functions

11.16 Show that the improper integral $\int_0^1 \frac{e^t - 1}{t\sqrt{1 - t}} \, dt$ converges.

11.17 Show that the improper integral $\displaystyle\int_1^{+\infty} \frac{t + \sin t}{t^2\sqrt{1+t}} \, dt$ converges and compute an upper bound for it.

11.18 Does the improper integral $\displaystyle\int_2^{+\infty} \frac{t^\alpha}{t^4 - 1} \, dt$ converge if either $\alpha = 3$ or $\alpha = 1$?

11.19 Show that the improper integral $\displaystyle\int_0^{+\infty} e^{-x} \arctan(1 - e^x) \, dx$ converges.

11.20 Determine if the improper integral $\displaystyle\int_e^{+\infty} \frac{\log x}{x + x \log^4 x} \, dx$ converges and, in the affirmative case, compute its value.

11.21 Consider the function $f(x) = \dfrac{x - 1}{x + 1 - \sqrt{x}}$.

(a) In which closed and bounded intervals is f integrable?
(b) Determine if the improper integral $\int_0^{+\infty} f(x) \, dx$ converges.

11.22 For which values of $\alpha > 0$ does $\displaystyle\int_{-\infty}^{+\infty} \frac{\sqrt{\pi^2 - 4 \arctan^2 t}}{|t^2 - 1|^\alpha} \, dt$ converge?

11.23 Determine for which $r \in \mathbb{R}$ the following improper integrals converge

(a) $\displaystyle\int_1^r \frac{\arctan t - t - t\sqrt[3]{t}}{\pi - 2 \arctan \frac{1}{1+t^2}} \, dt$ (b) $\displaystyle\int_{-2}^r \frac{t + \arctan |t|}{\sqrt[3]{(t + 3)^2}(\sqrt[3]{t - 3} - \sqrt[3]{2t - 3})^2} \, dt$.

11.24 Consider the function $f(x) = \displaystyle\int_0^x \frac{e^t - 1}{t\sqrt{1+t}} \, dt$.

(a) Find the domain of f and determine its behaviour at the boundary.
(b) Establish if f is invertible in its domain.

11.25 Consider the function $f(x) = \displaystyle\int_2^x \frac{1}{t^2 + 2t - 3} \, dt$.

(a) Draw the graph of f.
(b) Write an explicit analytic expression of f.

11.26 Consider the function $f(x) = \displaystyle\int_0^x \frac{1}{(t - 1)\sqrt{t + 4}} \, dt$.

(a) Find the domain of f and determine where f is continuous and where it is differentiable.
(b) Compute, if meaningful, $f(1/2)$.

11.27 Consider the function $f(x) = \int_2^x (\frac{\log t}{t} + \frac{t}{\log t}) \, dt$.

(a) Find the domain of f.
(b) Where is f monotonic?

11.28 Find the domain of $f(x) = \int_1^x \frac{\log(1+t) - t}{\log(e^{t^2} - t) + t} \, dt$.

11.29 Consider the function $f(x) = \int_a^x \frac{\sin t - |t|}{t^2 \sqrt{|t+1|}} \, dt$.

(a) Find the domain of f as a ranges in $\in R$.
(b) Draw the graph of f for $a = -1$.

11.30 Consider the function $f(x) = \int_{-x^2}^{2x} \frac{1}{(2t-1)\sqrt[3]{t+1}} \, dt$.

(a) Find the domain of f.
(b) Find an interval in which f is monotonic.

11.31 Consider the function $f(x) = \int_1^x \frac{\sin(t^2 - t) - t \log |t|}{\log(|t^2 - 1|) - t^2} \, dt$.

(a) Find the domain of f.
(b) Where is the restriction of f to $(2, +\infty)$ monotonic?
(c) Compute $\lim_{x \to 1} f(x)/|x - 1|^a$ for $a > 0$.

11.32 Consider the function $f(x) = \int_r^x \frac{(1 + \frac{1}{t})^t - 2}{\log(4t^2 - 3|t|)} \, dt$.

(a) Find the domain of f as r ranges in $\mathbb{R}$.
(b) Put $r = 3$. Compute the limits of f at the boundary of the domain, and determine where f is differentiable.
(c) Put $r = 3$. Determine where f is monotonic and draw the graph of f.
(d) Sia $r = 3$. Compute, if existing, $\lim_{x \to +\infty}(f(x) - \log x)$.

11.33 Consider the function $f(x) = \int_{-\infty}^x \frac{r + \log(|\arctan t|)}{t - \cos^2 t} \, dt$.

(a) Find the domain of f as r ranges in $\mathbb{R}$.
(b) Put $r = \log(2/\pi)$. Determine where f is differentiable.
(c) Sia $r = \log(2/\pi)$. Compute, if existing, $\lim_{x \to 0}(f(x) - f(0))/x^2$.

11.34 Consider the function $f(x) = \int_x^{x-1} \frac{1}{(t+1)\sqrt[3]{1-t}} \, dt$.

(a) Find the domain of f and where f is differentiable.
(b) Find an interval I for which the restriction $f|_I$ is invertible.

11.35 Consider the function $f(x) = \int_{x^2-x}^{+\infty} \frac{\sqrt{4t+1}}{t(t-2)} \, dt$.

(a) Find the domain of f.
(b) Compute the limits of f at the boundary of its domain.
(c) Determine a positive integer n such that $f(n) - f(n-1) < 10^{-1}$.
(d) Does the the improper integral $\int_2^3 f(x) \, dx$ converge?

11.36 Consider the function

$$f(x) = \begin{cases} -\frac{1}{x+1} & x < -1 \\ 0 & x = -1 \\ \frac{1}{\sqrt{x+1}} & -1 < x \leq 0 \\ x & x > 0. \end{cases}$$

(a) For which values of $a, b \in \mathbb{R}$ does the integral of f in $[a, b]$ converge in some sense?
(b) Find where $g(x) = \int_0^x f(t) \, dt$ is defined, where it is continuous and where it is differentiable.

Chapter 12
Numerical Series

In mathematics, by *series* it is meant an infinite sum, which is performed in a given order. Thus, numerical series are infinite, ordered sums of real numbers.

12.1 Convergence

Definition 12.1 (*Convergent series*) Given the sequence $(a_n)_{n\geq 0}$ of real numbers, consider the associated sequence of *partial sums* $(s_n)_{n\geq 0}$, where

$$s_n = \sum_{k=0}^{n} a_k.$$

The series

$$\sum_{n=0}^{+\infty} a_n,$$

is said to *converge* to the *sum* $S \in \mathbb{R}$ if the sequence of partial sums converges to S, and in this case one writes

$$\sum_{n=0}^{+\infty} a_n = \lim_n s_n = S.$$

If instead the sequence of partial sums diverges to either $+\infty$ or to $-\infty$, that is, if

$$\sum_{n=0}^{+\infty} a_n = \lim_n s_n = \pm\infty$$

then the series is called *divergent*, whereas if the limit $\lim_n s_n$ does not exist, then the series is called *indeterminate*.

© Springer International Publishing Switzerland 2016

M. Baronti et al., *Calculus Problems*, UNITEXT - La Matematica per il 3+2 101,
DOI 10.1007/978-3-319-15428-2_12

As for sequences, the terminology introduced in Definition 12.1 is not completely standard. While the notion of convergent series is agreed upon by the international community, the term "divergent series" is used for all the remaining cases, whether the limit of the partial sums is $\pm\infty$ or does not exist. The terminology chosen here is to comply with the Italian tradition.

To any sequence $(a_n)_{n\geq 0}$ of real numbers there corresponds a series. It is customary to call a_n the *general term* of the series. The choice of the first integer in the numbering of the sequence, namely $n = 0$ if one writes $(a_n)_{n\geq 0}$, is not important. It is equally possible to sum a sequence $(a_n)_{n\geq n_0}$ with $n_0 \geq 1$, in which case one writes

$$\sum_{n=n_0}^{+\infty} a_n$$

to mean the limit of the sequence of partial sums $s_n = a_{n_0} + \cdots + a_n$. As for integrals, the name of the dummy variable n is inessential, so that of course

$$\sum_{k=0}^{+\infty} a_k = \sum_{n=0}^{+\infty} a_n.$$

A basic example of converging series is the *geometric* series. For $q \in \mathbb{R}$ and $N \in \mathbb{N}$

$$\sum_{n=N}^{+\infty} q^n = \frac{q^N}{1-q} \qquad (12.1)$$

provided that $|q| < 1$. Indeed, the identity $(1-q)(1+q+\cdots+q^N) = 1 - q^{N+1}$ is equivalent, whenever $q \neq 1$, to

$$\sum_{n=0}^{N} q^n = \frac{1 - q^{N+1}}{1-q},$$

which tends to $1/(1-q)$ if $|q| < 1$. This implies that

$$\sum_{n=0}^{+\infty} q^n = \frac{1}{1-q}.$$

Formula (12.1) follows from this by subtracting off the first N terms. Other classical examples are the so-called *telescoping* series, namely those of the form

$$\sum_{n=0}^{+\infty} (b_{n+1} - b_n)$$

where $(b_n)_{n \geq 0}$ is a given sequence. Indeed, the associated partial sum is

$$\sum_{k=0}^{n} (b_{k+1} - b_k) = (b_1 - b_0) + (b_2 - b_1) + \cdots + (b_{n+1} - b_n) = b_{n+1} - b_0.$$

Hence, if $(b_n)_{n \geq 0}$ converges, then so does the telescoping series, for $b_n \to \ell$ implies

$$\sum_{n=0}^{+\infty} (b_{n+1} - b_n) = \ell - b_0.$$

An important example of telescoping series is the *Mengoli* series, namely

$$\sum_{n=1}^{+\infty} \frac{1}{n(n+1)} = \sum_{n=1}^{+\infty} \left(\frac{1}{n} - \frac{1}{n+1} \right) = 1.$$

Nonzero constant sequences give rise to diverging series, while an example of indeterminate series is the series associated with the sequence $(a_n)_{n \geq 0}$ where $a_n = (-1)^n$. Indeed, in this case $s_n = 1$ if n is even, and $s_n = 0$ if n is odd, and the limit does not exist.

Proposition 12.1 (Cauchy criterion) *The series $\sum_{n=0}^{+\infty} a_n$ converges if and only if for every $\varepsilon > 0$ there exists $N \in \mathbb{N}$ such that for every $p \geq N$ and every $q \in \mathbb{N}$*

$$\left| \sum_{n=p}^{p+q} a_n \right| < \varepsilon.$$

Specializing to the case $q = 0$ the above statement, the following holds:

Proposition 12.2 *If the series $\sum_{n=0}^{+\infty} a_n$ converges, then necessarily $a_n \to 0$.*

The converse statement of Proposition 12.2 is false. For example, the *harmonic series*

$$\sum_{n=1}^{+\infty} \frac{1}{n} = +\infty$$

although $1/n \to 0$. This is best seen by observing that

$$s_n = 1 + \frac{1}{2} + \frac{1}{3} + \cdots + \frac{1}{n} \geq \int_1^{n+1} \frac{1}{x} \, dx = \log(n+1).$$

The previous inequality is illustrated in Fig. 12.1, where it is shown that the partial sum may be interpreted as the area of the region between the x-axis and the piecewise constant function that has value $1/n$ in the interval $[n, n+1)$ (see Theorem 12.8).

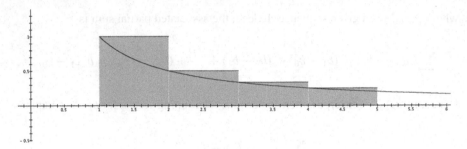

Fig. 12.1 The function $f(x) = 1/x$ is dominated by the piecewise constant function

Proposition 12.3 *Suppose that the sequences $(a_n)_{n\geq 0}$ and $(b_n)_{n\geq 0}$ are equal except for a finite number of values of n. Then*

(i) $\sum_{n=0}^{+\infty} a_n$ *converges if and only if* $\sum_{n=0}^{+\infty} b_n$ *converges;*
(ii) $\sum_{n=0}^{+\infty} a_n$ *diverges if and only if* $\sum_{n=0}^{+\infty} b_n$ *diverges;*
(iii) $\sum_{n=0}^{+\infty} a_n$ *is indeterminate if and only if* $\sum_{n=0}^{+\infty} b_n$ *is indeterminate.*

Definition 12.2 (*Absolute convergence*) The series $\sum_{n=0}^{+\infty} a_n$ is said to be *absolutely convergent* if the series $\sum_{n=0}^{+\infty} |a_n|$ is convergent.

Theorem 12.1 *If the series $\sum_{n=0}^{+\infty} a_n$ is absolutely convergent, then it is convergent. Moreover, in this case*

$$\left| \sum_{n=0}^{+\infty} a_n \right| \leq \sum_{n=0}^{+\infty} |a_n|.$$

The converse statement of Theorem 12.1 does not hold true: see the example after Theorem 12.10. In many cases, it is thus possible to show that a series converges by showing that it is actually absolutely convergent. This strengthens the motivation for studying series associated with non negative sequences.

12.2   Positive Series: Criteria

In this section it is assumed that the sequences to be summed $(a_n)_{n\geq 0}$ have non negative terms, that is $a_n \geq 0$ for all n. For this class of series, many results are available. First of all:

Theorem 12.2 *The series associated with a non negative sequence either converges or diverges to $+\infty$.*

Theorem 12.3 (Comparison) *Suppose that $(a_n)_{n\geq 0}$ and $(b_n)_{n\geq 0}$ are two non negative sequences and that eventually $a_n \leq b_n$. Then:*

(i) if $\sum_{n=0}^{+\infty} b_n$ converges, then $\sum_{n=0}^{+\infty} a_n$ converges;
(ii) if $\sum_{n=0}^{+\infty} a_n$ diverges, then $\sum_{n=0}^{+\infty} b_n$ diverges.

The comparison criterion can be used to show that the so-called *p-series*, namely

$$\sum_{n=1}^{+\infty} \frac{1}{n^p},$$

diverges when $0 < p < 1$. Indeed, for $p \in (0, 1]$ the inequality $n^{-p} \geq n^{-1}$ holds, and since the harmonic series diverges, so does the *p*-series. Below it is shown that the *p*-series with $p > 1$ converges.

Theorem 12.4 (Asymptotic equivalence) *Suppose that $(a_n)_{n \geq 0}$ and $(b_n)_{n \geq 0}$ are two non negative sequences and that $a_n \sim \lambda b_n$ for some $\lambda \in \mathbb{R} \setminus \{0\}$. Then $\sum_{n=0}^{+\infty} a_n$ converges if and only if $\sum_{n=0}^{+\infty} b_n$ converges.*

Theorem 12.5 *Suppose that $(a_n)_{n \geq 0}$ is a non negative sequence. If there exists a number $\ell \in (0, 1)$, such that eventually $\sqrt[n]{a_n} \leq \ell$ then $\sum_{n=0}^{+\infty} a_n$ converges. If $\sqrt[n]{a_n} \geq 1$ for infinitely many n, then $\sum_{n=0}^{+\infty} a_n$ diverges.*

Corollary 12.1 (Root test, or Cauchy test) *Suppose that $(a_n)_{n \geq 0}$ is a non negative sequence and suppose that*

$$\lim_n \sqrt[n]{a_n} = \ell$$

If $\ell < 1$, then $\sum_{n=0}^{+\infty} a_n$ converges, and if $\ell > 1$, then $\sum_{n=0}^{+\infty} a_n$ diverges.

Under the assumptions of the root test, if $\ell = 1$, then no conclusion may be drawn: there are cases in which the associated series converges and cases in which it diverges. Clearly, $\sqrt[n]{1/n} \to 1$ and the harmonic series diverges. For any $p > 1$ it holds that $\sqrt[n]{1/n^p} \to 1$ as well, but it will be shown below that the associated *p*-series converges. Using the root test, it is easy to prove that for $0 < p < 1$ and $\alpha \in \mathbb{R}$

$$\sum_{n=0}^{+\infty} p^n n^\alpha < +\infty.$$

Theorem 12.6 *Suppose that $(a_n)_{n \geq 0}$ is eventually positive. If there exists a number $\ell \in (0, 1)$, such that eventually $a_{n+1}/a_n \leq \ell$, then $\sum_{n=0}^{+\infty} a_n$ converges. If eventually $a_{n+1} \geq a_n$, then $\sum_{n=0}^{+\infty} a_n$ diverges.*

Corollary 12.2 (Ratio test, or D'Alembert test) *Suppose that $(a_n)_{n \geq 0}$ is eventually positive and suppose that*

$$\lim_n \frac{a_{n+1}}{a_n} = \ell$$

If $\ell < 1$, then $\sum_{n=0}^{+\infty} a_n$ converges, and if $\ell > 1$, then $\sum_{n=0}^{+\infty} a_n$ diverges.

As with the root test, the ratio test cannot be used when $\ell = 1$. Again, the ratios associated with the harmonic series and with any p-series converge to $\ell = 1$. By means of the ratio test it is possible to establish that the *exponential series*

$$\sum_{n=0}^{+\infty} \frac{x^n}{n!}$$

converges for every $x > 0$. By the criterion of absolute convergence (see Theorem 12.1 above), the exponential series actually converges for every $x \in \mathbb{R}$.

Theorem 12.7 (Condensation) *Suppose that $(a_n)_{n \geq 0}$ is a non negative, decreasing sequence. Then $\sum_{n=0}^{+\infty} a_n$ converges if and only if the condensed series $\sum_{n=0}^{+\infty} 2^n a_{2^n}$ converges.*

The condensation test can be applied to show at once that the p-series converges provided that $p > 1$. Indeed, the general term of the condensed series is in this case

$$2^n \frac{1}{(2^n)^p} = \left(\frac{1}{2^{p-1}}\right)^n$$

which, for $p > 1$ is the general term of a converging geometric series. The condensation criterion is particularly useful when logarithms appear.

12.3 Order, Series and Integrals

As indicated in the case of the harmonic series, there is a strong connection linking the theory of series and that of improper integrals. The idea behind is that a sequence can be interpreted as a piecewise constant function that is constant on the open intervals with extreme points at the integers, so that, for positive sequences, the associated series should represent the area below the piecewise constant function. The comparison test for integrals thus applies to infer, or use, the convergence of the series in relation to the convergence of the appropriate improper integral.

Theorem 12.8 (Integral test) *Let $N \in \mathbb{N}$ and suppose that $f : [N, +\infty) \to [0, +\infty)$ is a positive and decreasing function. Then the following are equivalent:*

(i) the series $\displaystyle\sum_{n=N}^{+\infty} f(n)$ converges;

(ii) the improper integral $\displaystyle\int_{N}^{+\infty} f(x)\, dx$ converges.

Furthermore,

$$\int_N^{+\infty} f(x)\,dx \le \sum_{n=N}^{+\infty} f(n) \le f(N) + \int_N^{+\infty} f(x)\,dx.$$

Notice that if the series converges to S, then the last inequality implies

$$0 < S - s_N = \sum_{n=N+1}^{+\infty} f(n) \le \int_N^{+\infty} f(x)\,dx.$$

The above result, in connection with the order criteria for improper integrals, calls for a sensible notion of order for sequences that tend to 0. When treating a series $\sum a_n$, the necessary condition that $a_n \to 0$ expressed in Proposition 12.2 is, as mentioned, not sufficient to assess its convergence. If a stronger requirement on the *order* with which this happens is satisfied, then the convergence of $\sum a_n$ is guaranteed. Definition 5.9 of Chap. 5 readily adapts to the case of a sequence $(a_n)_{n \ge 0}$.

Definition 12.3 (Order of a sequence) Let $\alpha > 0$ be fixed. The sequence $(a_n)_{n \ge 0}$ is said to tend to 0 of order α if there exists $\lambda \in \mathbb{R} \setminus \{0\}$ such that $a_n \sim \lambda n^{-\alpha}$, namely

$$\lim_n a_n n^\alpha = \lambda.$$

It is said to tend to 0 of order greater than α if

$$\lim_n a_n n^\alpha = 0$$

and smaller than α if

$$\lim_n a_n n^\alpha = \pm\infty.$$

Theorem 12.9 (Order criteria for series) *Suppose that $(a_n)_{n \ge 0}$ is a non negative sequence and assume that $a_n \to 0$. Then:*

(i) *if $a_n \to 0$ with an order that is greater than or equal to α for some $\alpha > 1$, then the series $\sum_{n=0}^{+\infty} a_n$ converges;*

(ii) *if $a_n \to 0$ with an order that is smaller than or equal to α for some $\alpha \le 1$, then the series $\sum_{n=0}^{+\infty} a_n$ diverges.*

12.4 Alternating Series

Among the series that have not been treated in Sect. 12.2 or in Sect. 12.3, of particular relevance are the *alternating series*, namely those of the form

$$\sum_{n=0}^{+\infty} (-1)^n a_n, \qquad a_n \geq 0,$$

in which the sign of the general term depends only on the parity of its index n.

Theorem 12.10 (Leibniz) *Suppose that $(a_n)_{n \geq 0}$ is positive, decreasing and tends to 0. Then the alternating series*

$$\sum_{n=0}^{+\infty} (-1)^n a_n$$

converges. Furthermore, the partial sums s_{2n} with even indices approximate from above the sum S and the partial sums s_{2n+1} with odd indices approximate from below the sum S. Finally, for every $n \geq 0$

$$|S - s_n| < a_{n+1}$$

that is, the remainder of the sum is controlled by the first term that is not summed.

A series that satisfies the assumptions of the above theorem is called a *Leibniz series*. It follows from Leibniz' theorem that the series

$$\sum_{n=1}^{+\infty} (-1)^n \frac{1}{n}$$

converges, an example of a converging series that is not absolutely convergent.

12.5   Guided Exercises on Numerical Series

12.1 Establish if the series $\displaystyle\sum_{n=1}^{+\infty} \frac{e^n}{n!(\sqrt{n+1} - \sqrt{n})}$ converges.

Answer. Consider the ratio a_{n+1}/a_n, namely

$$\frac{e^{n+1}}{(n+1)!(\sqrt{n+2} - \sqrt{n+1})} \cdot \frac{n!(\sqrt{n+1} - \sqrt{n})}{e^n} = \frac{e}{n+1} \cdot \frac{\sqrt{n+1} - \sqrt{n}}{\sqrt{n+2} - \sqrt{n+1}}.$$

Rationalizing, collecting $\sqrt{n}$ and taking the limit

$$\lim_n \frac{e}{n+1} \cdot \frac{\sqrt{1 + \frac{2}{n}} + \sqrt{1 + \frac{1}{n}}}{\sqrt{1 + \frac{1}{n}} + 1} = 0 < 1,$$

so that the series converges. This is an elementary application of the ratio test.

12.2 Establish when the series $\sum_{n=1}^{+\infty}\left(\dfrac{1}{n}+\sin\dfrac{1}{n}\right)n^\alpha$ converges, as α ranges in $\mathbb{R}$.

Answer. Since $\sin(1/n)$ is asymptotically equivalent to $1/n$, the given series converges if and only if so does the series

$$\sum_{n=1}^{+\infty}\frac{2}{n}n^\alpha$$

Now, $2n^{\alpha-1}$ is a multiple of a p-series term, and hence the series converges provided that $1-\alpha>1$, that is, $\alpha<0$.

This is an easy application of the criterion of asymptotic equivalence, which yields a p-series.

12.3 Establish for which $k>0$ the series $\sum_{n=0}^{+\infty}\left(k+\dfrac{1}{n+3}\right)^n\dfrac{n+1}{n^2+1}$ converges.

Answer. Taking nth roots gives

$$\lim_n \sqrt[n]{a_n}=\lim_{n\to+\infty}(k+\frac{1}{n+3})\sqrt[n]{\frac{n+1}{n^2+1}}=k.$$

Hence, by the root test, if $k>1$ the series diverges and if $0<k<1$ it converges. If $k=1$, then the series becomes

$$\sum_{n=0}^{+\infty}\left(1+\frac{1}{n+3}\right)^n\frac{n+1}{n^2+1}$$

Now, the necessary condition is satisfied because

$$\lim_n\left[\left(1+\frac{1}{n+3}\right)^{n+3}\right]^{\frac{n}{n+3}}\frac{n+1}{n^2+1}=0,$$

However,

$$\lim_n\frac{a_n}{1/n}=\lim_n\left[\left(1+\frac{1}{n+3}\right)^{n+3}\right]^{\frac{n}{n+3}}\frac{n(n+1)}{n^2+1}=e,$$

shows that $a_n\to0$ with order exactly 1, hence the series diverges.

Here the particular form of a_n calls for an application of the root test, which gives the answer in the case $k\neq1$ If $k=1$, then the sensible criterion to apply is the order criterion: indeed, for $k=1$ the general term is the product of a bounded convergent sequence and a sequence asymptotically equivalent to $1/n$.

12.4 Establish when the series $\displaystyle\sum_{n=1}^{+\infty}(-1)^n\frac{n}{n^2+1}$ converges.

Answer. This is an alternating series with

$$a_n = \frac{n}{n^2+1}$$

that is clearly positive and tends to 0. Furthermore

$$a_{n+1} - a_n = \frac{-n^2-n+1}{(n^2+1)((n+1)^2+1)} < 0$$

shows that the sequence is decreasing. By Leibniz' theorem, the series converges.
   This is a straightforward application of Leibniz' theorem.

12.5 Establish if the series $\displaystyle\sum_{n=0}^{+\infty}\frac{(n!)^n}{e^{n!}}$ converges.

Answer. The ratio a_{n+1}/a_n satisfies

$$\frac{((n+1)!)^{n+1}}{e^{(n+1)!}} \cdot \frac{e^{n!}}{(n!)^n} = \frac{n!(n+1)^{n+1}}{e^{n\cdot n!}}$$

$$\leq \frac{n^n(n+1)^{n+1}}{e^{n\cdot n!}}$$

$$\leq \frac{e^{n^2}e^{n^2+n}}{e^{n\cdot n!}}$$

$$= e^{2n^2+n-n\cdot n!},$$

because $n^n \geq n!$ and $e^n \geq n+1$. Since $e^{2n^2+n-n\cdot n!} \to 0$ the series converges.
   This is an application of the ratio test. In analyzing the ratio, some estimate on the various powers is necessary. The "convex" inequality $e^x \geq x+1$ is often very handy.

12.6 Consider a sequence $(a_n)_{n\geq 0}$ such that $a_n > 0$ and such that $a_n \to 0$ with order 3. Establish when the series

$$\sum_{n=1}^{+\infty} n^\alpha a_n$$

converges, as α ranges in $\mathbb{R}$.

Answer. By assumption, $a_n/n^{-3} \to \lambda \in \mathbb{R} \setminus \{0\}$. Then

$$\lim_n a_n n^\alpha = \begin{cases} 0 & 0 < \alpha < 3 \text{ (with order } 3\text{-}\alpha) \\ \lambda & \alpha = 3 \\ +\infty & \alpha > 3 \end{cases}$$

Write $b_n = n^\alpha a_n$. Now, if $\alpha < 3$, then $b_n \to 0$ with order $3 - \alpha > 1$ if $\alpha < 2$. Therefore, if $\alpha < 2$, then the series converges. If $2 \le \alpha < 3$, then $b_n \to 0$ with order ≤ 1 and the series diverges to $+\infty$. If $\alpha \ge 3$, then the series diverges to $+\infty$ because its general term is positive and does not tend to 0.

The standing assumption in this exercise is an information about the order with which a_n tends to 0, which translates at once into an information about the behaviour of $b_n = n^\alpha a_n$. As α increases, the order of $(b_n)_{n \ge 0}$ decreases exactly by α with respect to the order of $(a_n)_{n \ge 0}$, so the critical value for α is clearly 2.

12.7 Establish for which $k \in \mathbb{R}$ and $\alpha \in (0, +\infty)$ the series $\displaystyle\sum_{n=1}^{+\infty} \frac{1 - \cos(k/n)}{\sin(1/n^\alpha)}$ converges.

Answer. When $k = 0$, the general term is 0 for every $\alpha \in \mathbb{R}$, hence the series converges. For $k \ne 0$, and $\alpha > 0$ the general term b_n is

$$b_n = \frac{1 - \cos(k/n)}{\sin(1/n^\alpha)} = \frac{1 - \cos(k/n)}{k^2/n^2} \cdot \frac{1/n^\alpha}{\sin(1/n^\alpha)} \cdot \frac{k^2}{n^{2-\alpha}}$$

and hence

$$b_n \to \begin{cases} +\infty & \alpha > 2 \\ k^2/2 & \alpha = 2 \\ 0 & 0 < \alpha < 2. \end{cases}$$

Thus the series diverges to $+\infty$ if $k \ne 0$ and $\alpha \ge 2$. If $k \ne 0$ and $0 < \alpha < 2$, then $b_n \to 0$ with order $2 - \alpha$. Evidently, $2 - \alpha > 1$ only for $\alpha < 1$. Therefore, for $k \ne 0$ and $0 < \alpha < 1$ the series converges, while for $\alpha \ge 1$ it diverges.

This exercise is a very standard application of order criteria, in combination with asymptotic equivalence. Indeed, $(b_n)_{n \ge 1}$ has been written in a way that displays its asymptotic equivalence, up to non zero constants, with the sequence $(a_n)_{n \ge 1}$, where $a_n = k^2 n^{\alpha-2}$, and the latter is almost the prototypical scale of powers.

12.8 Establish for which $x \in \mathbb{R}$ and $\alpha \in (0, +\infty)$ the series $\displaystyle\sum_{n=1}^{+\infty} \frac{\log^n(1 + x^2)}{n^\alpha}$ converges.

Answer. Evidently,

$$a_n = \frac{\log^n(1+x^2)}{n^\alpha} \geq 0$$

for every $x \in \mathbb{R}$, and $a_n = 0$ for every $n \geq 1$ if and only if $x = 0$. In particular, if $x = 0$, then the series converges. If $x \neq 0$, then $a_n > 0$ for every $n \geq 1$ and the root test may be applied. Indeed,

$$\lim_n \frac{\log(1+x^2)}{(\sqrt[n]{n})^\alpha} = \log(1+x^2).$$

Hence, if $\log(1+x^2) < 1$, namely $|x| < \sqrt{e-1}$, the series converges, whereas if $\log(1+x^2) > 1$, namely $|x| > \sqrt{e-1}$, then it diverges to $+\infty$. If $|x| = \sqrt{e-1}$, then $a_n = n^{-\alpha}$ and $a_n \to 0$ with order α. Therefore, in the case when $x = \pm\sqrt{e-1}$, the series converges for $\alpha > 1$, and if $0 < \alpha \leq 1$ it diverges to $+\infty$.

This is a standard exercise on the root test. The critical case where the limit is 1 occurs for $x = \pm\sqrt{e-1}$ and any α, but then the series is really a *p*-series.

12.9 Let x be a real parameter, and consider the series $\displaystyle\sum_{n=1}^{+\infty} \frac{(4\cos x - 1)^{2n+1}}{n+9^n}$.

(a) Establish when the series converges, as x varies in $\mathbb{R}$.
(b) Put $x = \pi/4$. Find a partial sum that approximates the sum of the series with an error less than 10^{-2}.

Answer. (a) Observe that

$$\left|\frac{(4\cos x - 1)^{2n+1}}{n+9^n}\right| = \frac{|4\cos x - 1|((4\cos x - 1)^2)^n}{n+9^n} \leq 5\frac{((4\cos x - 1)^2)^n}{9^n},$$

and that the series

$$\sum_{n=1}^{+\infty} \left(\frac{(4\cos x - 1)^2}{9}\right)^n$$

is a geometric series that converges provided that $(4\cos x - 1)^2 < 9$, that is, if $|4\cos x - 1| < 3$. The solutions of these inequalities are the values of x for which $-1/2 < \cos x < 1$, namely for which $x \in A$, where

$$A := \bigcup_{m\in\mathbb{Z}} \{x \in \mathbb{R} : -2\pi/3 + 2m\pi < x < 2\pi/3 + 2m\pi, \ x \neq 2m\pi\}.$$

Thus, if $x \in A$, then the series is absolutely convergent, hence convergent. If $x \notin A$, then $(4\cos x - 1)^2 \geq 9$ and hence

$$\lim_n \frac{(4\cos x - 1)^{2n+1}}{n + 9^n} = \lim_n \frac{(4\cos x - 1)}{1 + \frac{n}{9^n}} \left(\frac{(4\cos x - 1)^2}{9}\right)^n$$

$$= \begin{cases} -\infty & \cos x < -1/2 \\ 4\cos x - 1 & \cos x = -1/2, \text{ or } \cos x = 1. \end{cases}$$

This shows that if $x \notin A$ then the general term of the series does not tend to 0, and hence the series does not converge.

(b) If $x = \pi/4$, then

$$a_n = \frac{(2\sqrt{2} - 1)^{2n+1}}{n + 9^n}.$$

Denote by S the sum of the series and by s_k the kth partial sum. Then

$$|S - s_k| = \sum_{n=k+1}^{+\infty} \frac{(2\sqrt{2} - 1)^{2n+1}}{n + 9^n} \leq (2\sqrt{2} - 1) \sum_{n=k+1}^{+\infty} \left(\frac{(2\sqrt{2} - 1)^2}{9}\right)^n$$

$$= (2\sqrt{2} - 1)\frac{1}{1 - \frac{(2\sqrt{2}-1)^2}{9}} \left(\frac{(2\sqrt{2} - 1)^2}{9}\right)^{k+1} \leq 10^{-2}$$

if $k \geq 5$. Hence s_3 approximates from below the sum S with an error that is smaller than 10^{-2}.

To answer question (a) the idea is to compare the given series to a geometric series because, after all, the behaviour of the denominator $n + 9^n$ is dictated by 9^n. When this estimate is done, then the series is dominated by a series of geometric type of the form $\sum |f(x)|^n$. When $|f(x)| < 1$ then the geometric series converges, and for the values of x for which $|f(x)| \geq 1$ the general term of the original series does not tend to 0. In question (b) the idea is to appeal to a nice property of geometric series, namely that the remainder, that is, $S - s_n$, has the explicit expression (12.1). The final step is to find the correct value of k that makes the complicated-looking expression less than 10^{-2}.

12.10 Consider the series:

$$\sum_{n=1}^{+\infty} \frac{\arcsin(1/n)}{n^2 + \log^5 n}$$

Prove that it converges and find a partial sum s_n that approximates the sum S with an error smaller than 10^{-2}, specifying if $s_n < S$ or $s_n > S$.

Answer. For $n \geq 1$

$$a_n = \frac{\arcsin(1/n)}{n^2 + \log^5 n} \leq \frac{\arcsin(1/n)}{n^2} \leq \frac{\pi}{2}\frac{1}{n^2} = b_n.$$

Thus, a simple application of the comparison theorem yields the convergence of the given series.

First of all, observe that $a_n > 0$, and hence it is $s_n < S$ for every $n \geq 1$. In order to estimate the sum, the idea is to apply the integral test. Consider f : $[1, +\infty) \to \mathbb{R}$ defined by $f(x) = \pi/2x^2$, which is clearly positive, continuous and strictly decreasing. Now, using the integral test

$$0 < S - s_n = \sum_{k=n+1}^{+\infty} a_n \leq \sum_{k=n+1}^{+\infty} b_n \leq \int_n^{+\infty} f(x)\,dx = \frac{\pi}{2} \int_n^{+\infty} \frac{1}{x^2}\,dx = \frac{\pi}{2n}.$$

Thus $0 < S - s_n < 10^{-2}$ provided that $n > 50\pi > 157$.

This exercise can be solved using a combination of comparison and integral tests, specially in its second part. The convergence can easily be established with simple order criteria, but the evaluation of the remainder is very well handled with an integral. To wit, the integral of the function x^{-2}.

12.11 Consider the series:

$$\sum_{n=1}^{+\infty} \frac{1}{n\sqrt{n+1}\log(n+2)}.$$

Establish if the series converges and, if so, find a partial sum s_n that approximates the sum S with an error smaller than 10^{-1}, specifying if $s_n < S$ or $s_n > S$.

Answer. Consider the function f defined for $x \geq 1$ by

$$f(x) = \frac{1}{x\sqrt{x+1}\log(x+2)},$$

so that for every positive integer n

$$a_n = \frac{1}{n\sqrt{n+1}\log(n+2)} = f(n).$$

By the integral test, the series converges if and only if the improper integral

$$\int_1^{+\infty} \frac{1}{t\sqrt{t+1}\log(t+2)}\,dt.$$

converges. Indeed, f is continuous, positive and decreasing in $[1, +\infty)$. Furthermore, for $t \geq 1$ it is $\log(t+2) \geq \log 3 \geq 1$ and hence

$$\frac{1}{t\sqrt{t+1}\log(t+2)} \leq \frac{1}{t\sqrt{t}}.$$

By order criteria, and by comparison

$$\int_1^{+\infty} \frac{1}{t\sqrt{t+1}\log(t+2)}dt \leq \int_1^{+\infty} \frac{1}{t\sqrt{t}}dt < +\infty$$

and, consequently, the series converges. Again by the integral test,

$$0 \leq S - s_n = \sum_{k=n+1}^{+\infty} \frac{1}{k\sqrt{k+1}\log(k+2)} \leq \sum_{k=n+1}^{+\infty} \frac{1}{\log(n+3)k\sqrt{k+1}}$$

$$\leq \frac{1}{\log(n+3)} \int_n^{+\infty} \frac{1}{x^{3/2}}dx = \frac{2}{\log(n+3)\sqrt{n}}.$$

For $n = 32$ it is $0 \leq S - s_n \leq 9.95 \times 10^{-2} < 10^{-1}$.

12.12 Let α denote a real parameter and consider the series $\sum_{n=1}^{+\infty}((1/n) - \sin(1/n))n^\alpha$.

(a) Establish when the series converges, as α varies in $\mathbb{R}$.
(b) Put $\alpha = 1$. Find a partial sum s_n that approximates the sum S with an error smaller than 10^{-1}.

Answer. (a) By the fourth order McLaurin expansion of the sine function with Lagrange's remainder

$$\frac{1}{n} - \sin\frac{1}{n} = \frac{1}{n^3}\left(\frac{1}{6} - \frac{\cos c}{120}\frac{1}{n^2}\right)$$

with $c \in [0, 1/n]$. Evidently, the general term of the series tends to 0 only for $\alpha < 3$ and, in this case, the order with which this happens is $3 - \alpha$. Therefore the series converges only for $\alpha < 2$.
(b) As just established

$$0 \leq \left(\frac{1}{n} - \sin\frac{1}{n}\right)n = \frac{1}{n^2}\left(\frac{1}{6} - \frac{\cos c}{120}\frac{1}{n^2}\right) \leq \frac{1}{n^2}\left(\frac{1}{6} + \frac{1}{120n^2}\right).$$

Hence, by the integral test

$$0 \leq S - s_n = \sum_{k=n+1}^{+\infty} \left(\frac{1}{k} - \sin\frac{1}{k}\right)k \leq \sum_{k=n+1}^{+\infty} \left(\frac{1}{6k^2} + \frac{1}{120k^4}\right)$$

$$\leq \int_n^{+\infty} \left(\frac{1}{6t^2} + \frac{1}{120t^4}\right)dt = \frac{1}{6n} + \frac{1}{360n^3} < \frac{1}{3n}.$$

The partial sum s_4 is thus sufficient.

This is a fairly standard use of the integral test in combination with comparison and order criteria. The novelty here is to derive an estimate by means of Taylor's expansions.

12.13 Establish for which values of $b \in \mathbb{R} \setminus \{-2\}$ the series $\displaystyle\sum_{n=0}^{+\infty} \frac{b^n}{b^n + 2^n}$ converges.

Answer. If $b = 0$, then the series trivially converges to 0. If $b \neq 0$, write

$$a_n = \frac{b^n}{b^n + 2^n} = \frac{1}{1 + (2/b)^n}.$$

If $|2/b| < 1$, then $a_n \to 1$ and the series diverges to $+\infty$. If $b = 2$, then $a_n = 1/2$ for every n and the series diverges to $+\infty$.

The cases where $|2/b| > 1$ must be studied according as $b > 0$ or $b < 0$. Suppose first $0 < b < 2$. Then $2/b > 1$ and $a_n \to 0$ with exponential order, hence with order greater than 2: the series converges. Assume finally $-2 < b < 0$, and write

$$a_n = \frac{1}{1 + (\frac{2}{b})^n} = (-1)^n \frac{1}{(-1)^n + |\frac{2}{b}|^n} = (-1)^n \frac{1}{(-1)^n + d^n}$$

where $d := |2/b| > 1$. The sequence $(1/((-1)^n + d^n))_{n \geq 1}$ is clearly positive and tends to 0. It is also eventually decreasing. Indeed consider the inequality

$$(-1)^n + d^n < (-1)^{n+1} + d^{n+1}.$$

If n is odd, it is equivalent to $d^n < d^{n+1} + 2$, which is true for every n since $d > 1$; if n is even, it is equivalent to $d^{n+1} > d^n + 2$, that is $d > 1 + 2d^{-n}$. The latter is eventually true because $d^{-n} \to 0$. Hence, by Leibniz' theorem, the series converges.

The series under consideration can possibly converge only when $|2/b| > 1$. In turn, for positive b the series is convergent, as can be seen either with order criteria or observing that it is dominated by a converging geometric series. The case of interest is when $b < 0$. It is then clear that in order to use Leibniz' theorem the only issue to be examined is the fact that the sequence is decreasing, because in the denominator there is the oscillating summand $(-1)^n$. This, however, turns out to be negligible with respect to the fast growing exponential d^n.

12.14 Suppose that $(a_n)_{n \geq 0}$ is positive and decreasing. Prove that if $\sum a_n$ converges, then $na_n \to 0$, and that the converse statement is false.

Answer. Denoting by $[n/2]$ the *integral part* of $n/2$ (the largest integer $\leq n/2$), it is

$$\sum_{k=[n/2]}^{n} a_k \geq (n - [n/2] + 1)a_n \geq \frac{1}{2}na_n.$$

By Cauchy's criterion, if the series converges, then the left hand side tends to 0. If $a_n = 1/(n \log n)$, then $na_n = 1/\log n \to 0$ but the series $\sum a_n$ does not converge, because by condensation

$$\sum_{n=1}^{+\infty} 2^n a_{2^n} = \sum_{n=1}^{+\infty} \frac{2^n}{2^n \log(2^n)} = \sum_{n=1}^{+\infty} \frac{1}{n \log 2} = +\infty.$$

This exercise has a theoretical flavour: it is asked to show that the general term of a converging series must tend to 0 with order greater than 1 (provided that $(a_n)_{n\geq 1}$ is positive and decreasing), but "greater than 1" is not enough. Observe the subtle, but crucial, difference between saying "order greater than 1" and saying "order greater than or equal to some α with $\alpha > 1$", as in the statement of Theorem 12.9. Indeed, the chosen example, where $a_n = 1/(n \log n)$, tends to zero with order greater than 1, but for any $\alpha > 1$ it is

$$\lim_n \frac{1/(n \log n)}{1/n^\alpha} = \lim_n \frac{n^{\alpha-1}}{\log n} = +\infty,$$

so that there is no $\alpha > 1$ for which $(a_n)_{n\geq 1}$ tends to 0 with order greater than or equal to α.

12.6 Problems on Series

12.15 Establish if the series $\sum a_n$ converges, where:

(a) $a_n = \dfrac{n + \log n}{(n - \log n)^3}$

(b) $a_n = e^{-\sqrt{n}}$

(c) $a_n = \dfrac{n + \log n}{n^{1/3}}$

(d) $a_n = e^{-n^2}$

(e) $a_n = (\sqrt{n+1} - \sqrt{n})^2$.

12.16 Establish if the series $\sum a_n$ converges, where:

(a) $a_n = \dfrac{1}{\sqrt{n^2 - n}(\sqrt{n+1} - \sqrt{n})}$

(b) $a_n = \dfrac{(n+2)e^n}{n!}$.

12.17 Establish if the series $\sum a_n$ converges, where:

(a) $a_n = \left(e^{1/n} + \dfrac{1}{n}\right)^n$

(b) $a_n = \dfrac{1}{n^2 - 21}$

(c) $a_n = \dfrac{n+1}{n^2+1}$

(d) $a_n = \dfrac{n^k + a}{n^k - b}$, $\quad a, b \in \mathbb{R}, k > 0$

(e) $a_n = \dfrac{1}{\sqrt{n^2 + n}}$.

12.18 Use the root, ratio and condensation tests to establish if the series $\sum a_n$ converges, where:

(a) $a_n = \dfrac{1}{\sqrt{n^3 - n}}$

(b) $a_n = \dfrac{\log(\log n)}{n(\log n)^2}$

(c) $a_n = \dfrac{\sin(1/n)}{\log n}$

(d) $a_n = \dfrac{\sqrt{n+1} - \sqrt{n}}{(\log n)^3}$

(e) $a_n = \dfrac{n^{103}}{n!}$

(f) $a_n = \dfrac{(n!)^2}{(2n)!}$

(g) $a_n = n^{-n^{1/3}}$

(h) $a_n = \left(\dfrac{\log(\log n)}{\log n}\right)^n$.

12.19 Establish if the series $\sum (-1)^n a_n$ converges, where:

(a) $a_n = \dfrac{\sqrt{n} - 1}{\sqrt{n} + 1}$

(b) $a_n = \dfrac{\log n}{(\log n)^2 - 1}$

(c) $a_n = \dfrac{(n!)^2}{n^2 (2n)!}$

(d) $a_n = \dfrac{\pi}{2} - \arctan n$.

12.20 Establish if the series $\sum a_n$ converges as b, c range in $\mathbb{R}$ and k in $\mathbb{N}$, where:

(a) $a_n = \dfrac{n^n - n!}{(bn)^n - e^n}$

(b) $a_n = \dbinom{n}{k} \dfrac{1}{n(n \log n)^c}$

(c) $a_n = \dfrac{n(\log n - b) - \log(n!)}{n(n+1)^{1/b}}$

(d) $a_n = (1 - \dfrac{1}{n^2})^{n^b}$

(e) $a_n = \left(\sin(b + \dfrac{c}{n})\right)^n$, $\quad 0 \le b < \pi/2$.

12.21 Establish or which values of $\alpha \in \mathbb{R}$ the series $\sum_{n=1}^{+\infty} \dfrac{(\sin \alpha)^n}{n + \log n}$ converges.

12.22 Consider the series $\sum_{n=1}^{+\infty} \left(e^{\frac{1}{n}} - \cos \dfrac{1}{n} - \sin \dfrac{\alpha}{n} \right)$.

(a) Establish when the series converges, as α ranges in $\mathbb{R}$.
(b) Put $\alpha = 1$. Find a partial sum that approximates the sum S of the series with an error smaller than 10^{-3}

12.23 Establish if the series $\sum_{n=1}^{+\infty} ne^{-n^2}$ converges and, if yes, find a partial sum that approximates the sum S of the series with an error smaller than 10^{-3}, specifying if $s_n < S$ or $s_n > S$.

12.24 Establish if the series

$$\sum_{n=1}^{+\infty} \frac{e^n}{e^{2n} + 1}$$

converges and, if yes, find a partial sum that approximates the sum S of the series with an error smaller than 10^{-1}, specifying if $s_n < S$ or $s_n > S$.

Chapter 13
Separation of Variables

13.1 Differential Equations

A large number of real-life problems considered in a variety of disciplines including engineering, physics, economics and biology, are modeled by *differential equations*. They are equations in which the unknown is a function, and, loosely speaking, involve the function itself together with its rates of change. If the function to be found is a function of a single real variable, then one speaks of an *ordinary* differential equation, whereas if the unknown function depends on several variables, then the equation is called a *partial* differential equation. In the jargon, these two very different classes of equations are often abbreviated as ODE and PDE, respectively. In this book, all differential equations are ordinary differential equations. Hence, from now on they will simply be referred to as differential equations.

The general form of a differential equation is

$$f(x, y, y^{(1)}, y^{(2)}, \ldots, y^{(n)}) = 0, \tag{13.1}$$

where $f : A \to \mathbb{R}$ is a function defined on some subset $A \subseteq \mathbb{R}^{n+2}$ and where the unknown y is a function $y : I \to \mathbb{R}$ defined on some interval I. The *order* of the differential equation is the highest order of differentiation involved in the equation, namely n in (13.1). The equation is said to be in *normal form* if it may be written as

$$y^{(n)} = g(x, y, y^{(1)}, y^{(2)}, \ldots, y^{(n-1)}), \tag{13.2}$$

where $g : B \to \mathbb{R}$ is a function defined on some subset $B \subseteq \mathbb{R}^{n+1}$. Thus, f in (13.1) or g in (13.2) express the relation between the function y and its derivatives. An important issue, as clarified in the definition below, is that the equation must be satisfied pointwise in the independent variable x.

© Springer International Publishing Switzerland 2016

M. Baronti et al., *Calculus Problems*, UNITEXT - La Matematica per il 3+2 101, DOI 10.1007/978-3-319-15428-2_13

Definition 13.1 A function $y : I \to \mathbb{R}$ with n derivatives in the interval I is called a *solution* of the differential equation (13.1) if

$$f\left(x, y(x), y^{(1)}(x), y^{(2)}(x), \ldots, y^{(n)}(x)\right) = 0$$

holds for every $x \in I$. The set of all solutions of (13.1) is called its *general solution*, sometimes referred to as *general integral*.

The general solution of a differential equation often consists of infinitely many functions. For this reason, in many problems additional requirements are considered.

Definition 13.2 The *Cauchy problem* associated with the differential equation (13.2) and with the *initial value* $(x_0, y_0, \ldots, y_{n-1}) \in \mathbb{R}^{n+1}$ is the problem

$$\begin{cases} y^{(n)} = g\left(x, y(x), y^{(1)}(x), y^{(2)}(x), \ldots, y^{(n-1)}(x)\right) \\ y(x_0) = y_0 \\ y^{(1)}(x_0) = y_1 \\ \vdots \\ y^{(n-1)}(x_0) = y_{n-1}. \end{cases} \tag{13.3}$$

A solution of the Cauchy problem (13.3) is a function y that has n derivatives in a neighborhood I of x_0 and is such that (13.3) holds for every $x \in I$.

13.2   The Method of Separation of Variables

Among the many interesting classes of differential equations, are those first order equations that have the normal form

$$y'(x) = a(x)b(y). \tag{13.4}$$

Here $a : I \to \mathbb{R}$ and $b : J \to \mathbb{R}$ are two functions, and both I and J are open intervals. The associated Cauchy problem is

$$\begin{cases} y'(x) = a(x)b(y) \\ y(x_0) = y_0. \end{cases} \tag{13.5}$$

If $b(y_0) = 0$, then the constant function $y(x) = y_0$ solves (13.5) because both sides of the equation vanish identically on I. Such solutions are called *singular solutions*.

The reason why equations like (13.4) are of particular relevance is twofold. Firstly, there are many important examples of scientific problems that can be modeled by such equations and, secondly, because much can be said about their solutions and how to go about finding them. Below are three basic theorems.

Theorem 13.1 *Suppose that I and J are open intervals, and take $x_0 \in I$ and $y_0 \in J$. If $a \in C^0(I)$ and if $b \in C^0(J)$ is such that $b(y_0) \neq 0$, then there exists a neighborhood V of x_0 on which a unique solution $y : V \to \mathbb{R}$ to the Cauchy problem (13.5) is defined.*

Uniqueness really means that if $z : V \to \mathbb{R}$ is another solution defined in the same set V, then $y(x) = z(x)$ for every $x \in V$. The above theorem is of theoretical relevance but does not explain neither the name of the method nor how to find the solution. Here is some heuristics. Consider the Cauchy problem (13.5). If $b(y_0) \neq 0$, then

$$\frac{dy}{dx} = a(x)b(y)$$

calls for separating of variables, in the sense that one is led to write

$$\int a(x)\, dx = \int \frac{1}{b(y)}\, dy, \qquad (13.6)$$

and the right hand side makes sense in a neighborhood of y_0 where b does not vanish. This informal argument can be made precise. Suppose that y is a solution of (13.4) and that B is a primitive function of $1/b$ in an open interval containing y_0. Then

$$\frac{d}{dx}B(y(x)) = B'(y(x))y'(x) = \frac{1}{b(y(x))}y'(x) = a(x),$$

so that $B(y(x))$ is a primitive function of a. If A is any primitive function of a on I, then $B(y(x)) = A(x) + c$. This clarifies (13.6). Finally, since in a neighborhood of y_0

$$\frac{1}{b(y)} = \frac{dB}{dy} \neq 0,$$

the function B is strictly monotonic, hence invertible. Now, $B(y(x)) = A(x) + c$, yields $y(x) = B^{-1}(A(x) + c)$ and $c = B(y_0) - A(x_0)$ so that

$$y(x) = B^{-1}(A(x) + B(y_0) - A(x_0)). \qquad (13.7)$$

Upon applying B to both sides, the above is a restatement of

$$\int_{x_0}^{x} a(x)\, dx = \int_{x_0}^{x} \frac{y'(x)}{b(y(x))}\, dx, \qquad (13.8)$$

which is a precise version of (13.6). This is how the solutions are practically found. Notice that formulae (13.7) or (13.8) involve a final inversion of B, a problem that does not always lead to an explicit solution, as studied in Sect. 2.2.

It is not hard to find situations where $b \in C^0(J)$ is not enough to guarantee uniqueness if $b(y_0) = 0$. For example, the Cauchy problem

$$\begin{cases} y'(x) = \sqrt[3]{y} \\ y(0) = 0 \end{cases} \tag{13.9}$$

has at least the three solutions defined in $I = \mathbb{R}$:

$$y_1(x) = 0, \quad y_2(x) = \begin{cases} (2x/3)^{3/2} & x \geq 0 \\ 0 & x < 0 \end{cases}, \quad y_3(x) = -y_2(x).$$

If the function b is more regular, then the assumption $b(y_0) \neq 0$ in Theorem 13.1 can actually be removed, and the following theorem holds.

Theorem 13.2 *Suppose that I and J are open intervals, and take $x_0 \in I$ and $y_0 \in J$. If $a \in C^0(I)$ and $b \in C^1(J)$, then there exists a neighborhood V of x_0 and a unique solution $y : V \to \mathbb{R}$ to the Cauchy problem* (13.5).

The third useful result deals in more detail with the case when $b(y_0)$ vanishes. Observe that in the example (13.9), the function $b(y) = \sqrt[3]{y}$ vanishes of order $1/3$ at the origin.

Theorem 13.3 *Suppose that I and J are open intervals, and take $x_0 \in I$ and $y_0 \in J$. Assume that $a \in C^0(I)$ and suppose that $b \in C^0(J)$ is such that $b(y_0)$ vanishes with order α as $x \to x_0$. Then*

(i) if $\alpha \geq 1$, the constant solution $y(x) = y_0$ is the unique solution of (13.5);
(ii) if $\alpha < 1$, the constant solution $y(x) = y_0$ is not the unique solution of (13.5).

The dychotomy treated in Theorem 13.3 stems from the fact that the integral in the right hand side of (13.8) can be interpreted in the improper sense, and will either diverge, if $\alpha \geq 1$, or converge, if $\alpha < 1$. In the latter case, the improper integral represents a legitimate solution, in the former it does not, and hence only the constant solution is indeed available. It must be noticed that Theorem 13.3 extends to cover cases where $J = [y_0, y_1)$.

13.3 Guided Exercises on Separation of Variables

13.1 Consider the Cauchy problem

$$\begin{cases} y'(x) = \dfrac{1 + y^2(x)}{\sqrt{x}} \\ y(1) = 1. \end{cases}$$

(a) Discuss local existence and uniqueness of solutions.
(b) Find the solution, if and where it exists.

Answer. (a) The differential equation is of the form $y'(x) = a(x)b(y(x))$, where $a(x) = 1/\sqrt{x}$ is continuous on $(0, +\infty)$ and $b \in C^1(\mathbb{R})$. By Theorem 13.2 a local solution exists and is unique.
(b) Separating the variables,

$$\int_1^x \frac{1}{\sqrt{t}}\, dt = \int_1^x \frac{y'(t)}{1+y^2(t)}\, dt \quad \Longrightarrow \quad 2\sqrt{x}-2 = \int_1^{y(x)} \frac{du}{1+u^2} = \arctan y(x) - \frac{\pi}{4}.$$

Therefore

$$\arctan y(x) = 2\sqrt{x} + \frac{\pi}{4} - 2$$

for $x \geq 0$ and

$$-\frac{\pi}{2} < 2\sqrt{x} + \frac{\pi}{4} - 2 < \frac{\pi}{2}.$$

Hence the solution is

$$y(x) = \tan\left(2\sqrt{x} + \frac{\pi}{4} - 2\right)$$

with domain $(0, ((\pi + 8)/8)^2)$.

This exercise is a straightforward application of the method of separation of variables. In particular, the hypotheses of Theorem 13.2 are immediately checked and the needed integrals are very easy to compute.

13.2 Consider the Cauchy problem

$$\begin{cases} y'(x) = e^{y(x)+x-1} \\ y(0) = y_0. \end{cases}$$

(a) Discuss local existence and uniqueness of solutions as y_0 ranges in $\mathbb{R}$.
(b) Find the values of $y_0 \in \mathbb{R}$, if any, for which the solution is defined in $\mathbb{R}$.

Answer. (a) The differential equation is of the form $y'(x) = a(x)b(y(x))$, where $a(x) = e^{x-1}$ is continuous on $\mathbb{R}$ and $b(y) = e^y$ is of class C^1 on $\mathbb{R}$. By Theorem 13.2 a local solution exists and is unique for every $y_0 \in \mathbb{R}$.
(b) Separating the variables,

$$\int_0^x e^{t-1}\, dt = \int_0^x \frac{y'(t)}{e^{y(t)}}\, dt \quad \Longrightarrow \quad e^{x-1} - e^{-1} = \int_{y_0}^{y(x)} e^{-u}\, du = -e^{-y(x)} + e^{-y_0}.$$

Therefore

$$e^{-y(x)} = e^{-y_0} + e^{-1} - e^{x-1}.$$

The above equality implies that $e^{x-1} < e^{-1} + e^{-y_0}$. Thus, for no $y_0 \in \mathbb{R}$ there is a solution defined on $\mathbb{R}$.

Again, Theorem 13.2 is applicable. In order to asses the existence everywhere defined solutions, however, the explicit expression given by formula (13.8) is needed.

13.3 Consider the Cauchy problem

$$\begin{cases} y'(x) = \dfrac{x^2}{\sqrt{y(x) - 1}} \\ y(0) = k. \end{cases}$$

(a) Discuss local existence and uniqueness of solutions in a neighborhood of $x_0 = 0$.
(b) Put $k = 10$. Find the domain of the solution.
(c) Put $k = 10$. Compute the limit of $y(x)/x^2$ as $x \to +\infty$.
(d) Put $k = 10$. Is there x_0 in the domain of $y(x)$ such that $y(x) \to 0$ as $x \to x_0$?

Answer. (a) The differential equation is of the form $y'(x) = a(x)b(y(x))$, where $a(x) = x^2$ is continuous on $\mathbb{R}$ and $b(y) = 1/\sqrt{y-1}$ is of class $C^1((1, +\infty))$. By Theorem 13.2, for $k \le 1$ there are no solutions, whereas for $k > 1$ a local solution exists and is unique in a neighborhood of $x_0 = 0$.

(b) Separating the variables,

$$\int_0^x \sqrt{y(t) - 1} \, y'(t) \, dt = \int_0^x t^2 \, dt.$$

The substitution $y(t) - 1 = u$ yields

$$\int_9^{y(x)-1} \sqrt{u} \, du = \frac{x^3}{3} \quad \Longrightarrow \quad \frac{2}{3}\sqrt{(y(x) - 1)^3} - 18 = \frac{x^3}{3}.$$

Therefore

$$y(x) = 1 + \sqrt[3]{\left(\frac{x^3 + 54}{2}\right)^2}$$

with domain $(-3\sqrt[3]{2}, +\infty)$.

(c) Evidently,

$$\lim_{x \to +\infty} \frac{y(x)}{x^2} = \lim_{x \to +\infty} \frac{1}{x^2} + \sqrt[3]{\left(\frac{x^3 + 54}{2x^3}\right)^2} = \frac{1}{\sqrt[3]{4}}.$$

(d) No, because the solution is necessarily differentiable, hence continuous. If any such x_0 existed, then $\lim_{x \to x_0} y(x) = y(x_0) = 0$, while $y(x) > 1$.

The explicit expression of the solution is needed in order to find the domain and also for the behaviour as $x \to +\infty$, hence formula (13.8) is necessary. For question (a) Theorem 13.2 is enough, while a simple a priori argument looking at the form of $b(y)$ answers question (d).

13.4 Consider the Cauchy problem

$$\begin{cases} y'(x) = \dfrac{xy(x)}{y(x) - 4} \\ y(0) = y_0. \end{cases}$$

(a) Determine for which values of $y_0 \in \mathbb{R}$ if any, the solution exists and is unique.
(b) Put $y_0 = 1$. Write the second order McLaurin polynomial of the solution.
(c) Put $y_0 = 1$. Find some analytic formula for the solution.

Answer. (a) The differential equation is of the form $y'(x) = a(x)b(y(x))$, where $a(x) = x$ is continuous on $\mathbb{R}$ and $b(y) = y/(y - 4)$ is in $C^1 (\mathbb{R} \setminus \{4\})$. Therefore, for any $y_0 \neq 4$ Theorem 13.2 applies and guarantees that a solution exists and is unique.

(b) If $y(0) = y_0 = 1$, then $y'(0) = 0$. Differentiating both sides of the differential equation, which is possible because the right hand side is a ratio of differentiable functions, gives $y''(x)$ in terms of x, $y(x)$ and $y'(x)$. The explicit calculation is

$$y''(x) = \frac{(y(x) + xy'(x))(y(x) - 4) - xy(x)y'(x)}{(y(x) - 4)^2}.$$

Hence $y''(0) = -1/3$ and $p_2(x) = 1 - x^2/6$.
(c) Separating the variables, it follows

$$\int_0^x \frac{y(t) - 4}{y(t)} y'(t)\, dt = \int_0^x t\, dt$$

and hence

$$\frac{x^2}{2} = \int_1^{y(x)} \frac{u - 4}{u}\, du = y(x) - 4 \log y(x) - 1.$$

Question (a) is answered by Theorem 13.2, and sets the stage for (b) and (c). As for question (b), Taylor polynomials at the point where the initial datum is given can be computed using the differential equation itself, inasmuch as its right hand side is differentiable. It might be tedious, but it is doable if enough regularity is available. Question (c) is the tricky one. As discussed in the beginning of this chapter, the primitive function B of $b(y)$, in this case $B(y) = y - 4 \log y - 1$, is strictly monotonic in a neighborhood of $y(0) = 1$, hence invertible. But the explicit formula for its inverse is not to be found explicitly. This is what is meant by "find some formula". As it often happens, the method of separation of variables cannot be carried out in full, as the inversion process cannot be completed explicitly. So the most reasonable formula is of the kind $B(y(x)) = A(x) + B(y_0) - A(x_0)$ (see Sect. 13.2).

13.5 Consider the Cauchy problem

$$\begin{cases} y'(x) = \dfrac{x+|x|}{y+1} \\ y(x_0) = y_0. \end{cases}$$

(a) Discuss local existence and uniqueness of solutions as x_0 and y_0 range in $\mathbb{R}$.
(b) Put $x_0 = 0$ and $y_0 = -2$. Find all the solutions of the problem.
(c) Put $x_0 = y_0 = 0$. Does the second order McLaurin polynomial of $y(x)$ exist?

Answer. (a) The differential equation is of the form $y'(x) = a(x)b(y(x))$, where $a(x) = x+|x|$ and $b(y) = 1/(y+1)$. Evidently, $a \in C^0(\mathbb{R})$ and $b \in C^1(\mathbb{R}\setminus\{-1\})$. By Theorem 13.2, for every $x_0 \in \mathbb{R}$ and every $y_0 \in \mathbb{R}\setminus\{-1\}$ there exists a neighborhood of x_0 in which the solution exists and is unique.

(b) As observed in (a), there exists a neighborhood V of $x_0 = 0$ on which the solution exists and is unique. Separating the variables and integrating,

$$\int_0^x y'(t)(y(t)+1)\,dt = \int_0^x (t+|t|)\,dt =: g(x),$$

where evidently

$$g(x) = \begin{cases} x^2 & x \geq 0 \\ 0 & x < 0. \end{cases}$$

Now, under the substitution $z = y(x)$, the first indicated integral is computed, so that

$$g(x) = \int_{y(0)}^{y(x)} (z+1)\,dz = \frac{y^2(x)}{2} + y(x).$$

This second degree equation in $y(x)$ has a solution provided that $\Delta = 1+2g(x) \geq 0$, which is true for every $x \in \mathbb{R}$. Its solutions are

$$y_1(x) = -1 - \sqrt{1+2g(x)}$$
$$y_2(x) = -1 + \sqrt{1+2g(x)}$$

but only the first one satisfies the condition $y(0) = -2$. It follows that the unique solution is $y(x) = -1 - \sqrt{1+2g(x)}$.

(c) Take a neighborhood W of $x_0 = 0$ where the solution exists and is unique. The second order McLaurin polynomial of y exists provided that y is twice differentiable at $x_0 = 0$. Now, $y'(0) = 0$ and if $x \in W$ satisfies $x > 0$, then

$$\frac{y'(x) - y'(0)}{x} = \frac{2x}{x}\frac{1}{y(x)+1} \implies \lim_{x\to 0^+} \frac{y'(x)-y'(0)}{x} = 2$$

whereas if $x \in W$ satisfies $x < 0$, then $y'(x) = 0$ because of the differential equation, and hence $y''_-(0) = 0$. It follows that $y(x)$ is not twice differentiable at $x_0 = 0$ and hence that the second order McLaurin polynomial of y does not exist.

Once again, question (a) is just a way to understand when the Cauchy problem at hand is meaningful. Question (b) is about inverting locally a second order polynomial, so it is about finding the correct branch of the parabola. Finally, question (c) is about finding a McLaurin polynomial of the solution. As observed earlier, this is possible depending on the regularity of the right hand side of the differential equation. Now, the presence of $|x|$ must be alerting, and indeed the shape of the function g, which is identically zero to the left of the origin, but not to the right, reveals that the task has a negative answer.

13.6 Consider the differential problem

$$\begin{cases} y'(1 - x^2) = 1 - y(x) \\ y(0) = 0. \end{cases}$$

(a) Show that the problem has one and only one solution.
(b) Find the solution, indicating explicitly its domain.
(c) Does the solution admit an extension to $(0, +\infty)$?

Answer. (a) In the interval $I = (-1, 1)$ the given problem is equivalent to

$$\begin{cases} y'(x) = \dfrac{1}{1 - x^2}(1 - y(x)) \\ y(0) = 0. \end{cases}$$

The differential equation is of the form $y'(x) = a(x)b(y(x))$ with $a(x) = \frac{1}{1-x^2}$ and $b(y) = 1 - y$. Now, $a \in C^0(\mathbb{R} \setminus \{-1, 1\})$ and $b \in C^1(\mathbb{R})$. Hence there exists a neighborhood V of 0 on which a solution exists and is unique.

(b) In V the solution satisfies

$$\frac{y'(t)}{1 - y(t)} = \frac{1}{1 - t^2}$$

for every $t \in V$. Integrating

$$\int_0^x \frac{y'(t)}{1 - y(t)} \, dt = \int_0^x \frac{1}{1 - t^2} \, dt$$
$$= \frac{1}{2} \int_0^x \frac{1}{1 - t} \, dt + \frac{1}{2} \int_0^x \frac{1}{1 + t} \, dt$$
$$= \frac{1}{2} \log(1 + x) - \frac{1}{2} \log(1 - x)$$

whenever $x \in V$. Finally, the substitution $y(t) = z$ yields

$$\int_0^{y(x)} \frac{1}{1-z}\, dz = -\log(1 - y(x)) = \frac{1}{2}\log\left(\frac{1+x}{1-x}\right)$$

for $x \in V$. it follows that

$$\log\left(\frac{1}{1 - y(x)}\right) = \log\left(\sqrt{\frac{1+x}{1-x}}\right) \implies \frac{1}{1 - y(x)} = \sqrt{\frac{1+x}{1-x}}$$

$$\implies \quad y(x) = 1 - \sqrt{\frac{1-x}{1+x}}$$

which is the unique solution in $(-1, 1)$.

(c) Observe that the above solution is defined at 1 but is not differentiable at 1. Furthermore, $\lim_{x \to 1^-} y'(x) = +\infty$. Thus, there cannot be a function

$$\bar{y}(x) = \begin{cases} y(x) & x \in (-1, 1) \\ \bar{y}(x) & x \geq 1 \end{cases}$$

which solves the problem, because it cannot be differentiable at 1.

In item (a) of this problem, a preliminary step is required, namely that of writing the given differential equation in normal form. This amounts to dividing up by $1 - x^2$, which seems to pose some limitation but has the advantage of enabling separation of variables. A posteriori, though, the limitation is not really there. In item (b) the method is applied to its very end, meaning that the final inversion process does lead to an explicit answer. In part (c) it is asked whether it is possible to go across the point 1, where all the previously found formulae clearly break down. Technically, the obstruction is in the lack of differentiability, which cannot be circumvented.

13.7 Consider the Cauchy problem

$$\begin{cases} y'(x) = (x + 1)(|y(x)| - 2) \\ y(0) = y_0. \end{cases}$$

(a) Discuss local existence and uniqueness of solutions as y_0 ranges in $\mathbb{R}$.
(b) Put $y_0 = 0$. Establish if the solution is twice differentiable in $x_0 = 0$ and draw a local graph of the solution around $x_0 = 0$.
(c) Put $y_0 = 0$. Find the explicit analytic expression of the solution in a suitable neighborhood of $x_0 = 0$.

Answer. (a) The differential equation is of the form $y'(x) = a(x)b(y(x))$, where $a(x) = x + 1$ is in $C^0(\mathbb{R})$ and $b(y) = |y| - 2$ in $C^1(\mathbb{R} \setminus \{0\})$. Therefore, by Theorem 13.2, for all $y_0 \in \mathbb{R} \setminus \{0\}$ there exists a neighborhood of $x_0 = 0$ where a solution is defined and unique. Furthermore, $b \in C^0(\mathbb{R})$ and $b(0) = -2 \neq 0$ so that, by Theorem 13.1, also for $y_0 = 0$ there exists a neighborhood of $x_0 = 0$ where a solution is defined and unique.

(b) Notice that $y'(0) = -2$. Since $y'(x) = (x+1)(|y(x)| - 2)$, in a neighborhood V of $x_0 = 0$ also $y' \in C^0(V)$ and by permanence of sign $y'(x) < 0$ in a neighborhood of $x_0 = 0$, so that y is strictly decreasing in that neighborhood. Since $y(0) = 0$, it follows then that $y(x) > 0$ in a left neighborhood of $x_0 = 0$ and $y(x) < 0$ in a right neighborhood of $x_0 = 0$. Now,

$$\frac{y'(x) - y'(0)}{x} = \frac{y'(x) + 2}{x} = (x+1)\frac{|y(x)|}{x} - 2,$$

which, taking into account that $\lim_{x\to 0} y(x)/x = y'(0) = -2$, implies

$$\lim_{x\to 0^+} \frac{y'(x) - y'(0)}{x} = 0$$

$$\lim_{x\to 0^-} \frac{y'(x) - y'(0)}{x} = -4.$$

In conclusion, y is not twice differentiable at $x_0 = 0$. A local graph of the solution is in Fig. 13.1.

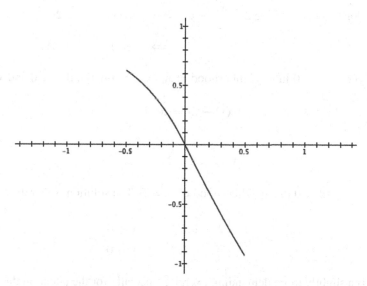

Fig. 13.1 Local graph of the solution around $x_0 = 0$, Exercise 13.7

(c) For t in a neighborhood V of $x_0 = 0$ it is $y'(t) = (t+1)(|y(t)|-2)$. Separation of variables and integration yields

$$\int_0^x \frac{y'(t)}{|y(t)|-2}\,dt = \int_0^x (t+1)\,dt = \frac{x^2}{2}+x$$

and by substitution

$$\int_{y(0)}^{y(x)} \frac{1}{|z|-2}\,dz = \frac{x^2}{2}+x.$$

Now, if $x > 0$, then $y(x) < 0$, so that if $x \in V$ and $x > 0$

$$-\log|y(x)+2| + \log 2 = \frac{x^2}{2}+x.$$

Analogously, if $x \in V$ and $x < 0$ it follows that $y(x) > 0$, so that

$$\log|y(x)-2| - \log 2 = \frac{x^2}{2}+x.$$

In conclusion, if $x > 0$ and $x \in V$, then

$$\log|y(x)+2| = \log 2 - \frac{x^2}{2} - x \quad \Longrightarrow \quad |y(x)+2| = 2e^{-\frac{x^2}{2}-x}$$

$$\Longrightarrow \quad y(x) = -2 + 2e^{-\frac{x^2}{2}-x}$$

because $y(x)+2 > 0$ in a neighborhood of $x_0 = 0$. Similarly, if $x < 0$ and $x \in V$, then

$$|y(x)-2| = 2e^{\frac{x^2}{2}+x}$$

and

$$y(x) = 2 - 2e^{\frac{x^2}{2}+x}$$

because $y(x)-2 < 0$ in a neighborhood of $x_0 = 0$. The solution is therefore

$$y(x) = \begin{cases} 2 - 2e^{\frac{x^2}{2}+x} & x \le 0 \\ -2 + 2e^{-\frac{x^2}{2}-x} & x > 0. \end{cases}$$

This is a slightly more demanding exercise, specially for the attention that needs to be paid to sign issues. Question (a) is almost standard: apply the stronger theorem, Theorem 13.2, whenever possible, and otherwise the weaker one, Theorem 13.1 As for question (b), clearly one has to consider the difference quotient of y' at the origin and infer as much as possible from the differential equation. After writing things down, one has primarily to deal with $|y(x)|/x$ near zero. Knowing the sign of $y'(0)$, this is doable. This is the main issue for item (b). Indeed, drawing a local graph only requires knowing something about y', an information that is available from the differential equation. Finally, question (c) involves a careful integration followed by a careful inversion process. Neither is really hard, but requires bookkeeping of the sign, and the size, of $y(x)$ around the origin, which is facilitated by the local drawing that was sketched in (b).

13.8 Consider the Cauchy problem

$$\begin{cases} y'(x) = \dfrac{(1 + y^2(x)) \arctan y(x)}{x} \\ y(1) = y_0. \end{cases}$$

(a) For which $y_0 \in \mathbb{R}$ does there exist a unique solution in a neighborhood of $x_0 = 1$?
(b) Put $y_0 = 0$. Decide if a solution exist and, if yes, specify its maximal domain.
(c) Put $y_0 = 1$. Draw a graph of the solution, if it exists, in a neighborhood of $x_0 = 1$.
(d) Put $y_0 = \sqrt{3}$. Decide if a solution exist and, if yes, specify its maximal domain.

Answer. (a) The differential equation is of the form $y'(x) = a(x)b(y(x))$, where $a(x) = 1/x$ is continuous in $\mathbb{R} \setminus \{0\}$ and and $b(y) = (1 + y^2) \arctan y$ is in $C^1(\mathbb{R})$. By Theorem 13.2 there exists a unique solution in a neighborhood of $x_0 = 1$ for every $y_0 \in \mathbb{R}$.

(b) Observe that $b(y_0) = b(0) = 0$, and hence $y(x) = 0$ is a solution of the problem in $(0, +\infty)$. This is actually the only one, as established in (a).

(c) By (a), the unique solution defined in a neighborhood of $x_0 = 1$ satisfies

$$y(1) = 1, \qquad y'(1) = 2 \arctan 1 = 2\frac{\pi}{4} = \frac{\pi}{2}.$$

Being a composition of differentiable functions, $y'(x)$ is clearly differentiable, and

$$y''(x) = \frac{(y'(x) + 2y(x)y'(x) \arctan y(x))x - (1 + y^2(x)) \arctan y(x)}{x^2}$$

$$y''(1) = \frac{\pi}{2} + \frac{\pi^2}{4} - \frac{\pi}{2} = \frac{\pi^2}{4}.$$

Since $y''(x)$ is continuous, by permanence of sign $y''(x) > 0$ in a neighborhood of $x_0 = 1$. Hence a local graph is as in Fig. 13.2.

The above graph is actually the restriction to $\{x : |x - 1| < 1/4\}$ of the graph of the second order Taylor polynomial of the solution at $x = 1$, which is

$$P_2(x) = 1 + \frac{\pi}{2}(x - 1) + \frac{\pi^2}{8}(x - 1)^2.$$

Fig. 13.2 Local graph of the
solution around $x_0 = 1$,
Exercise 13.8

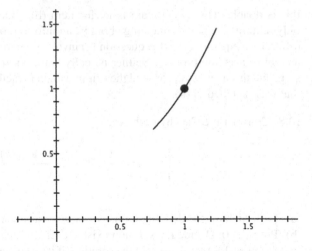

(d) Put $y_0 = \sqrt{3}$. Separating the variables and integrating gives

$$\int_1^x \frac{y'(t)}{(1+y^2(t))\arctan y(t)}\,dt = \int_1^x \frac{1}{t}\,dt = \log x$$

for $x > 0$. Hence

$$\log x = \int_{\sqrt{3}}^{y(x)} \frac{du}{(1+u^2)\arctan u} = \log(\arctan u)\Big|_{\sqrt{3}}^{y(x)} = \log\left(\frac{\arctan y(x)}{\frac{\pi}{3}}\right)$$

with $y(x) > 0$. Finally,

$$\frac{\arctan y(x)}{\frac{\pi}{3}} = x \quad \Longrightarrow \quad y(x) = \tan\left(\frac{\pi}{3}x\right)$$

with $0 < \pi x/3 < \pi/2$, namely $0 < x < 3/2$.

Question (a) is a straightforward application of Theorem 13.2. Question (b) is just
the observation that the zero solution on $(0, +\infty)$ is unique and cannot be extended
any further because the denominator of the right hand side of the differential equation
is x. Question (c) calls for a Taylor expansion, which is easily obtained up to order
two and thus gives a reasonable approximation of a local graph. Question (d) is a
plain application of the technique of separation of variables followed by inversion.

13.9 Consider the Cauchy problem

$$\begin{cases} y'(x) = 1 + \dfrac{1}{2\sqrt{y+1}-1} \\ y(x_0) = y_0. \end{cases}$$

(a) Discuss local existence and uniqueness of solutions as x_0 and y_0 range in $\mathbb{R}$.
(b) Put $x_0 = -1$ and $y_0 = 0$. Find the solution of the problem, specifying its domain.
(c) Put $x_0 = 0$, $y_0 = -4/5$. Show that the solution of the problem is bounded.

Answer. (a) The differential equation is of the form $y'(x) = a(x)b(y(x))$, where $a(x) = 1$ is obviously continuous on $\mathbb{R}$ and

$$b(y) = 1 + \frac{1}{2\sqrt{y+1} - 1}$$

is in $C^1((-1, +\infty) \setminus \{-3/4\})$. Hence, by Theorem 13.2, for every $x_0 \in \mathbb{R}$ and every $y_0 \in (-1, +\infty) \setminus \{-3/4\}$ there exists a neighborhood of x_0 where a solution exists and is unique. If $y_0 = -1$, then the constant function $y(x) = -1$ is a solution, but b vanishes of order $1/2 < 1$ as $y \to -1$, and hence, by an appropriate extension of Theorem 13.3 (see comment below), the constant function $y(x) = -1$ is not the unique solution.

(b) In this case, there exists a solution in a neighborhood U of $x_0 = -1$ and is unique. From

$$y'(t) = 1 + \frac{1}{2\sqrt{y(t) + 1} - 1}$$

it follows that

$$y'(t)\left(1 - \frac{1}{2\sqrt{y(t) + 1}}\right) = 1.$$

Integrating,

$$\int_{-1}^{x} y'(t)\left(1 - \frac{1}{2\sqrt{y(t) + 1}}\right) dt = \int_{-1}^{x} 1\, dt - x + 1.$$

and substituting

$$\int_{0}^{y(x)} \left(1 - \frac{1}{2\sqrt{z + 1}}\right) dz = x + 1,$$

whence

$$y(x) - \sqrt{y(x) + 1} + 1 = x + 1.$$

Upon setting $q(x) = \sqrt{y(x) + 1}$, this equation is $q^2(x) - q(x) - x - 1 = 0$, and has positive discriminant provided that $x \geq -5/4$. From $q(-1) = 1$ it follows that in $[-5/4, +\infty)$ the equation has the solution

$$q(x) = \frac{1 + \sqrt{1 + 4(x + 1)}}{2}.$$

Hence the solution of the problem is

$$y(x) = -1 + \frac{1}{4}\left(1 + \sqrt{1+4(x+1)}\right)^2, \qquad x \in \left(-\frac{5}{4}, +\infty\right).$$

(c) From what has been established in (a), there is a neighborhood U of $x_0 = 0$ in which a solution exists and is unique, that is a function $y : U \to \mathbb{R}$ differentiable in U and such that

$$y'(x) = 1 + \frac{1}{2\sqrt{y(x)+1} - 1}$$

for every $x \in U$. Hence, it must be $y(x) \geq -1$ for all $x \in U$. Furthermore, there cannot be $x \in U$ for which $2\sqrt{y(x)+1} - 1 = 0$, namely, such that $y(x) = -3/4$. Finally, since

$$y(0) = -\frac{4}{5} \in \left[-1, -\frac{3}{4}\right)$$

and since y is continuous in U, it must be $-1 \leq y(x) < -3/4$ for $x \in U$, which shows that y is bounded.

The new ingredient of this exercise is a case where the solution is not unique because b vanishes of order less than 1. However, in the case at hand, Theorem 13.3 cannot be applied as it has been stated, because the natural domain J of the function b is $J = [-1, -3/4) \cup (3/4, +\infty)$, which is neither open nor an interval. Nevertheless, Theorem 13.3 is valid for the interval $J = [-1, -3/4)$, as remarked after the statement of the theorem.

The remaining part of the exercise involves the standard technique of separation of variables. The inversion entails the correct choice of the branch of a parabola. Finally, the differential equation itself, in particular the form of $b(y)$, strongly suggests that the solution must satisfy a priori bounds in order for $b(y)$ to make sense at all.

13.10 Consider the Cauchy problem

$$\begin{cases} y'(x) = \dfrac{\sqrt{1+y(x)}}{xy(x)} \\ y(1) = y_0. \end{cases}$$

(a) Establish for which $y_0 \in \mathbb{R}$ there exists a unique local solution.
(b) Sia $y_0 = 3$. Draw a graph of the solution.
(c) Sia $y_0 = 3$. Find some formula for the solution.

Answer. (a) The differential equation is of the form $y'(x) = a(x)b(y(x))$, where $a(x) = 1/x$ is continuous in $\mathbb{R} \setminus \{0\}$ and

$$b(y) = \frac{\sqrt{1+y}}{y}$$

is continuous in $[-1, 0) \cup (0, +\infty)$. If $y_0 \in (-1, 0) \cup (0, +\infty)$, $b(y_0) \neq 0$ then by Theorem 13.1 there exists a unique solution in a neighborhood of $x_0 = 1$. If $y_0 = -1$, then $b(y_0) = 0$ and the constant function $y(x) = -1$ in $(0, +\infty)$ is a solution, which is not unique in view of (the extension of) Theorem 13.3 because b vanishes of order $1/2$ as $y \to -1^+$.

(b) Put $y_0 = 3$. As shown in (a), the solution is unique around $x_0 = 1$. Furthermore, $y'(1) = 2/3$ and $y'(x)$ is differentiable because it is a composition of differentiable functions. It is thus easy to compute y'' as in Exercise 13.8, and hence to infer that $y''(1) = -41/54$. Hence a local graph is as in Fig. 13.3.

This is derived by restricting the graph of the second order Taylor polynomial at $x = 1$ of the solution, which has been computed to be

$$P_2(x) = 3 + \frac{2}{3}(x - 1) - \frac{41}{108}(x - 1)^2.$$

(c) Separating the variables and integrating yields

$$\int_1^x \frac{y(t)}{\sqrt{1 + y(t)}} y'(t)dt = \int_1^x \frac{1}{t}dt$$

whence, for $x > 0$. Substituting $1 + u = t^2$ with $t > 0$

$$\log x = \int_3^{y(x)} \frac{u}{\sqrt{1 + u}}du = \int_2^{\sqrt{1+y(x)}} 2(t^2 - 1)dt$$

$$= 2\left(\frac{(1 + y(x))^{\frac{3}{2}}}{3} - (1 + y(x))^{\frac{1}{2}} - \frac{2}{3}\right).$$

Fig. 13.3 Local graph of the solution for which $y_0 = 3$, Exercise 13.10

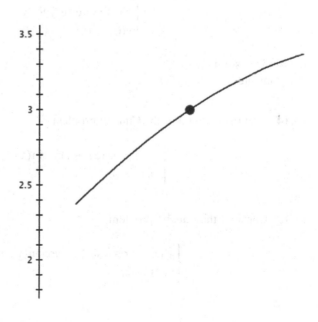

Another instance of non uniqueness for $y_0 = -1$. In question (b), the issue is the computation of a low order Taylor polynomial at $x_0 = 1$, which is doable. Again, the method of separation of variables is not carried out in full, and the requested formula is of the kind $B(y(x)) = A(x) + B(y_0) - A(x_0)$ (see Sect. 13.2).

13.4   Problems on Separation of Variables

13.11 Find the solution of the Cauchy problem

$$\begin{cases} y'(x) = e^{y(x)}(x+2) \\ y(0) = 0. \end{cases}$$

13.12 Find the solution of the Cauchy problem

$$\begin{cases} y'(x) = xe^{y(x)} \\ y(0) = 0 \end{cases}$$

and decide if it is invertible.

13.13 Consider the Cauchy problem

$$\begin{cases} y'(x) = xe^{x^2}e^{-y(x)} \\ y(0) = 0. \end{cases}$$

(a) Find the solution.
(b) Compute $\lim_{x \to +\infty} y(x)/x^2$.

13.14 Find the solution of the Cauchy problem

$$\begin{cases} y'(x) = (y(x) - 1)^2 x \sin(x^2) \\ y(0) = 0. \end{cases}$$

13.15 Consider the Cauchy problem

$$\begin{cases} y'(x) = x(y^2(x) + 1) \arctan(y(x)) \\ y(0) = y_0. \end{cases}$$

(a) Solve the problem for $y_0 = 1$.
(b) Solve the problem for $y_0 = 0$.

13.16 Consider the Cauchy problem

$$\begin{cases} y'(x) = \sqrt{x+1}(1+y^2(x)) \\ y(x_0) = y_0. \end{cases}$$

(a) Discuss local existence and uniqueness of solutions as x_0 and y_0 range in $\mathbb{R}$.
(b) Put $x_0 = y_0 = 0$. Determine the order with which $y(x)$ vanishes as $x \to 0^+$.
(c) Put $x_0 = y_0 = 0$. Find the solution of the problem.

13.17 Consider the Cauchy problem

$$\begin{cases} y'(x) = \dfrac{y(x)(1+y(x))}{x} \\ y(1) = y_0. \end{cases}$$

(a) Discuss local existence and uniqueness of solutions as y_0 ranges in $\mathbb{R}$.
(b) Put $y_0 = 0$. Find the solution of the problem.
(c) Put $y_0 = 1$. Find the solution of the problem.

13.18 Find the solution of the Cauchy problem

$$\begin{cases} y'(x) = \dfrac{x+2}{|y(x)-1|} \\ y(0) = 2. \end{cases}$$

13.19 Consider the Cauchy problem

$$\begin{cases} x^2 y' \tan y = 1 \\ y(1) = \frac{\pi}{4}. \end{cases}$$

(a) Discuss local existence and uniqueness of solutions.
(b) Find the solution in the interval in which it is defined.

13.20 Consider the Cauchy problem

$$\begin{cases} y'(x) = \dfrac{\sin y(x)}{\sqrt{4-y^2(x)}} \\ y(0) = \frac{\pi}{2}. \end{cases}$$

(a) Compute $\lim_{x \to 0^+} \left(y(x) - \dfrac{\pi}{2} \right)^k / \sin^3 x$ as k ranges in $\mathbb{R}$.

(b) Is there $x_0 \in \mathbb{R}$ such that $y(x_0) = -1$?

13.21 Consider the differential equation $y'(x) = -xy - y - y' \log y(x)$.

(a) Is the set V of solutions of the differential equation a vector space?

(b) Consider the Cauchy problem

$$\begin{cases} y'(x) = \dfrac{-xy(x) - y(x)}{1 + \log y(x)} \\ y(x_0) = y_0. \end{cases}$$

For which $x_0, y_0 \in \mathbb{R}$ is there a unique solution?

(c) Put $x_0 = 0$, $y_0 = e^{-4}$. Find the solution of the Cauchy problem, specifying its domain.

13.22 Consider the Cauchy problem

$$\begin{cases} y'(x) = \dfrac{\tan y}{x} \\ y(1) = \frac{\pi}{4}. \end{cases}$$

(a) Discuss local existence and uniqueness of solutions.

(b) Draw the graph of the solution in a neighborhood of $x_0 = 1$.

(c) Write the linearization of $y = y(x)$ at $x_0 = 1$.

(d) Find $a \in \mathbb{R}$ such that $f(x) = \arcsin(ax)$ is a solution of the problem in a neighborhood of $x_0 = 1$.

13.23 Consider the Cauchy problem

$$\begin{cases} y'(x) = \dfrac{1 + y^2(x)}{y(x) \log(1 + y^2(x))} \\ y(0) = -1. \end{cases}$$

(a) Does the solution exist, and is it unique?

(b) Write the solution, specifying the interval in which it is defined.

13.24 Consider the Cauchy problem

$$\begin{cases} y'(x) = xy(x)\sqrt{e^{x^2}(1 - \log^2 y(x))} \\ y(0) = y_0. \end{cases}$$

(a) For which $y_0 \in \mathbb{R}$ does there exist a unique solution?
(b) Is the solution bounded?
(c) Put $y_0 = \sqrt{e}$. Determine the domain of the solution.

13.25 Consider the Cauchy problem

$$\begin{cases} y'(x) = \dfrac{2y - y^2}{x - 1} \\ y(0) = 1. \end{cases}$$

(a) Does the solution exist, and is it unique?
(b) Is the solution bounded?
(c) Does the solution have local extreme points? Is it monotonic?
(d) Is the solution twice differentiable in its domain? Write, if it exists, the second order McLaurin polynomial of the solution.
(e) Draw the graph of the solution in a neighborhood of 0.

13.26 Write all solutions of the Cauchy problem

$$\begin{cases} y'(x) = xy\sqrt{\log y} \\ y(0) = 1. \end{cases}$$

13.27 Consider the Cauchy problem

$$\begin{cases} y'(x) = ky^2(x) + |y(x)| \\ y(0) = k. \end{cases}$$

(a) For which k does there exist a unique solution?
(b) Find the solution for $k = -1$.

13.28 Consider the Cauchy problem

$$\begin{cases} y(x)y'(x) = \dfrac{y^4(x) + 1}{\sqrt{1 - x^2}} \\ y(0) = 1. \end{cases}$$

(a) Does there exist a unique solution?
(b) Write, if it exists, the second order McLaurin polynomial of the solution.
(c) Determine if the solution is monotonic in its domain.
(d) Determine the solution in the interval in which it is defined.

13.29 Assume that there exist an open interval I containing 1 and a solution $y \in C^0(I)$ of the equation

$$y(x) = 1 + \int_1^x \frac{1}{y(t)\sqrt{t}}\, dt.$$

Find y and the interval I.

Chapter 14
First Order Linear Differential Equations

14.1 First Order Linear Equations with Continuous Coefficients

A very important class of differential equations is the class of *linear* differential equations. These are the (ordinary) differential equations for which the function f that appear in (13.1), or the function g in (13.2), of Chap. 13 are linear in the unknown function y and its derivatives. In this chapter, the focus is on *first order linear* equations. In normal form, they can be written as

$$y' = a(x)y + b(x) \tag{14.1}$$

where the standing assumption is that the functions a and b are defined in a common interval I on which they are continuous. For this reason the full name that they should be given is: first order linear differential equations with continuous coefficients. Evidently, the associated Cauchy problem is

$$\begin{cases} y' = a(x)y + b(x) \\ y(x_0) = y_0, \end{cases} \tag{14.2}$$

where $x_0 \in I$ and $y_0 \in \mathbb{R}$ is the initial value. One of the reasons of their importance is that they are one of the not too many classes for which a full satisfactory theory is available.

Theorem 14.1 *Let I be an interval containing x_0 and suppose that $a, b \in C^0(I)$. If $\overline{y} : I \to \mathbb{R}$ is any solution of (14.1), then the general solution of (14.1) consists of all the functions $y : I \to \mathbb{R}$ of the form*

$$y(x) = \overline{y}(x) + c e^{\int_{x_0}^x a(t)\,dt} \tag{14.3}$$

as c ranges in $\mathbb{R}$, independently of the choice of $x_0 \in I$.

© Springer International Publishing Switzerland 2016
M. Baronti et al., *Calculus Problems*, UNITEXT - La Matematica per il 3+2 101,
DOI 10.1007/978-3-319-15428-2_14

As explicitly stated in the theorem, the representation of any solution to (14.1) contains the exponentiated integral function $e^{\int_{x_0}^{x} a(t)\,dt}$ which depends on the chosen $x_0 \in I$. Notice that, however, if z_0 is any other point in the interval I, then the corresponding exponentiated integral function differs from the previous one by a multiplicative constant, so that in the end the general integral, as a set, is independent of x_0. The second element on which the general integral is built is the so-called *particular solution*, namely the solution $\overline{y}$ which is supposed to be given. Such a particular solution, in fact, can always be written and, in principle, computed, in the sense that there is actually a formula for it, namely

$$\overline{y}(x) = e^{\int_{x_0}^{x} a(t)\,dt} \int_{x_0}^{x} e^{-\int_{x_0}^{t} a(s)\,ds} b(t)\,dt. \tag{14.4}$$

The fact that $\overline{y}$ does indeed solve (14.1) is a straightforward computation, based, among other things, on the assumption that both a and b are continuous. From Theorem 14.1 and the above formula the next result follows at once.

Theorem 14.2 *Let I be an interval containing x_0 and suppose that $a, b \in C^0(I)$. The general solution of (14.1) consists of all the functions $y : I \to \mathbb{R}$ of the form*

$$y(x) = e^{\int_{x_0}^{x} a(t)\,dt} \left(c + \int_{x_0}^{x} e^{-\int_{x_0}^{t} a(s)\,ds} b(t)\,dt \right) \tag{14.5}$$

as c ranges in $\mathbb{R}$, independently of the choice of $x_0 \in I$. Among these,

$$y(x) = e^{\int_{x_0}^{x} a(t)\,dt} \left(y_0 + \int_{x_0}^{x} e^{-\int_{x_0}^{t} a(s)\,ds} b(t)\,dt \right) \tag{14.6}$$

is the unique solution of the Cauchy problem (14.2).

Notice that, in particular, Theorem 14.2 asserts that solutions do exist in I, that is, in any interval in which a and b are continuous.

14.2   Guided Exercises on Linear First Order Equations

14.1 Consider the Cauchy problem

$$\begin{cases} y'(x) = -\dfrac{1}{x+1}y(x) + e^{x^2} \\ y(0) = 0. \end{cases}$$

(a) Show that the problem admits a unique solution and determine its domain.
(b) Compute the solution explicitly.

Answer. (a) Denote by $a(x) = -1/(x+1)$ and $b(x) = e^{x^2}$. Then $a \in C^0((-1, +\infty))$ and $b \in C^0(\mathbb{R})$. Thus, there exists a unique solution in $(-1, +\infty)$.
(b) Using (14.6), the solution is

$$y(x) = e^{\int_0^x -\frac{1}{t+1}dt} \int_0^x e^{\int_0^t \frac{1}{s+1}ds} e^{t^2} dt$$

$$= e^{-\log(x+1)} \int_0^x e^{\log(t+1)} e^{t^2} dt$$

$$= \frac{1}{x+1} \int_0^x (t+1)e^{t^2} dt.$$

This is a straightforward application of Theorem 14.2. One observation, though, is in order. The final formula cannot be made fully explicit, because e^{t^2} does not admit primitive functions that can be written in elementary form. The function te^{t^2}, however, is the derivative of $e^{t^2}/2$, so the final formula above could be partially unravelled.

14.2 Consider the Cauchy problem

$$\begin{cases} y'(x) = y(x)\log x + e^{-x} \\ y(1) = 0. \end{cases}$$

(a) Compute $\lim_{x \to +\infty} y(x)$.
(b) Determine the concavity of the solution in a neighborhood of 1.

Answer. (a) Using (14.6), the solution is

$$y(x) = e^{\int_1^x \log t\, dt} \int_1^x e^{-t} e^{-\int_1^t \log u\, du} dt$$

$$= e^{x(\log x - 1)} \int_1^x e^{-t} e^{-t\log t + t} dt$$

$$= e^{x(\log x - 1)} \int_1^x e^{-t\log t} dt$$

per $x > 0$. Observe that as $t \to +\infty$, the function $e^{-t\log t}$ vanishes with an order that is larger than any positive number, and hence $\int_1^{+\infty} e^{-t\log t} dt = l > 0$. This implies that $y(x) \to +\infty$ as $x \to +\infty$.
(b) The solution is certainly twice differentiable in $(0, +\infty)$ because $y'(x)$ is a sum of differentiable functions. Now,

$$y''(x) = \frac{y(x)}{x} + (\log x)y'(x) - e^{-x} \quad \Longrightarrow \quad y''(1) = -e^{-1} < 0.$$

Furthermore, y'' is a continuous function, and hence, by permanence of sign, $y'' < 0$ in a neighborhood of 1. Therefore the function is concave in a neighborhood of 1.

Again, a direct application of Theorem 14.2, and again the integral given by formula 14.6 cannot be solved in full. The relevant fact that it actually converges as improper integral as $x \to \infty$, however, can be inferred anyhow. Notice that knowing the behaviour at $+\infty$ of the integral function is important, because it clarifies the behaviour of one of the two factors in the product under investigation. Question (b) can be answered knowing the sign of $y''(1)$, which is easy to compute by differentiating the differential equation and substituting the value 1.

14.3 Consider the Cauchy problem

$$\begin{cases} y'(x) = \sqrt{x+1}(1+y(x)) \\ y(0) = 0. \end{cases}$$

(a) Determine existence and uniqueness of solutions.
(b) Determine the order with which the solution $y(x)$ vanishes as $x \to 0^+$.
(c) Find the solution.

Answer. (a) Since $a(x) = b(x) = \sqrt{x+1}$ are continuous in $[-1, +\infty)$, by Theorem 14.2 the solution exists and is unique in $[-1, +\infty)$.
(b) Evidently, $y(x) \to y(0) = 0$ as $x \to 0^+$, and $y'(0) = 1 + y(0) = 1 \neq 0$. By McLaurin's formula, $y(x) = x + o(x)$ so that $y(x)$ vanishes with order 1.
(c) From formula (14.6)

$$y(x) = e^{\int_0^x \sqrt{t+1}\,dt} \int_0^x (t+1)^{\frac{1}{2}} e^{-\int_0^t \sqrt{u+1}\,du}\,dt$$

$$= e^{\frac{2}{3}(x+1)^{\frac{3}{2}} - \frac{2}{3}} \int_0^x (t+1)^{\frac{1}{2}} e^{\frac{2}{3} - \frac{2}{3}(t+1)^{\frac{3}{2}}}\,dt$$

with $x \geq -1$.

All questions call for an application of the basic Theorem 14.2. The expression of the solution is again given as an integral function. Although it is not the point of the exercise, this can actually be solved with the substitution $z = (t+1)^{3/2}$ and yields

$$y(x) = -1 + e^{-2/3} e^{\frac{2}{3}(x+1)^{3/2}}.$$

14.4 Consider the Cauchy problem

$$\begin{cases} y'(x) = (x+1)y(x) + e^x \\ y(0) = -1 \end{cases}$$

(a) Does the solution have in 0 a local extreme point?
(b) Compute

$$\lim_{x \to 0^+} \frac{(y(x)+1)^k}{x \sin x}$$

as k ranges in $\mathbb{R}$.

Answer. (a) By Theorem 14.2 the solution exists and is unique in $\mathbb{R}$. Furthermore, it belongs to $C^\infty(\mathbb{R})$. Indeed, y' is differentiable because it is a sum of differentiable functions in $\mathbb{R}$. Hence y'' exists, and is itself a sum of differentiable functions. By induction, the solution is actually indefinitely differentiable. Now,

$$y''(x) = y(x) + (x+1)y'(x) + e^x$$
$$y'''(x) = y'(x) + y'(x) + (x+1)y''(x) + e^x,$$

whence $y(0) = -1$, $y'(0) = 0$, $y''(0) = 0$ e $y'''(0) = 1$. Thus the third order McLaurin expansion of the solution is $y(x) = -1 + x^3/6 + o(x^3)$, which exhibits 0 as an inflection point.
(b) From McLaurin's formula

$$\lim_{x \to 0^+} \frac{(y(x)+1)^k}{x \sin x} = \lim_{x \to 0^+} \frac{(\frac{x^3}{6} + o(x^3))^k}{x(x + o(x))} = \lim_{x \to 0^+} \begin{cases} 0 & k > \frac{2}{3} \\ (\frac{1}{6})^{\frac{2}{3}} & k = \frac{2}{3} \\ +\infty & k < \frac{2}{3}. \end{cases}$$

This exercise is all about deriving the third order McLaurin expansion of the solution, without delving into using formula (14.6).

14.5 Consider the Cauchy problem

$$\begin{cases} y'(x) = \dfrac{2}{x} y(x) + \dfrac{1}{\log(2x^2 + x)} \\ y(x_0) = 0. \end{cases}$$

(a) Discuss existence and uniqueness of solutions as x_0 ranges in $\mathbb{R}$, specifying their domains.
(b) Put $x_0 = 1$. Write the solution in integral form.
(c) Put $x_0 = 1$. Is the improper integral $\int_1^{+\infty} y(x)^{-1} \, dx$ convergent?

Answer. (a) Put

$$a(x) = \frac{2}{x}, \qquad b(x) = \frac{1}{\log(2x^2 + x)}.$$

Clearly, $a \in C^0(\mathbb{R} \setminus \{0\})$ and $b \in C^0((\text{Dom}(b)))$, where

$$\text{Dom}(b) = (-\infty, -1) \cup \left(-1, -\frac{1}{2}\right) \cup \left(0, \frac{1}{2}\right) \cup \left(\frac{1}{2}, +\infty\right).$$

Hence, for every $x_0 \in \mathrm{Dom}(b)$ there is a solution, defined and unique in the maximal subinterval of $\mathrm{Dom}(b)$ to which x_0 belongs.

(b) In this case the solution exists and is unique in $(\frac{1}{2}, +\infty)$, and is given by

$$
\begin{aligned}
y(x) &= e^{\int_1^x \frac{2}{t}\, dt} \int_1^x e^{-\int_1^t \frac{2}{s}\, ds}\, \frac{1}{\log(2t^2 + t)}\, dt \\
&= e^{2\log x} \int_1^x e^{-2\log t}\, \frac{1}{\log(2t^2 + t)}\, dt \\
&= x^2 \int_1^x \frac{1}{t^2 \log(2t^2 + t)}\, dt.
\end{aligned}
$$

(c) As just shown, $y(x) = x^2 g(x)$, where

$$
g(x) = \int_1^x h(t)\, dt, \qquad h(t) = \frac{1}{t^2 \log(2t^2 + t)}.
$$

Clearly, $h \in C^0((1/2, +\infty))$ and $h(t) > 0$ for $t > 1/2$. By the fundamental theorem of calculus, $g \in C^1((1/2, +\infty))$ and $g'(x) = h(x) > 0$ for $x > 1/2$. Hence g is strictly increasing in $(1/2, +\infty)$ and since $g(1) = 0$, it follows that $g(x) > 0$ for $x > 1$. Therefore, since $y(x) = x^2 g(x)$, the function $x \to 1/y(x)$ is defined and continuous in $(1, +\infty)$ and $1/y(x) \to +\infty$ as $x \to 1^+$. The given improper integral is convergent provided that for any fixed $\bar{x} \in (1, +\infty)$ both integrals

$$
\int_1^{\bar{x}} \frac{1}{y(x)}\, dx, \qquad \int_{\bar{x}}^{+\infty} \frac{1}{y(x)}\, dx
$$

converge. Since $g'(1) = h(1) = (\log 3)^{-1} \neq 0$, the function g vanishes with order 1 as $x \to 1^+$. Therefore, $1/y(x) \to +\infty$ with order 1 and the first integral diverges. Since $y(x) > 0$ for every $x > 1$, the full improper integral on $(1, +\infty)$ diverges.

Part (a) of this exercise requires to realize that the domain of the function b is a disjoint union of intervals, and hence that a solution can only be defined in one and only one of them. Part (b) is a direct application of the usual formula (14.6), and is really the setup for part (c), which is an application of the techniques of Chap. 11.3. What matters here is the qualitative behaviour of the solution as $x \to 1^+$, and as $x \to +\infty$. Now, the solution $y(x)$ vanishes with critical order 1 as x approaches 1 from the right and this is enough to conclude that the improper integral of its reciprocal does not converge. Since the solution is actually positive in the whole interval, this permits to infer that the improper integral diverges.

14.6 Consider the Cauchy problem

$$
\begin{cases}
y'(x) = a(x)y(x) + x \\
y(0) = 1
\end{cases}
$$

and assume that $a : \mathbb{R} \to \mathbb{R}$ is differentiable.

(a) Write the second order McLaurin polynomial of the solution if $a(0) = 2$ and $a'(0) = 1$.
(b) Put $a(x) = x$. Write the solution.
(c) Put $a(x) = x$. Does the improper integral $\int_0^{+\infty} y(x)^{-1} \, dx$ converge?

Answer. (a) First of all $y'(0) = 2$ and $y(0) = 1$. From the differential equation, y' is differentiable and

$$y''(x) = a'(x)y(x) + a(x)y'(x) + 1$$

whence $y''(0) = 6$. Therefore

$$p_2(x) = y(0) + y'(0)x + \frac{y''(0)}{2}x^2 = 1 + 2x + 3x^2.$$

(b) The Cauchy problem is

$$\begin{cases} y'(x) = xy(x) + x \\ y(0) = 1 \end{cases}$$

so that formula (14.6) gives

$$y(x) = e^{\int_0^x t \, dt}\left(1 + \int_0^x t e^{-\int_0^t u \, du} \, dt\right)$$

$$= e^{\frac{x^2}{2}}\left(1 + \int_0^x t e^{-\frac{t^2}{2}} \, dt\right)$$

$$= e^{\frac{x^2}{2}}\left(1 + 1 - e^{-\frac{x^2}{2}}\right)$$

$$= 2e^{\frac{x^2}{2}} - 1.$$

(c) By what has just been established, $y(x) = 2e^{\frac{x^2}{2}} - 1$. Hence $\text{Dom}(y) = \mathbb{R}$, and $y(x) > 1$ for $x \in \mathbb{R}$. Furthermore, $y(x) \to +\infty$ as $x \to +\infty$ with an order that is bigger than any positive real power. Hence $y(x)^{-1}$ is continuous in $\mathbb{R}$ and the improper integral $\int_0^{+\infty} y(x)^{-1} \, dx$ converges.

In part (a) formula (14.6) plays no role: the differential equation, together with the information on a, is enough. On the contrary, in part (b) formula (14.6) gives an explicit solution, which, in turn, leads to an easy answer to (c).

14.7 Consider the Cauchy problem

$$\begin{cases} y'(x) = a(x)y(x) + x + 4 \\ y(0) = 0. \end{cases}$$

(a) Compute, if existing,

$$\lim_{x \to 0^+} \frac{y(x) - 4x}{e^{x^2} - 1},$$

assuming that $a(x)$ is a continuous function defined in $\mathbb{R}$ and that $a(0) = -3$.

(b) Take $a(x) = 1/(x+5)$. Write the explicit solution, specifying its domain.

Answer. (a) Put $b(x) = x + 4$. Evidently, $a, b \in C^0(\mathbb{R})$. The Cauchy problem has thus a unique solution in $\mathbb{R}$. Since $y'(0) = 4$, it is

$$\lim_{x \to 0} \frac{y'(x) - y'(0)}{x} = \lim_{x \to 0} \frac{y'(x) - 4}{x} = \lim_{x \to 0} \left(a(x) \frac{y(x)}{x} + 1 \right) = -3y'(0) + 1 = -11.$$

Hence y is differentiable in $\mathbb{R}$ and twice differentiable at 0. It is therefore legitimate to use the second order McLaurin's expansion of $y(x)$, which reads

$$y(x) = 4x - \frac{11}{2}x^2 + x^2 \omega(x)$$

for every $x \in \mathbb{R}$, where $\omega(x) \to 0$ as $x \to 0$. Therefore

$$\lim_{x \to 0} \frac{y(x) - 4x}{e^{x^2} - 1} = \lim_{x \to 0} \frac{x^2(-\frac{11}{2} + \omega(x))}{e^{x^2} - 1} = -\frac{11}{2}.$$

(b) The function $a(x) = 1/(x+5)$ is in $C^0((-5, +\infty))$, and $b \in C^0(\mathbb{R})$. Therefore there exists one and only one solution in $(-5, +\infty)$, and is given by

$$y(x) = e^{\int_0^x \frac{1}{t+5} dt} \int_0^x e^{-\int_0^t \frac{1}{s+5} ds}(t+4) dt$$

$$= e^{\log(x+5)-\log 5} \int_0^x e^{-\log(t+5)+\log 5}(t+4) dt$$

$$= (x+5) \int_0^x \frac{t+4}{t+5} dt$$

$$= (x+5) \int_0^x 1 - \frac{1}{t+5} dt$$

$$= (x+5)(x - \log(x+5) + \log 5).$$

The most interesting part is question (a), where the required limit calls for a second order expansion, as the first order term is subtracted off in the numerator. This can be achieved using the differential equation alone. Part (b) boils down to carrying out the integration of a rational function.

14.8 Consider the Cauchy problem

$$\begin{cases} y'(x) = -y(x) \log(|x|) + \log(x^2 + x) \\ y(x_0) = y_0 \end{cases}$$

(a) Discuss existence and uniqueness of solutions as x_0 and y_0 range in $\mathbb{R}$, specifying their domains.
(b) Write the solution in integral form if $x_0 = 1$ and $y_0 = 0$.
(c) Put $x_0 = 1$ and $y_0 = 0$. Compute, if existing, $\lim_{x\to+\infty} y(x)$.

Answer. (a) Put $a(x) = \log(|x|)$ and $b(x) = \log(x^2 + x)$. Evidently, $a \in C^0(\mathbb{R} \setminus \{0\})$ and $b \in C^0((-\infty, -1) \cup (0, +\infty))$. By Theorem 14.2, for every $x_0 \in (-\infty, -1) \cup (0, +\infty)$ and every $y_0 \in \mathbb{R}$ a solution exists and is unique. If $x_0 \in (-\infty, -1)$, the solution is unique in $(-\infty, -1)$, whereas if $x_0 \in (0, +\infty)$, the solution is unique in $(0, +\infty)$.
(b) As discussed in (a), for $x_0 = 1$ and $y_0 = 0$ the solution exists and is unique in $(0, +\infty)$, and is given by

$$y(x) = e^{-\int_1^x \log t\, dt} \int_1^x e^{\int_1^t \log s\, ds} \log(t^2 + t)\, dt$$

$$= e^{-x\log x + x - 1} \int_1^x e^{t\log t - t + 1} \log(t^2 + t)\, dt$$

$$= \frac{e^x}{x^x} \int_1^x \frac{t^t}{e^t} \log(t^2 + t)\, dt.$$

(c) Observe that $y(x) = h(x)/(x^x e^{-x})$, where

$$h(x) = \int_1^x g(t)\, dt, \qquad g(t) = \frac{t^t}{e^t} \log(t^2 + t).$$

Now, since $g(t) \to +\infty$ as $t \to +\infty$, it follows that $h(x) \to +\infty$ as $x \to +\infty$ as well. Furthermore, $g \in C^0((0, +\infty))$ so that by the fundamental theorem of calculus it is $h \in C^1((0, +\infty))$ and $h'(x) = g(x)$. Finally, by the de l'Hôpital theorem, since

$$\lim_{x\to+\infty} \frac{\frac{d}{dx}h(x)}{\frac{d}{dx}(x^x e^{-x})} = \lim_{x\to+\infty} \frac{2\log x + \log(1 + \frac{1}{x})}{\log x} = 2,$$

it follows that

$$\lim_{x\to+\infty} y(x) = \lim_{x\to+\infty} \frac{h(x)}{x^x e^{-x}} = 2.$$

Again, the limiting behaviour of $y(x)$ is inferred from a careful use of the de l'Hôpital theorem. Parts (a) and (b) are standard.

14.9 Consider the Cauchy problem:

$$\begin{cases} y' = |x|y(x) + \dfrac{x}{x^2 - 1} \\ y(x_0) = y_0. \end{cases}$$

(a) Discuss existence and uniqueness of solutions as x_0 and y_0 range in $\mathbb{R}$, specifying their domains.
(b) Put $x_0 = 0$. Compute, if existing, $y''(0)$ as y_0 ranges in $\mathbb{R}$.
(c) Put $y(-2) = 0$. Compute, if existing, $\lim_{x \to -\infty} y(x)$.
(d) Are there solutions $y = y(x)$ such that $\lim_{x \to 1^+} y(x) \in \mathbb{R}$?

Answer. (a) Put $a(x) = |x|$ and $b(x) = x/(x^2 - 1)$. Evidently, $a \in C^0(\mathbb{R})$ and $b \in C^0(\mathbb{R} \setminus \{-1, 1\})$. Hence, for every $x_0 \in \mathbb{R} \setminus \{-1, 1\}$ and for every $y_0 \in \mathbb{R}$ there exists one and only one solution. More precisely: if $x_0 < -1$, the solution exists and is unique in $(-\infty, -1)$; if $x_0 \in (-1, 1)$ the solution exists and is unique in $(-1, 1)$; if $x_0 > 1$ the solution exists and is unique in $(1, +\infty)$.
(b) If $x_0 = 0$ the solution exists and is unique in $(-1, 1)$ and $y'(0) = 0$. Therefore

$$\frac{y'(x) - y'(0)}{x} = \frac{|x|}{x} y(x) + \frac{1}{x^2 - 1}.$$

Now, since $y(x)$ tends to y_0 as $x \to 0$,

$$\lim_{x \to 0} \frac{y'(x) - y'(0)}{x} = \begin{cases} -1 & y_0 = 0 \\ \nexists & y_0 \neq 0. \end{cases}$$

Thus, $y''(0)$ exists, and is equal to -1, if and only if $y_0 = 0$.
(c) If $x_0 = -2$ and $y_0 = 0$ the solution exists and is unique in $(-\infty, -1)$, and is given by

$$y(x) = e^{\int_{-2}^{x} |t| \, dt} \int_{-2}^{x} e^{-\int_{-2}^{t} |s| \, ds} \frac{t}{t^2 - 1} \, dt$$

$$= e^{-\frac{x^2}{2} + 2} \int_{-2}^{x} e^{\frac{t^2}{2} - 2} \frac{t}{t^2 - 1} \, dt$$

$$= e^{-\frac{x^2}{2}} \int_{-2}^{x} e^{\frac{t^2}{2}} \frac{t}{t^2 - 1} \, dt.$$

Thus, $y(x) = h(x)/e^{\frac{x^2}{2}}$, where

$$h(x) = \int_{-2}^{x} g(t) dt, \qquad g(t) = e^{\frac{t^2}{2}} \frac{t}{t^2 - 1}.$$

Now, $g \in C^0((-\infty, -1))$ and by the fundamental theorem of calculus $h \in C^1((-\infty, -1))$ and $h'(x) = g(x)$ for $x < -1$. Further, $g(t) \to -\infty$ as $t \to -\infty$ an hence $h(x) \to +\infty$. Therefore, $y(x)$ is the ratio of two functions that diverge as $x \to -\infty$. The ratio of the corresponding derivatives is

$$\frac{\frac{d}{dx}h(x)}{\frac{d}{dx}e^{\frac{x^2}{2}}} = \frac{e^{\frac{x^2}{2}}\frac{x}{x^2-1}}{xe^{\frac{x^2}{2}}} = \frac{1}{x^2-1}$$

and tends to 0 as $x \to -\infty$. Hence, by the de l'Hôpital theorem, $\lim_{x\to-\infty} y(x) = 0$.
(d) There are no solutions $y = y(x)$ such that $\lim_{x\to 1+} y(x) \in \mathbb{R}$. Indeed, if any such did exist, then for any fixed $\overline{x} > 1$ and any $x \in (1, \overline{x})$ it would hold that

$$\int_x^{\overline{x}} y'(t)\,dt = \int_x^{\overline{x}} ty(t)\,dt + \int_x^{\overline{x}} \frac{t}{t^2-1}\,dt,$$

whence

$$y(\overline{x}) - y(x) = \int_x^{\overline{x}} ty(t)\,dt + \frac{1}{2}\log(\overline{x}^2 - 1) - \frac{1}{2}\log(x^2 - 1).$$

But then $t \to ty(t)$ would have a continuous extension at 1, and it would follow that

$$\lim_{x\to 1+} \left(y(\overline{x}) - y(x) - \int_x^{\overline{x}} ty(t)dt - \frac{1}{2}\log(\overline{x}^2 - 1) \right) \in \mathbb{R}$$

whereas $\lim_{x\to 1+} -\log(x^2 - 1)/2 = +\infty$, a contradiction.

Questions (a), (b) and (c) are, at this point, standard. Question (d) is instead a rather tricky one. The analysis carried out previously clearly indicates that there is a delicate issue at the point 1, due to the diverging behaviour of $b(x)$. The crucial observation is that not only does b diverge as $x \to 1^+$, but it does so in a non integrable fashion. Hence, an improper integral in some $(1, \overline{x})$, say for example in $(1, 2)$, should reveal something. Informally speaking, if $y(x) \to \ell \in \mathbb{R}$, then the equality

$$\int_1^2 y'(t)\,dt = \int_1^2 ty(t)\,dt + \int_1^2 \frac{t}{t^2-1}\,dt,$$

would contain two converging integrals, the first two, and a diverging one, the third.

14.3   Problems on Linear First Order Equations

14.10 Solve the Cauchy problem

$$\begin{cases} y'(x) = y(x) - 4x^2 \\ y(-1) = 0. \end{cases}$$

14.11 Solve the Cauchy problem

$$\begin{cases} y'(x) = \dfrac{1}{x-1}y(x) + 1 \\ y(0) = 0. \end{cases}$$

14.12 Consider the Cauchy problem

$$\begin{cases} y'(x) = \sqrt{x+1} + y(x) \\ y(x_0) = y_0. \end{cases}$$

(a) Discuss existence and uniqueness of solutions as x_0 and y_0 range in $\mathbb{R}$.
(b) Write the solution if $x_0 = y_0 = 0$.

14.13 Consider the Cauchy problem

$$\begin{cases} xy'(x) + y(x) = x \\ y(1) = 0. \end{cases}$$

(a) Determine if the solution exists and is unique.
(b) Find $A, B, C \in \mathbb{R}$ such that $f(x) = Ax + B/x + C$ solves the problem in $(0, +\infty)$.

14.14 Consider the Cauchy problem

$$\begin{cases} y'(x) = xy(x) + x^2 \\ y(0) = k. \end{cases}$$

Find, if existing, $k \in \mathbb{R}$ such that y has a local minimum at 0.

14.15 How many times is the solution of the Cauchy problem

$$\begin{cases} y'(x) = y(x) + |x| \\ y(0) = 0 \end{cases}$$

differentiable at 0?

14.16 Solve the Cauchy problem

$$\begin{cases} y''(x) - xy'(x) = 0 \\ y(0) = 1, \ y'(0) = 1 \end{cases}$$

and compute $\lim\limits_{x \to 0^+} (y(x) - x - 1)/(\sin x)^k$ as k varies in $\mathbb{R}$.

14.17 Consider the Cauchy problem

$$\begin{cases} y''(x) = -xy'(x) \\ y(0) = 0, \ y'(0) = 1. \end{cases}$$

(a) Find the solutions.
(b) Compute $\lim\limits_{x \to +\infty} y(x)$.

14.18 Consider the Cauchy problem

$$\begin{cases} y'(x) = y(x) \cos x + 1 \\ y(0) = 0. \end{cases}$$

(a) Solve the problem.
(b) Compute $\lim\limits_{x \to 0^+} (y(x) - e^{\sin x} x)/(e^x - 1)$.

14.19 Consider the differential equation $xy' - 2y = x^3 e^x$.

(a) Determine all the solutions of the equation in $(0, +\infty)$.
(b) Determine all the solutions of the equation in $\mathbb{R}$.

14.20 Consider the Cauchy problem

$$\begin{cases} y'(x) = ky(x) + x \\ y(0) = 1. \end{cases}$$

(a) Compute $\lim\limits_{x \to 0^+} (y(x) - x - 1)^2/(1 - \cos x)$.
(b) Put $k = 0$. Compute $\lim\limits_{x \to +\infty} y(x)$ and $\lim\limits_{x \to -\infty} y(x)$.

14.21 Consider the Cauchy problem

$$\begin{cases} y'(x) = -3x^2 y(x) + \sqrt{2x^4 + 1} \\ y(x_0) = y_0. \end{cases}$$

(a) Discuss existence and uniqueness of solutions as x_0 and y_0 range in $\mathbb{R}$.
(b) Put $x_0 = y_0 = 0$. Compute $\lim\limits_{x \to +\infty} y(x)$.
(c) Put $x_0 = y_0 = 0$. Compute the second order McLaurin polynomial of the solution.

14.22 Consider the Cauchy problem

$$\begin{cases} y'(x) = \left(\dfrac{1-2x}{x^2-x}\right) y(x) + \dfrac{1}{x+\arctan x} \\ y(x_0) = y_0. \end{cases}$$

(a) Discuss existence and uniqueness of solutions as x_0 and y_0 range in $\mathbb{R}$, specifying their domains.
(b) Put $x_0 = -1$, $y_0 = 0$. Determine, if existing, the solution.
(c) Compute, if existing, $\lim\limits_{x\to-\infty} y(x)$.

14.23 Consider the problem

$$\begin{cases} y''(x) + \dfrac{y'(x)}{x} = kx^2 \\ y(2) = 0. \end{cases}$$

(a) Discuss existence and uniqueness of solutions, specifying their domains.
(b) For which $k \in \mathbb{R}$ is the set of solution a vector space?
(c) Find all solutions if $k = 1$.

14.24 Consider the differential equation $x^2 y' + xy = 1$.

(a) Determine all solutions in $(0, +\infty)$ and in $(-\infty, 0)$.
(b) Is there a solution of the equation defined in $\mathbb{R}$?
(c) Determine the solution of the equation that satisfies $y(1) = 0$.
(d) Find the third order Taylor polynomial centered at 1 of the solution in (c).

14.25 Consider the problem

$$\begin{cases} y(x)y'(x) = \dfrac{y^2(x)}{2x} + \dfrac{\log^3 x}{2} \\ y(1) = 0. \end{cases}$$

Find all solutions, specifying the interval in which they are defined.

14.26 Consider the problem

$$\begin{cases} (x^2 - x)y'(x) = xy(x) + (x-1)^2 \\ y(1) = 0. \end{cases}$$

Find all solutions, specifying the interval in which they are defined.

14.27 Consider the differential equation $y'(x) = \dfrac{4x^2 - 2}{x(x^2 - 1)}y(x) + \dfrac{1}{\arctan(x - 1)}$.

(a) Discuss existence and uniqueness of the solutions $y(x)$ such that $y(x_0) = y_0$, as x_0 and y_0 range in $\mathbb{R}$.

(b) Let $y(x)$ be the solution for which $y(2) = 0$. Does $\lim\limits_{x \to +\infty} y(x)/x^4$ converge?

(c) Let $y(x)$ be the solution for which $y(2) = 0$. Compute, if existing, $\lim\limits_{x \to 1^+} y(x)$.

(d) Let $y(x)$ be the solution for which $y(2) = 0$. Compute $\lim\limits_{x \to 2^+} y(x)/(x - 2)^k$, if existing, as k ranges in $\mathbb{R}$.

14.28 Consider the equation $(2x^2 + 1)y'(x) - (4x^3 - 6x)y(x) = 4x^4 - 8x^3 - 1$.

(a) Are there solutions that are increasing in a neighborhood of 0?

(b) Find all solutions.

(c) Are there solutions that vanish as $x \to +\infty$?

14.29 Consider the differential equation $xy'(x) = y(x) + \log(1 + x^2)$.

(a) Are there solutions such that $y(0) = 1$?

(b) Find all solutions defined in $(0, +\infty)$ such that $\lim\limits_{x \to +\infty} y(x) = -\infty$.

Chapter 15
Constant Coefficient Linear Differential Equations

15.1 Linear Equations with Constant Coefficients

As pointed out in Chap. 14, one of the most important classes of differential equations is the class of linear differential equations and, among these, the simplest and perhaps best understood is that of *constant coefficient* linear differential equations. In normal form (see (13.2) of Chap. 13), an order n equation of this type is

$$y^{(n)} = \sum_{k=0}^{n-1} a_k y^{(k)} + b(x), \tag{15.1}$$

where $a_0, \ldots, a_{n-1}$ are real numbers, whence the name "constant coefficient", and where $b : I \to \mathbb{R}$ is some given function defined in some open interval I, and is referred to as the *source term*. Also, as usual, $y^{(0)}$ stands for y itself. As in (13.3) of Chap. 13, the associated Cauchy problem is

$$\begin{cases} y^{(n)} = \displaystyle\sum_{k=0}^{n-1} a_k y^{(k)} + b(x) \\ y(x_0) = y_0 \\ y^{(1)}(x_0) = y_1 \\ \vdots \\ y^{(n-1)}(x_0) = y_{n-1}. \end{cases} \tag{15.2}$$

If $b = 0$, the equation is called *homogeneous*, otherwise it is called *nonhomogeneous*. Constant coefficient equations are completely understood, and can always be solved, at least in principle.

© Springer International Publishing Switzerland 2016
M. Baronti et al., *Calculus Problems*, UNITEXT - La Matematica per il 3+2 101,
DOI 10.1007/978-3-319-15428-2_15

Theorem 15.1 *Suppose that I is an open interval, take $x_0 \in I$ and fix an initial value $(y_0, y_1, \ldots, y_{n-1}) \in \mathbb{R}^n$. If $b \in C^0(I)$, then there exists a unique solution $y : I \to \mathbb{R}$ of the Cauchy problem* (15.2).

15.2   The Homogeneous Equation

The basic step in the theory consists in considering the homogeneous case first, where the equation is simply

$$y^{(n)} = \sum_{k=0}^{n-1} a_k y^{(k)}. \tag{15.3}$$

Here is the fundamental structural result.

Theorem 15.2 (Structure theorem, I) *The general solution Y_0 of the order n homogeneous differential equation* (15.3) *is an n dimensional vector subspace of $C^0(\mathbb{R})$.*

Recall that by general solution it is meant the set of all solutions of the differential equation under consideration. It follows immediately from this theorem that the general solution Y_0 of (15.3) is completely described by any set of n *linearly independent solutions*. Recall that the continuous functions $y_1, \ldots, y_n$ are linearly independent in $C^0(\mathbb{R})$ if whenever for some real numbers $c_1, \ldots, c_n$ the identity

$$c_1 y_1(x) + \cdots + c_n y_n(x) = 0$$

holds for every $x \in \mathbb{R}$, then $c_1 = \cdots = c_n = 0$. This is the usual notion of linear independence, formulated for the real vector space $C^0(\mathbb{R})$.

One of the remarkable facts in the theory of constant coefficient equations is that a basis of solutions of (15.3) can be determined by looking at the *characteristic polynomial* associated with (15.3), namely the n degree polynomial

$$P(\lambda) = \lambda^n - \sum_{k=0}^{n-1} a_k \lambda^k. \tag{15.4}$$

In particular, the roots of the algebraic equation

$$P(\lambda) = 0, \tag{15.5}$$

known as the *characteristic equation* of (15.3), play a fundamental role, as clarified by the theorem that follows.

Theorem 15.3 *Denote by $\lambda_1, \ldots, \lambda_r$ the distinct (complex) roots of the characteristic equation* (15.5) *with multiplicities $m_1, \ldots, m_r$, so that $m_1 + \cdots + m_r = n$. Then the n distinct non zero functions among the functions*

$$\varphi_{jk}(x) = x^k \mathfrak{R}\left(e^{\lambda_j x}\right)$$
$$\psi_{jk}(x) = x^k \mathfrak{I}\left(e^{\lambda_j x}\right)$$

where $j = 1, \ldots, r$ and $k = 0, \ldots, m-1$, form a basis of solutions of the differential equation (15.3), that is, a basis of the vector space Y_0.

Some comments are in order. First of all, $\mathfrak{R}(z)$ and $\mathfrak{I}(z)$ stand for the real and imaginary parts of the complex number z, respectively. To wit, if $z = a + ib$, then $\mathfrak{R}(z) = a$ and $\mathfrak{I}(z) = b$.

Secondly, the theorem states that with each (complex) root λ there are associated some functions (the φ's and the ψ's) that will collectively give rise to a basis of solutions. It is important to understand how the case of a real root differs from the case of a complex root.

Suppose that λ is a real root with multiplicity m of the characteristic equation. Then the solutions that correspond to λ are precisely the m functions

$$e^{\lambda x}, \; x e^{\lambda x}, \; x^2 e^{\lambda x}, \ldots, \; x^{m-1} e^{\lambda x}.$$

Indeed, in this case $\mathfrak{R}\left(e^{\lambda x}\right) = e^{\lambda x}$ for every $x \in \mathbb{R}$, whereas $\mathfrak{I}\left(e^{\lambda x}\right) = 0$, so that in the real case only the functions φ appear, and none of the ψ's.

Suppose next that $\lambda = \alpha + i\beta$ is a complex root, that is, suppose that $\beta \neq 0$ and denote again by m its multiplicity. In this case, using the basic property of complex exponentials, $e^{u+v} = e^u e^v$, and Euler's formula $e^{it} = \cos t + i \sin t$, it follows

$$\mathfrak{R}\left(e^{\lambda x}\right) = \mathfrak{R}\left(e^{\alpha x}(\cos(\beta x) + i \sin(\beta x))\right) = e^{\alpha x} \cos(\beta x)$$
$$\mathfrak{I}\left(e^{\lambda x}\right) = \mathfrak{I}\left(e^{\alpha x}(\cos(\beta x) + i \sin(\beta x))\right) = e^{\alpha x} \sin(\beta x).$$

Hence Theorem 15.3 associates with λ the $2m$ distinct functions

$$e^{\alpha x} \cos(\beta x), \; x e^{\alpha x} \cos(\beta x), \; \ldots, \; x^{m-1} e^{\alpha x} \cos(\beta x)$$
$$e^{\alpha x} \sin(\beta x), \; x e^{\alpha x} \sin(\beta x), \; \ldots, \; x^{m-1} e^{\alpha x} \sin(\beta x).$$

Contrary to intuition, these are not too many. The point is that since the characteristic equation has real coefficients, also the conjugate complex number $\bar{\lambda} = \alpha - i\beta$ is a root, with the same multiplicity m. With this second distinct root, the theorem associates, up to sign, exactly the same $2m$ functions as those associated with λ: precisely the same involving cosines, the φ's, and minus those involving sines, the ψ's. Therefore, to the pair $(\lambda, \bar{\lambda})$ the theorem assigns collectively $2m$ functions, accounting for the correct cardinality of the basis. The functions that are defined in Theorem 15.3 are sometimes referred to as *fundamental solutions*.

In conclusion, the general solution Y_0 of (15.3) is a vector space of dimension n, and a basis, say $\{y_1, y_2, \ldots, y_n\}$, is found by solving the characteristic polynomial and hence by associating to its roots the fundamental solutions, as prescribed by Theorem 15.3. An arbitrary solution of (15.3) is therefore of the form

$$y(x) = c_1 y_1(x) + c_2 y_2(x) + \cdots + c_n y_n(x) \tag{15.6}$$

where $c_1, \ldots, c_n \in \mathbb{R}$.

15.3   The Nonhomogeneous Equation

The second step consists in bringing into the picture the source term $b(x)$, that is, in looking at the full differential equation (15.1). In this case, the structure Theorem 15.2 must be refined, and becomes:

Theorem 15.4 (Structure theorem, II) *The general solution Y of the order n nonhomogeneous differential equation (15.1) is an n dimensional affine subspace of $C^0(\mathbb{R})$. More precisely, it is given by*

$$Y = Y_0 + \overline{y}, \tag{15.7}$$

where Y_0 is the general solution of the homogeneous equation (15.3) associated with (15.1), and $\overline{y}$ is any solution of (15.1).

According to Theorem 15.4, the general solution of the nonhomogeneous equation is obtained, geometrically speaking, by translating inside $C^0(\mathbb{R})$ the vector space Y_0 by any given solution $\overline{y}$ of the nonhomogeneous equation. In practice, in order to describe Y, the procedure is the following. First one looks at the homogeneous equation (15.3) associated with (15.1), simply by ignoring the source term $b(x)$, that is, by setting it to zero. The solution of the characteristic equation in combination with Theorem 15.3 yields a basis of Y_0, say $\{y_1, y_2, \ldots, y_n\}$. Then, any solution of the nonhomogeneous equation is

$$y(x) = c_1 y_1(x) + c_2 y_2(x) + \cdots + c_n y_n(x) + \overline{y}, \tag{15.8}$$

where $c_1, \ldots, c_n \in \mathbb{R}$ and where $\overline{y}$ is any *particular solution* of the nonhomogeneous equation.

The final task to achieve is the determination of a particular solution. First of all, it must be clear that any such solution will do the job, for (15.7), or, equivalently, (15.8) holds true with any such $\overline{y}$. There are several techniques to obtain an expression for $\overline{y}$. In many practical situations, the following theorem is of great help.

Theorem 15.5 (Similitude method) *Consider the nonhomogeneous differential equation (15.1) and suppose that the source term is of the form*

$$b(x) = e^{\alpha x}\left(P(x)\cos(\beta x) + Q(x)\sin(\beta x)\right), \tag{15.9}$$

where $\alpha, \beta \in \mathbb{R}$ and where P and Q are polynomials of degrees d_1 and d_2, respectively. Then there exists a particular solution $\overline{y}$ of (15.1) of the form

$$\overline{y} = x^m e^{\alpha x} \left(\overline{P}(x) \cos(\beta x) + \overline{Q}(x) \sin(\beta x) \right), \tag{15.10}$$

where $\overline{P}$ and $\overline{Q}$ are polynomials of degree at most $d = \max\{d_1, d_2\}$ and m is the multiplicity of the root $\alpha \pm i\beta$ of the characteristic polynomial (here $m = 0$ if $\alpha \pm i\beta$ is not a root).

As illustrated below in the exercises, the way to use the above theorem is to assume that a solution of the form (15.10) exists, to write it with the correct number of unknown parameters for $\overline{P}$ and $\overline{Q}$ and then to compute its derivatives up to order n. Requiring $\overline{y}$ to satisfy (15.1) is equivalent to solving a system for the parameters, the solution of which, in the end, singles out $\overline{y}$.

A complete answer to the problem of finding a particular solution is given by the method of *variation of constants*, that derives its name from the formula (15.12) below that inspires its construction.

The starting point is a given basis $\{y_1, y_2, \dots, y_n\}$ of Y_0. Out of this basis, the so-called *Wronskian matrix* is built, namely

$$W(x) = \begin{pmatrix} y_1(x) & y_2(x) & \cdots & y_n(x) \\ y_1'(x) & y_2'(x) & \cdots & y_n'(x) \\ \vdots & \vdots & & \vdots \\ y_1^{(n-1)}(x) & y_2^{(n-1)}(x) & \cdots & y_n^{(n-1)}(x) \end{pmatrix} \tag{15.11}$$

which is defined for every $x \in \mathbb{R}$. A crucial property of $W(x)$ is that is an invertible matrix for every $x \in \mathbb{R}$.

Theorem 15.6 *Suppose that $\{y_1, y_2, \dots, y_n\}$ is a basis of the general solution Y_0 of the homogeneous equation associated with the nonhomogeneous equation (15.1) and let $W(x)$ denote its Wronskian matrix. Then the function*

$$\overline{y}(x) = c_1(x)y_1(x) + c_2(x)y_2(x) + \cdots + c_n(x)y_n(x) \tag{15.12}$$

is a particular solution of (15.1), defined for $x \in I$, where the functions $c_j : I \to \mathbb{R}$ are

$$\begin{pmatrix} c_1(x) \\ c_2(x) \\ \vdots \\ c_{n-1}(x) \\ c_n(x) \end{pmatrix} = \int_{x_0}^{x} W^{-1}(t) \begin{pmatrix} 0 \\ 0 \\ \vdots \\ 0 \\ b(t) \end{pmatrix} dt, \tag{15.13}$$

and where $x_0 \in I$ is chosen arbitrarily.

A particularly important consequence of Theorem 15.6 is that it asserts that a particular solution of the nonhomogeneous equation actually exists, and is defined in the interval in which the source term b is continuous. In combination with Theorem 15.4, this says that the general solution of the nonhomogeneous equation (15.1) is non-empty whenever b is continuous, and consists of functions that are defined in the interval I in which b is continuous.

In its full generality Theorem 15.6 is rather hard to apply, but in low dimensional cases it is actually reasonable. We illustrate this method in the simple case of a second order equation. Suppose that we have found a basis $\{y_1(x), y_2(x)\}$ for the homogeneous equation $y'' = a_1 y' + a_2 y$ so that the general solution is $y(x) = c_1 y_1(x) + c_2 y_2(x)$. The method assumes that the nonhomogeneous equation $y'' = a_1 y' + a_2 y + b(x)$ admits a solution is of the form (15.12), namely

$$\overline{y}(x) = c_1(x)y_1(x) + c_2(x)y_2(x)$$

where $c_1(x)$ and $c_2(x)$ are functions to be found. Now, its derivative is

$$\overline{y}'(x) = \{c_1'(x)y_1(x) + c_2'(x)y_2(x)\} + c_1(x)y_1'(x) + c_2(x)y_2'(x).$$

If the term inside curly brackets vanishes, then the second derivative is

$$\overline{y}''(x) = c_1'(x)y_1'(x) + c_2'(x)y_2'(x) + c_1(x)y_1''(x) + c_2(x)y_2''(x),$$

and an easy computation, taking into account that y_1 and y_2 solve the homogeneous equation, gives

$$\overline{y}''(x) - a_1\overline{y}'(x) - a_2\overline{y}(x) = c_1'(x)y_1'(x) + c_2'(x)y_2'(x).$$

The latter must be equal to $b(x)$. In summary, the functions $c_1(x)$ and $c_2(x)$ are such that $\overline{y}$ solves the nonhomogeneous equation if they satisfy

$$\begin{cases} c_1'(x)y_1(x) + c_2'(x)y_2(x) = 0 \\ c_1'(x)y_1' + c_2'(x)y_2'(x) = b(x). \end{cases} \tag{15.14}$$

In matrix notation, recalling the definition of $W(x)$ in (15.11), this is

$$\begin{pmatrix} y_1(x) & y_2(x) \\ y_1'(x) & y_2'(x) \end{pmatrix} \begin{pmatrix} c_1'(x) \\ c_2'(x) \end{pmatrix} = \begin{pmatrix} 0 \\ b(x) \end{pmatrix} \implies \begin{pmatrix} c_1'(x) \\ c_2'(x) \end{pmatrix} = W(x)^{-1} \begin{pmatrix} 0 \\ b(x) \end{pmatrix}$$

and it is therefore solved by (15.13). As illustrated below in Exercise 15.6, a system like (15.14) can be dealt with in a simpler way than appealing to (15.13).

15.4 Guided Exercises on Constant Coefficient Differential Equations

15.1 Consider the differential equation $y''(x) + 6y'(x) + (k + 5)y(x) = 0$, with $k \in \mathbb{R}$.

(a) Find $k \in \mathbb{R}$ in such a way that $y(x) = e^{-3x}\sin x$ is a solution.
(b) Solve the equation for $k = 4$.

Answer. (a) The equation $y''(x)+6y'(x)+(k+5)y(x) = 0$ has characteristic equation $\lambda^2+6\lambda+k+5 = 0$. In order for $e^{-3x}\sin x$ to be a solution of the differential equation, it is necessary and sufficient that $\lambda = -3-i$ solves the characteristic equation, which happens if and only if $4 - k = -1$, namely $k = 5$.

(b) Put $k = 4$. The characteristic equation is $\lambda^2 + 6\lambda + 9 = 0$ that has the only root $\lambda = -3$ with multiplicity 2. Therefore, the general integral of the differential equation is $y(x) = e^{-3x}(c_1 + c_2 x)$ with $c_1, c_2 \in \mathbb{R}$.

Part (a) of this basic exercise requires to see that the function $e^{-3x}\sin x$ is a basis function provided that $-3 \pm i$ are roots of the characteristic equation. Part (b) is an elementary application of Theorem 15.3. It is worthwhile observing that "solve the equation" means "find the general solution".

15.2 Consider the differential equation $ay''(x) - 2y'(x) + 4y(x) = f(x)$, where $a \geq 0$.

(a) Take $f(x) = 0$. Are there non-trivial solutions that vanish as $x \to +\infty$?
(b) Take $f(x) = 2e^x$ and $a = 1$. Solve the differential equation.

Answer. (a) The characteristic equation associated with the homogeneous equation is $a\lambda^2 - 2\lambda + 4 = 0$, and has discriminant $4(1 - 4a)$.

If $a = 0$, then $\lambda = 2$ and all solutions are of the form $y(x) = ce^{2x}$ with $c \in \mathbb{R}$. In this case, only when $c = 0$ does the solution $y(x)$ vanish as $x \to \infty$.

Suppose next that $0 < a < 1/4$. Then the roots are $(1 \pm \sqrt{1 - 4a})/a$, and are both positive. Then

$$y(x) = c_1 e^{\frac{1+\sqrt{1-4a}}{a}x} + c_2 e^{\frac{1-\sqrt{1-4a}}{a}x}$$

with $c_1, c_2 \in \mathbb{R}$. Only when $c_1 = c_2 = 0$ does the solution vanish as $x \to \infty$.

Finally suppose $a > 1/4$. Then

$$y(x) = e^{\frac{1}{a}x}\left(c_1 \sin\left(\frac{\sqrt{4a-1}}{a}x\right) + c_2 \cos\left(\frac{\sqrt{4a-1}}{a}x\right)\right)$$

with $c_1, c_2 \in \mathbb{R}$. Again only the trivial solution ($c_1 = c_2 = 0$) vanishes as $x \to \infty$.

In conclusion, there are no non-trivial solutions that vanish as $x \to +\infty$.

(b) The homogeneous equation associated with $y''(x) - 2y'(x) + 4y(x) = 2e^x$ has general solution $y(x) = e^x(c_1 \sin\sqrt{3}x + c_2 \cos\sqrt{3}x)$. By Theorem 15.5, a particular solution of the kind $\overline{y}(x) = Ae^x$ exists. Taking derivatives, this is a solution provided that the equation $Ae^x - 2Ae^x + 4Ae^x = 2e^x$ is satisfied, which happens for $A = 2/3$. Therefore, the general solution of the equation is

$$y(x) = e^x \left(c_1 \sin \sqrt{3}x + c_2 \cos \sqrt{3}x \right) + \frac{2}{3}e^x.$$

Part (a) requires first to solve a second order characteristic equation, and then a careful analysis of the behaviour at $+\infty$ of the fundamental solutions $e^{\lambda_j(a)x}$ where $\lambda_1(a) = (1 + \sqrt{1 - 4a})/a$ and $\lambda_2(a) = (1 - \sqrt{1 - 4a})/a$ may well be complex as a varies. This requires basic inspection of complex exponentials. Part (b) is a straightforward application of Theorem 15.5.

15.3 Consider the Cauchy problem

$$\begin{cases} y''(x) + 2y'(x) + ky(x) = 0 \\ y(0) = 1 \\ y'(0) = 0. \end{cases}$$

(a) Does the solution have at 0 a local maximum or minimum, as k varies in $\mathbb{R}$?
(b) Find the solution if $k = 4$.

Answer. (a) The solution exists and is unique in $\mathbb{R}$. Now,

$$0 = y''(0) + 2y'(0) + ky(0) = y''(0) + k \quad \Longrightarrow \quad y''(0) = -k.$$

If $k < 0$, then $y''(0) > 0$ and $y(x)$ has at 0 a local minimum, whereas if $k > 0$, then $y''(0) < 0$ and $y(x)$ has at 0 a local maximum. If $k = 0$, the equation becomes $y''(x) + 2y'(x) = 0$ and its solution is $y(x) = 1$ for every $x \in \mathbb{R}$.

(b) If $k = 4$, the equation is $y''(x) + 2y'(x) + 4y(x) = 0$ and has characteristic equation $\lambda^2 + 2\lambda + 4 = 0$, with solutions $-1 \pm \sqrt{3}i$. The general solution is thus

$$y(x) = e^{-x}(c_1 \sin \sqrt{3}x + c_2 \cos \sqrt{3}x),$$

so that if $y(0) = 1$, then $c_2 = 1$. A computation gives

$$y'(x) = -e^{-x}(c_1 \sin \sqrt{3}x + \cos \sqrt{3}x) + e^{-x}(c_1\sqrt{3}\cos \sqrt{3}x - \sqrt{3}\sin \sqrt{3}x),$$

whence $y'(0) = 0$ implies $c_1 = \sqrt{3}/3$. Hence the solution is

$$y(x) = e^{-x}\left(\frac{\sqrt{3}}{3} \sin \sqrt{3}x + \cos \sqrt{3}x\right).$$

In part (a) it is known that the origin is a critical point. Some information on $y''(0)$ should assess the nature of it, and it does, except when $k = 0$ in which case

y is clearly constant. Part (b) is a completely standard method for solving a Cauchy problem: first find the general solution and then tune the constants to fit the initial conditions. This is done by imposing them, and then solving for c_1 and c_2.

15.4 (a) Write a homogeneous, constant coefficient linear differential equation of minimal order such that the functions $y_1(x) = e^x$, $y_2(x) = e^{-x}$ and $y_3(x) = \sin x$ solve it.

(b) Write all the solutions of the equation found in (a) that diverge as $x \to +\infty$.

Answer. (a) The characteristic equation must have among its roots $1, -1, i$. The lowest order characteristic equation with this property is $(\lambda - 1)(\lambda + 1)(\lambda^2 + 1) = 0$, that is $\lambda^4 - 1 = 0$. The equation to be found is then $y^{(4)} - y = 0$.

(b) The general solution of $y^{(4)} - y = 0$ is

$$y(x) = c_1 e^x + c_2 e^{-x} + c_3 \sin x + c_4 \cos x,$$

with $c_1, c_2, c_3, c_4 \in \mathbb{R}$. The diverging solutions as $x \to +\infty$ are those for which $c_1 \in \mathbb{R} \backslash \{0\}$, for any choice of $c_2, c_3, c_4 \in \mathbb{R}$.

This is a basic exercise in which the problem addressed is somehow reversed: given some solutions, find an equation. This is of course achieved using the characteristic equation. The request on minimal order is reasonable: if $P(\lambda) = 0$, then of course $Q(\lambda)P(\lambda) = 0$ for any polynomial Q, so that the order can be arbitrarily increased. Question (b) is a standard analysis of exponentials, sines and cosines, unbounded and bounded functions.

15.5 Consider the differential equation $y'''(x) + y'(x)(1 - a^2) - ay(x) = bx^2$ and put

$$V = \left\{ y : \mathbb{R} \to \mathbb{R} : y \text{ solves the differential equation and } \lim_{x \to +\infty} y(x) = 0 \right\}.$$

$$W = \left\{ y : \mathbb{R} \to \mathbb{R} : y \text{ solves the differential equation and is bounded} \right\}.$$

(a) Put $b = 0$. Compute the dimension of V as a ranges in $\mathbb{R}$.

(b) Put $b = 0$. Compute the dimension of W as a ranges in $\mathbb{R}$.

(c) Put $a = b = 1$. Compute the general solution.

Answer. (a) The characteristic equation is

$$\lambda^3 + \lambda(1 - a^2) - a = (\lambda - a)(\lambda^2 + a\lambda + 1) = 0.$$

Evidently, $\lambda = a$ is a root, and the corresponding solution $y_1(x) = e^{ax}$ tends to 0 as $x \to +\infty$ only if $a < 0$. The solutions associated with the roots of $\lambda^2 + a\lambda + 1$ are, in general, complex. More precisely, if $z \in \mathbb{C}$ is a complex root, then $a = -2\Re(z)$. Therefore the real part of the roots $\{z, \bar{z}\}$ is negative if and only if $a > 0$. If the roots of $\lambda^2 + a\lambda + 1$ are real, hence if $z_1, z_2 \in \mathbb{R}$, then $z_1 z_2 = 1$ and $z_1 + z_2 = -a$, so that they have the same sign and are negative if and only if $a > 0$. Therefore,

$$\dim V = \begin{cases} 2 & a > 0 \\ 0 & a = 0 \\ 1 & a < 0. \end{cases}$$

(b) The solution $y_1(x) = e^{ax}$ is bounded if and only if $a = 0$, whereas the solutions associated to complex roots of $\lambda^2 + a\lambda + 1$ are bounded if and only if their real part is 0. Thus, in the case of complex roots, there are bounded solutions if and only if $a = 0$. In the case of real roots z_1 and z_2, the equality $z_1 z_2 = 1$ implies $z_1 \neq 0$ and $z_2 \neq 0$, so that the corresponding solutions are not bounded. Finally, observe that if $a = 0$, then the roots of $\lambda^2 + a\lambda + 1$ are actually complex. Therefore

$$\dim W = \begin{cases} 0 & a \neq 0 \\ 3 & a = 0. \end{cases}$$

(c) The characteristic equation of the homogeneous part of $y''' - y = x^2$ is

$$\lambda^3 - 1 = (\lambda - 1)(\lambda^2 + \lambda + 1) = 0$$

with roots $1, (-1 \pm i\sqrt{3})/2$. Therefore, the general solution of the homogeneous equation is

$$y_0(x) = c_1 e^x + c_2 e^{-\frac{1}{2}x} \cos \frac{\sqrt{3}}{2}x + c_3 e^{-\frac{1}{2}x} \sin \frac{\sqrt{3}}{2}x$$

with $c_1, c_2, c_3 \in \mathbb{R}$. The final step is the determination of a particular solution $\overline{y}$. Since the source term is a second degree polynomial, Theorem 15.5 indicates to put $\overline{y}(x) = Ax^2 + Bx + C$. Substituting $\overline{y}$ and its derivatives in the equation yields $-Ax^2 - Bx - C = x^2$, whence $A = -1$, $B = C = 0$. Therefore $\overline{y}(x) = -x^2$ is a particular solution. The general solution of the nonhomogeneous equation is thus

$$y(x) = c_1 e^x + c_2 e^{-\frac{1}{2}x} \cos \frac{\sqrt{3}}{2}x + c_3 e^{-\frac{1}{2}x} \sin \frac{\sqrt{3}}{2}x - x^2.$$

Parts (a) and (b) of this exercise ask to find the dimensions of two vector subspaces of the vector space Y_0 of solutions of a homogeneous (constant coefficient, linear) differential equation. The fact that V and W are indeed vector spaces is more or less trivial, because linear combinations of elements of V are clearly in V, and the zero function is in V. Analogous considerations hold for W. The question amounts to a discussion of the properties of the fundamental solutions and hence, ultimately, on the roots of the characteristic equation. Decay at $+\infty$ happens only for negative real part of the root, be it real or complex, whereas boundedness means that the real part is zero. Question (c) is about a standard nonhomogeneous equation.

15.6 Determine the general solution of the equation $y''(x) + 4y(x) = 1/\cos 2x$.

Answer. The characteristic equation associated to $y''(x) + 4y(x) = 0$ is $\lambda^2 + 4 = 0$, with roots $\lambda = \pm 2i$. Therefore the general solution of the homogeneous equation is

$$y(x) = c_1 \cos 2x + c_2 \sin 2x,$$

with $c_1, c_2 \in \mathbb{R}$. Next, the particular solution will be computed with the method of variations of constants. The candidate solution has the form $\bar{y}(x) = c_1(x)y_1(x) + c_2 y_2(x)$, where $y_1(x) = \cos 2x$, $y_2(x) = \sin 2x$ and $c_1(x)$ and $c_2(x)$ are differentiable functions that must solve the system (15.14), that is

$$\begin{cases} c_1'(x)y_1(x) + c_2'(x)y_2(x) = 0 \\ c_1'(x)y_1'(x) + c_2'(x)y_2'(x) = (\cos 2x)^{-1}. \end{cases}$$

The first equation yields $c_1'(x) = -c_2'(x) \sin 2x / \cos 2x$, and plugging it in the second

$$2c_2'(x)\left(\frac{\sin^2 2x}{\cos 2x} + \cos 2x\right) = \frac{1}{\cos 2x},$$

namely $2c_2'(x) = 1$, whence $c_2(x) = x/2$. Finally,

$$c_1'(x) = -\frac{1}{2}\frac{\sin 2x}{\cos 2x} \qquad \Longrightarrow \qquad c_1(x) = \frac{1}{4}\log(|\cos 2x|).$$

Hence, a particular solution is $\bar{y}(x) = \frac{1}{4}\log(|\cos x|)\cos 2x + \frac{x}{2}\sin 2x$, so that

$$y(x) = c_1 \cos 2x + c_2 \sin 2x + \frac{1}{4}\log(|\cos x|)\cos 2x + \frac{x}{2}\sin 2x$$

is the general solution, with $c_1, c_2 \in \mathbb{R}$.

This exercise is a standard application of the method of variations of constants to a second order equation.

15.7 Consider the differential equation $y'''(x) + y''(x) = 4$ and the set

$$V = \left\{y : \mathbb{R} \to \mathbb{R} : y \text{ solves the differential equation and } \lim_{x \to 0} y(x) = 0 \text{ with order } 2\right\}.$$

(a) Is V a vector space?
(b) Describe V.

Answer. (a) V is not a vector space because $y = 0$ is not in V, since the equation is nonhomogeneous.

(b) A solution y belongs to V if and only if

$$\begin{cases} y'''(x) + y''(x) = 4 \\ y'(0) = y(0) = 0 \\ y''(0) \neq 0. \end{cases}$$

The characteristic equation associated with $y''' + y'' = 0$ is $\lambda^3 + \lambda^2 = 0$, which has roots -1 with multiplicity 1 and 0 with multiplicity 2. Therefore the general solution of the homogeneous equation is

$$y(x) = c_1 e^{-x} + c_2 + c_3 x$$

with $c_1, c_2, c_3 \in \mathbb{R}$. A particular solution can be found with the aid of Theorem 15.5, taking $\overline{y}(x) = Ax^2$ with $A \in \mathbb{R}$. Now,

$$\overline{y}'''(x) + \overline{y}''(x) = 2A = 4$$

hence $A = 2$. Therefore $\overline{y}(x) = 2x^2$ and the general solution of the nonhomogeneous equation is

$$y(x) = c_1 e^{-x} + c_2 + c_3 x + 2x^2$$

with $c_1, c_2, c_3 \in \mathbb{R}$. In order to impose the required conditions, the first two derivatives must be computed, namely

$$y'(x) = -c_1 e^{-x} + c_3 + 4x, \qquad y''(x) = c_1 e^{-x} + 4.$$

The conditions are then

$$y(0) = c_1 + c_2 = 0$$
$$y'(0) = -c_1 + c_3 = 0$$
$$y''(0) = c_1 + 4 \neq 0.$$

In conclusion

$$V = \left\{ y : \mathbb{R} \to \mathbb{R} : y(x) = c_1(e^{-x} - 1 + x) + 2x^2, c_1 \in \mathbb{R} \backslash \{-4\} \right\}.$$

The point of this exercise is to solve a differential problem that is not a Cauchy problem, but almost, in the sense that first one must find the general solution of a nonhomogeneous linear third order equation with constant coefficients, and then impose some constraints on the free coefficients.

15.8 Consider the differential equation

$$y''(x) + 2y'(x) + y(x) = axe^{-x} + b \log x,$$

with $a, b \in \mathbb{R}$.

(a) Put $a = 1$ and $b = 0$. Find the general solution of the differential equation.

(b) Put $a = 1$ and $b = 0$. Are there solutions $y = y(x)$ of the differential equation such that the improper integral $\int_1^{+\infty} e^{-x} y(x)^{-1} \, dx$ converges?

(c) Put $a = 0$ and $b = 1$. Write the fourth order Taylor polynomial centered at 1 of the solution $y = y(x)$ of the differential equation such that $y(1) = y'(1) = 0$.

Answer. (a) Since the characteristic equation is $(\lambda + 1)^2 = 0$, the general solution of the associated homogeneous equation is given by $y(x) = c_1 e^{-x} + c_2 x e^{-x}$, with $c_1, c_2 \in \mathbb{R}$. In order to find a particular solution, observe that the source term is $e^{\alpha x}(P(x)\cos(\beta x) + Q(x)\sin(\beta x))$ with $\alpha = -1$, $\beta = 0$ and $P(x) = x$, so that Theorem 15.5 applies. Now, $\alpha = -1$ is a root with multiplicity 2 of the characteristic equation. Hence a particular solution will have the form

$$\overline{y}(x) = e^{-x} x^2 (Ax + B) = e^{-x}(Ax^3 + Bx^2)$$

where A an B are real numbers to be found. Now, $\overline{y}$ is a solution provided that for every $x \in \mathbb{R}$

$$\overline{y}''(x) + 2\overline{y}'(x) + \overline{y}(x) = x e^{-x}.$$

Since

$$\overline{y}'(x) = -e^{-x}(Ax^3 + Bx^2) + e^{-x}(3Ax^2 + 2Bx) = e^{-x}(-Ax^3 + x^2(3A - B) + 2Bx)$$

$$\overline{y}''(x) = -e^{-x}(-Ax^3 + x^2(3A - B) + 2Bx) + e^{-x}(-3Ax^2 + 2x(3A - B) + 2B)$$

$$= e^{-x}(Ax^3 + x^2(B - 6A) + x(6A - 4B) + 2B)$$

the equality $\overline{y}''(x) + 2\overline{y}'(x) + \overline{y}(x) = x e^{-x}$ implies that

$$x e^{-x} = e^{-x}(6Ax + 2B)$$

for every $x \in \mathbb{R}$. It follows that $A = 1/6$ and $B = 0$. Therefore, a particular solution is $\overline{y}(x) = x^3 e^{-x}/6$ and the general solution of the nonhomogeneous equation is

$$y(x) = c_1 e^{-x} + c_2 x e^{-x} + \frac{1}{6} x^3 e^{-x}.$$

(b) By what has been proved in (a),

$$\frac{e^{-x}}{y(x)} = \frac{1}{c_1 + c_2 x + \frac{1}{6} x^3} =: \frac{1}{q(x)}.$$

Now, if $c_2 > 0$, then q is strictly increasing in $[1, +\infty)$ and if $q(1) = c_1 + c_2 + 1/6 > 0$, then the function to be integrated is continuous in $[1, +\infty)$. Furthermore, $q(x) \to +\infty$ with order $3 > 1$ as $x \to +\infty$ and hence the improper integral converges. Thus, if $c_2 > 0$ and $c_1 + c_2 + 1/6 > 0$ the corresponding solutions $y = y(x)$ are such that the desired improper integral converges.

(c) Observe that now the source term b is in $C^0((0, +\infty))$ and that the given Cauchy problem has one and only one solution in $(0, +\infty)$. The required Taylor polynomial exists because the solution, which by definition is twice differentiable in $(0, +\infty)$, is actually (at least) four times differentiable in $(0, +\infty)$, as the following bootstrap argument shows. Indeed, $y'' = -2y' - y + \log x$, and since the right hand side is differentiable in $(0, +\infty)$, such is also y''. Thus, $y''' = -2y'' - y' + x^{-1}$ and since the right hand side is differentiable in $(0, +\infty)$, such is also y''', and $y^{(4)} = -2y^{(3)} - y'' - x^{-2}$. Thus $y^{(4)} \in C^0((0, +\infty))$ because the right hand side is such, and in conclusion $y \in C^4((0, +\infty))$. Finally, since $y(1) = y'(1) = y''(1) = 0$ while $y'''(1) = 1$ and $y^{(4)}(1) = -3$, it follows that

$$P_4(x) = \frac{1}{6}(x - 1)^3 - \frac{1}{8}(x - 1)^4.$$

Part (a) displays the standard procedure for the determination of the general solution of a nonhomogeneous linear differential equation with constant coefficients. Part (b) is an aside tour and simply requires to find some integrable cases, which is not hard since the function to be integrated is the reciprocal of a third degree polynomial. Part (c) is the interesting one, and shows an example of a typical bootstrap argument that is often used to prove regularity of solutions. In the case at hand, it is clear that the argument could be applied indefinitely, thereby proving that the solution is in fact of class C^∞. In this way it is possible to derive the value of as many derivatives at the point 1 as one wishes, hence to write Taylor polynomials.

15.9 Consider the Cauchy problem

$$\begin{cases} y''(x) + y(x) = \log x \\ y(\pi) = y'(\pi) = 0. \end{cases}$$

(a) In which intervals containing π does there exist a solution?
(b) Find an explicit analytic expression for the solution.

Answer. (a) Since the source term of the equation is defined and continuous only in $(0, +\infty)$, the problem has a unique solution in $(0, +\infty)$.

(b) The characteristic equation of the homogeneous equation is $\lambda^2 + 1 = 0$. The general integral of the homogeneous equation is $y(x) = c_1 \sin x + c_2 \cos x$ with $c_1, c_2 \in \mathbb{R}$. The method of variations of constants applies, and there is a particular solution of the form $\bar{y}(x) = c_1(x) \sin x + c_2(x) \cos x$, where the unknown functions c_1 and c_2 must satisfy

$$\begin{cases} c_1'(x) \sin x + c_2'(x) \cos x = 0 \\ c_1'(x) \cos x - c_2'(x) \sin x = \log x. \end{cases}$$

Next, multiplying the first by $\sin x$, the second by $\cos x$ and summing, and, similarly multiplying the first by $\cos x$, the second by $\sin x$ and subtracting, it follows

$$\begin{cases} c_1'(x) = \cos x \log x \\ c_2'(x) = -\sin x \log x. \end{cases}$$

It is then possible to choose

$$c_1(x) = \int_\pi^x \cos t \log t \, dt, \qquad c_2(x) = -\int_\pi^x \sin t \log t \, dt$$

which leads to the particular solution $\bar{y}(x) = \sin x \int_\pi^x \cos t \log t \, dt - \cos x \int_\pi^x \sin t \log t \, dt$.

The general solution is thus

$$y(x) = c_1 \sin x + c_2 \cos x + \sin x \int_\pi^x \cos t \log t \, dt - \cos x \int_\pi^x \sin t \log t \, dt.$$

From $y(\pi) = 0$ it follows $c_2 = 0$. Hence, differentiating

$$y'(x) = c_1 \cos x + \cos x \int_\pi^x \cos t \log t \, dt + \sin x \int_\pi^x \sin t \log t \, dt.$$

Finally, $y'(\pi) = 0$ implies $c_1 = 0$. The solution of the Cauchy problem is therefore

$$y(x) = \int_\pi^x \sin(x - t) \log t \, dt.$$

Part (b) is non-trivial. A careful look reveals that one can actually substitute throughout $\log x$ with any differentiable function φ defined in an open neighborhood of π and the very same argument yields the analogous answer $y(x) = \int_\pi^x \sin(x - t)\varphi(t) \, dt$. Notice that the variation of constants gives integral formulas for $c_1(x)$ and $c_2(x)$.

15.10 Consider the differential equation $y''(x) + 2y'(x) + y(x) = ke^x$, with $k \in \mathbb{R}$.

(a) Find the solutions, as k ranges in $\mathbb{R}$.
(b) Put $k = 0$ and define

$$V = \left\{ y : \mathbb{R} \to \mathbb{R} : y \text{ solves the equation and } \int_0^{+\infty} y(x) \, dx = 0 \right\}.$$

Prove that V is a vector space and find a basis for it.
(c) Put $k = 1$. Take the solution y of the differential equation that satisfies $y(0) = y'(0) = 0$. Compute, if existing,

$$\lim_{x \to 0^+} \frac{y(x) - \frac{x^2}{2}}{x - \sin x}.$$

Answer. (a) The characteristic equation has the root -1 with multiplicity 2, that give rise to the basis functions $y_1(x) = e^{-x}$ and $y_2(x) = xe^{-x}$. The general integral is thus $y(x) = c_1 e^{-x} + c_2 x e^{-x}$, with $c_1, c_2 \in \mathbb{R}$.

Assume now $k \neq 0$. The source term is of the type $e^{ax}(P(x)\cos bx + Q(x)\sin bx)$, with $a = 1$, $b = 0$, $P(x) = k$ and $Q(x) = 0$. Since $\lambda = 1$ is not a root of the characteristic equation, a particular solution can be taken of the form $\overline{y}(x) = Ae^x$, where $A \in \mathbb{R}$ is to be found. In this case, it must be

$$\overline{y}''(x) + 2\overline{y}'(x) + \overline{y}(x) = Ae^x + 2Ae^x + Ae^x = ke^x$$

for every $x \in \mathbb{R}$, which happens only if $A = k/4$. Therefore $\overline{y}(x) = \frac{k}{4}e^x$ and the general solution of the nonhomogeneous equation (hence $k \neq 0$) is

$$y(x) = c_1 e^{-x} + c_2 x e^{-x} + \frac{k}{4} e^x$$

with $c_1, c_2 \in \mathbb{R}$.

(b) If $k = 0$, the differential equation is homogeneous and its solutions form a vector space Y_0. Clearly the null function is in V, while if $\int_0^{+\infty} y_1(x)\,dx = \int_0^{+\infty} y_2(x)\,dx = 0$, then also $\int_0^{+\infty}(\lambda_1 y_1 + \lambda_2 y_2)\,dx = 0$, by linearity of the integral. Hence, V is a vector subspace of Y_0. When $k = 0$, the general integral computed in (a) is $y(x) = c_1 e^{-x} + c_2 x e^{-x}$. The integral condition then reads

$$\begin{aligned} 0 &= \int_0^{+\infty} y(x)\,dx \\ &= \lim_{y \to +\infty} \int_0^y (c_1 e^{-x} + c_2 x e^{-x})\,dx \\ &= \lim_{y \to +\infty} [(c_1 + c_2)(-e^{-y} + 1) + c_2(-ye^{-y})] \\ &= c_1 + c_2 \end{aligned}$$

namely $c_1 = -c_2$. Hence the elements of V are all of the form

$$y(x) = c_1 e^{-x}(1 - x)$$

which entails that V is one-dimensional, with basis $\{e^{-x}(1 - x)\}$.

(c) By what was shown in (a), $y \in C^\infty(\mathbb{R})$, so that a third order McLaurin expansion of y can be considered. From the differential equation it follows

$$y''(x) = e^x - 2y'(x) - y(x), \qquad y'''(x) = e^x - 2y''(x) - y'(x)$$

and from the initial conditions it follows $y''(0) = 1$ and $y'''(0) = -1$. Therefore, the third order McLaurin expansion of y is

$$y(x) = \frac{x^2}{2} - \frac{x^3}{6} + x^3 w(x)$$

with $w(x) \to 0$ as $x \to 0$. Next, from the third order McLaurin expansion of the sine function, namely, $\sin x = x - \frac{x^3}{3!} + x^3 w_1(x)$, where $w_1(x) \to 0$ as $x \to 0$, it follows

$$\lim_{x \to 0} \frac{y(x) - \frac{x^2}{2}}{x - \sin x} = \lim_{x \to 0} \frac{x^3(-\frac{1}{6} + w(x))}{x^3(\frac{1}{6} - w_1(x))} = -1.$$

This exercise requires first a standard computation of the general integral, but then only the homogeneous case is used in part (b). The particularly simple form of the basis of Y_0 allows to infer the explicit conditions that single out the elements of V, which is one-dimensional. Finally, in part (c) it is quite clear that McLaurin expansions are the natural tool, and actually easy to derive.

15.11 Consider the Cauchy problem

$$\begin{cases} y''(x) + 2y'(x) - 3y(x) = f(x) \\ y(0) = y'(0) = 0 \end{cases}$$

with $f : \mathbb{R} \to \mathbb{R}$ a continuous function.

(a) Find the solution, specifying its domain.
(b) Suppose that $\lim_{x \to +\infty} f(x) = 1$ and $f(x) \geq 0$ for every $x \in \mathbb{R}$ and compute $\lim_{x \to +\infty} y(x)$ for the solution $y(x)$ of the Cauchy problem.

Answer. (a) By Theorem 15.6, a solution exists and is unique in $\mathbb{R}$. The characteristic equation of the homogeneous equation $y''(x) + 2y'(x) - 3y(x) = 0$ is $\lambda^2 + 2\lambda - 3 = 0$, and has the real roots $-3, 1$. Therefore, the general solution is $y(x) = c_1 e^x + c_2 e^{-3x}$, with $c_1, c_2 \in \mathbb{R}$. The variation of constants method for the determination of a particular solution applies. Hence, the differentiable functions $c_1(x), c_2(x)$ are such that $\bar{y}(x) - c_1(x)e^x + c_2(x)e^{-3x}$ is a particular solution provided that they satisfy

$$\begin{cases} c_1'(x)e^x + c_2'(x)e^{-3x} = 0 \\ c_1'(x)e^x - 3c_2'(x)e^{-3x} = f(x) \end{cases}$$

or, equivalently

$$\begin{cases} c_1'(x) = \frac{1}{4} f(x) e^{-x} \\ c_2'(x) = -\frac{1}{4} f(x) e^{3x}. \end{cases}$$

It is then possible to select

$$\begin{cases} c_1(x) = \frac{1}{4} \int_0^x f(t)e^{-t}\, dt \\ c_2(x) = -\frac{1}{4} \int_0^x f(t)e^{3t}\, dt \end{cases}$$

thereby obtaining the general solution in the form

$$y(x) = c_1 e^x + c_2 e^{-3x} + \frac{1}{4} e^x \int_0^x f(t)e^{-t}\, dt - \frac{1}{4} e^{-3x} \int_0^x f(t)e^{3t}\, dt$$

with $c_1, c_2 \in \mathbb{R}$. Imposing the initial conditions gives $c_1 = c_2 = 0$, and hence the solution of the Cauchy problem is

$$y(x) = \frac{1}{4} e^x \int_0^x f(t)e^{-t}\, dtt - \frac{1}{4} e^{-3x} \int_0^x f(t)e^{3t}\, dt.$$

(b) Since $\lim_{t \to +\infty} f(t)e^{-t} = 0$ with an order bigger than any real power, the improper integral $\int_0^{+\infty} f(t)e^{-t}\, dt$ converges and is non-negative. Hence

$$\lim_{x \to +\infty} \frac{1}{4} \int_0^x f(t)e^{-t}\, dt = \ell > 0$$

As for the second integral in the formula for $y(x)$, observe that $f(t)e^{3t} \to +\infty$ as $t \to +\infty$, so that the improper integral $\int_0^{+\infty} f(t)e^{3t}\, dt$ diverges to $+\infty$. Write

$$h(x) = \int_0^x f(t)e^{3t}\, dt,$$

so that the behaviour of the ratio $h(x)/e^{3x}$ must be understood, because it takes on an indeterminate form as $x \to +\infty$. Since f is continuous, $h \in C^1(\mathbb{R})$ by the fundamental theorem of calculus and the ratio of derivatives is

$$\frac{\frac{d}{dx} h(x)}{\frac{d}{dx} e^{3x}} = \frac{f(x)e^{3x}}{3e^{3x}} = \frac{f(x)}{3} \to \frac{1}{3}$$

as $x \to +\infty$. Applying the de l'Hôpital theorem

$$\lim_{x \to +\infty} \frac{1}{4} e^{-3x} \int_0^x f(t)e^{3t}\, dt = \frac{1}{12}.$$

Therefore

$$\lim_{x \to +\infty} y(x) = +\infty.$$

Given the generality of the source term f, only a general solution involving integral formulae can be expected, as the method of variation of constants indicates. Once the initial conditions are plugged in, the solution is in fact a difference of two integral functions, one involving f and a decaying exponential, the other involving f and a diverging exponential. Clearly, under the assumptions on f that it is positive and with limit 1 at infinity, one of them converges and the other does not.

15.12 Consider the differential equation $y''(x) + 2y'(x) + a = \int_0^{bx} e^{-t^2} \, dt$.

(a) For which values of the parameters $a, b \in \mathbb{R}$ is the set of solutions a vector space?
(b) Put $a = 1$ and $b = 0$. Find the general solution of the equation.
(c) Put $a = b = 1$. Are there solutions that are convex and increasing in a neighborhood of 0?
(d) Put $a = 0$ e $b = -1$ and take the solution for which $y(0) = y'(0) = 0$. Draw its graph in a neighborhood of 0.

Answer. (a) The set of solutions is a vector space only if the equation is homogeneous, that is, if $\int_0^{bx} e^{-t^2} \, dt = a$ for every $x \in \mathbb{R}$, that is, if the function $g(x) = \int_0^{bx} e^{-t^2} \, dt$ is constant. Since $g \in C^1(\mathbb{R})$, by the fundamental theorem of calculus $g'(x) = be^{-b^2x^2}$ and hence g' is identically 0 if and only if $b = 0$. Hence, the set of solutions is a vector space if and only if $a = 0$ and $b = 0$.

(b) The equation is $y'' + 2y' = -1$, and its characteristic equation is $\lambda^2 + 2\lambda = 0$, with roots 0 and -2. The general solution of the homogeneous equation is therefore $y_0(x) = c_1 + c_2 e^{-2x}$ with $c_1, c_2 \in \mathbb{R}$. Consider next the nonhomogeneous equation. Since the source term is a constant, a particular solution will be of the form $\overline{y}(x) = Ax + B$. Hence $\overline{y}'(x) = A$ and $\overline{y}''(x) = 0$, whence $2A = -1$. The general solution is thus

$$y(x) = c_1 + c_2 e^{2x} - \frac{x}{2}.$$

(c) In this case the equation is $y''(x) + 2y'(x) = \int_0^x e^{-t^2} \, dt - 1$, and $y''(0) + 2y'(0) = -1$. Observe that $y \in C^2(\mathbb{R})$, actually $y \subset C^\infty(\mathbb{R})$, because

$$y'' = -2y' + \int_0^x e^{-t^2} \, dt - 1$$

and the right hand side is continuous in $\mathbb{R}$. Since $y''(0) + 2y'(0) = -1 < 0$ it must be $y''(0) < 0$ or $y'(0) < 0$ and by permanence of sign there is a neighborhood I of 0 such that $y''(x) < 0$ or $y'(x) < 0$ for every $x \in I$. Hence the solution is either concave or strictly decreasing in a neighborhood of 0, and the answer is no.

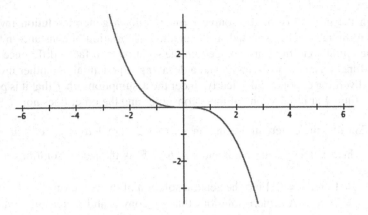

Fig. 15.1 Local graph of the solution in a neighborhood of 0, Exercise 15.12

(d) The Cauchy problem

$$\begin{cases} y''(x) + 2y'(x) = \int_0^{-x} e^{-t^2}\, dt \\ y(0) = y'(0) = 0 \end{cases}$$

has a unique solution in $\mathbb{R}$. The differential equation in normal form gives that $y''(0) = -2y'(0) = 0$ and, since the right hand side is differentiable in $\mathbb{R}$, it implies that such is also y'' and $y''' = -2y'' - e^{-x^2}$. Therefore $y'''(0) = -1$ and $y''' \in C^0(\mathbb{R})$. By permanence of sign there is a neighborhood I of 0 such that $y''' < 0$ for every $x \in I$. Hence y'' is strictly decreasing in I and $y''(0) = 0$ implies that $y''(x) > 0$ if $x \in I$ and $x < 0$, whereas $y''(x) < 0$ if $x \in I$ and $x > 0$. Thus y' it is strictly increasing in $I \cap (-\infty, 0)$ and strictly decreasing in $I \cap (0, +\infty)$. Finally, since $y'(0) = 0$ it follows that $y'(x) < 0$ for $x \in I - \{0\}$ and y is strictly decreasing in I. The local graph in a neighborhood of 0 is therefore as in Fig. 15.1, where it is actually depicted the graph of the third order McLaurin polynomial of the solution, namely $P_3(x) = -x^3/6$.

The interesting parts of exercise, (c) and (d), are based on the idea that local information on the solution can be inferred from the (second order) differential equation together with some knowledge at the origin, be it available from the differential equation, as in part (b), or as a priori knowledge, as in part (d). In part (c) the information is $y''(0) + 2y'(0) = -1$, which, carefully analized, says that near the origin the solution cannot be simultaneously convex and increasing, for otherwise the first and second derivatives would be non negative. Part (d) is about unraveling again the local information, which is encoded in the third order McLaurin polynomial because the lower order ones vanish.

15.5   Problems on Constant Coefficient Differential Equations

15.13 Find all solutions of $y'''(x) - 2y''(x) + y(x) = 0$.

15.14 Determine the general solution of the following differential equations:

(a) $9y''(x) - 9y'(x) + 2y(x) = 18xe^{\frac{x}{3}}$.
(b) $y'''(x) - y''(x) + 2y'(x) - 2y(x) = e^{2x}$.
(c) $y'''(x) + 3y''(x) - 2y'(x) - 2y(x) = 2xe^x + 1$.

15.15 Consider the differential equation $ky'''(x) + y''(x) - 2y'(x) - 1 = 0$.

(a) Determine the general solution as k ranges in $\mathbb{R}$.
(b) Put $k = 0$. Find all the solutions such that $y(0) = y(1)$.

15.16 Consider the differential equation $y^{(4)}(x) + ky(x) = 0$.

(a) For which $k \in \mathbb{R}$ is $y(x) = e^{3x}$ a solution?
(b) Put $k = 1$. Find all solutions.

15.17 Consider the problem

$$\begin{cases} y^{(5)}(x) + y^{(1)}(x) + 1 = \sin \frac{\sqrt{2}}{2}x \\ y(0) = 0. \end{cases}$$

(a) Find all solutions.
(b) Is the set of solutions a vector space?

15.18 Find the general solution of $y^{(4)}(x) + 3y^{(2)}(x) = x$.

15.19 Find all solutions of the problem

$$\begin{cases} y'''(x) - 2y''(x) + y'(x) = 3x^2 + x \\ y(0) = 0, \ y'(0) = 1. \end{cases}$$

15.20 Suppose that $y_1(x) = x$ and $y_2(x) = xe^x$ are solutions of a homogeneous, constant coefficient differential equation.

(a) Can this differential equation be of order 2?
(b) Write the minimal order homogeneous, constant coefficient differential equation that has y_1 and y_2 among its solutions.

15.21 Consider the differential equation: $y''(x) + y(x) = e^x(2\alpha^2 - 2) + e^{2x}(5\alpha - 5)$.

(a) Find the general solution of the differential equation.
(b) For which $\alpha \in \mathbb{R}$ are there bounded solutions?

15.22 Is there a linear, homogeneous, constant coefficient differential equation of order 2 that has among its solutions $y_1(x) = x$ and $y_2(x) = \sin x$?

15.23 Consider the Cauchy problem

$$\begin{cases} y''(x) + ky'(x) - y(x) = 1 \\ y(0) = y'(0) = 0. \end{cases}$$

(a) Write the third order McLaurin polynomial of the solution.
(b) Compute $\lim\limits_{x \to 0^+} y(x)/(1 - \cos x)$ as k ranges in $\mathbb{R}$.

15.24 Consider the differential equation: $y''(x) - y(x) = |x|$.

(a) Are the solutions three times differentiable in $\mathbb{R}$?
(b) Find all the solutions that vanish at 0.

15.25 Find a linear, constant coefficient, homogeneous differential equation that has $y_1(x) = x$ and $y_2(x) = x^2$ among its solutions.

15.26 Consider the functions $f(x) = xe^x$ and $g(x) = e^x \sin x$, $h(x) = x$.

(a) Find a linear, constant coefficient, homogeneous differential equation with minimal order of which f, g and h are solutions.
(b) Find the general solution of that equation.

15.27 Consider the functions $f(x) = x^2$, $g(x) = x + 1$, $h(x) = e^{2x} \sin x$.

(a) Find a linear, constant coefficient, homogeneous differential equation with minimal order of which f, g and h are solutions.
(b) Find the general solution of that equation.

15.28 Find a linear, constant coefficient, homogeneous differential equation that has $y_1(x) = \sin 2x$, $y_2(x) = 1$ among its solutions.

15.29 Find a linear, constant coefficient, homogeneous differential equation that has $y_1(x) = e^x$, $y_2(x) = x^2 e^x$ among its solutions.

15.30 Find the general solution of the equation $y''(x) + y'(x) = \log x$.

Chapter 16
Miscellaneous

This final chapter is devoted to a set of problems that are not designed with a specific issue in mind but rather require a variety of techniques, and should perhaps be addressed as a final check on the global preparation. The reader will notice that some of the problems that were earlier solved with a possibly laborious technique can be worked out in a more efficient way with more powerful tools, such as derivatives or Taylor expansions, that are usually introduced at a later stage of the theoretical presentation. The problems presented here are not necessarily more complex, but rather involve the whole apparatus that has been developed in the other chapters, and are thus designed to have an overlook of the subject matter. Most of the problems of this chapter were actually assigned in written examinations. Intentionally, there is no ordering according to increasing or decreasing difficulty and no solutions are given.

16.1 Problems

16.1 Let $h : [-1, 2] \to \mathbb{R}$ be the function whose graph is in depicted in Fig. 16.1. Draw the graphs of the functions:

$$h_1(x) = h(-x) \qquad h_2(x) = -h(x) \qquad h_3(x) = h(2x) \qquad h_4(x) = 2h(x)$$
$$h_5(x) = |h(x-3)| \qquad h_6(x) = |h(x) - 1| \qquad h_7(x) = h(1-x).$$

16.2 Find, if possible, a neighborhood of $x_0 = 0$ on which $f(x) = 2\arctan x - x^2$ is invertible.

16.3 For each $k \in \mathbb{R}$ consider the function $f_k(x) = \dfrac{x}{x^3 + x + k}$.

(a) Determine the values of $k \in \mathbb{R}$, if any, for which f_k is defined on $\mathbb{R}$.
(b) Determine the values of $k \in \mathbb{R}$, if any, for which f_k is continuous in $[0, +\infty)$.
(c) Let $k = 1$ and put $F(x) = f_1(|x|)$. Is F bounded on its domain?

© Springer International Publishing Switzerland 2016
M. Baronti et al., *Calculus Problems*, UNITEXT - La Matematica per il 3+2 101,
DOI 10.1007/978-3-319-15428-2_16

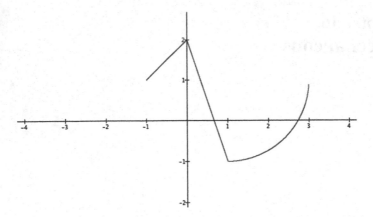

Fig. 16.1 Graph of the function h in Problem 16.1

16.4 A function $f : I \to \mathbb{R}$, where $I \subseteq \mathbb{R}$, is said to be *uniformly continuous* on I if for every $\varepsilon > 0$, there exists $\delta > 0$ for which whenever $x, y \in I$ are such that $|x - y| < \delta$, then $|f(x) - f(y)| < \varepsilon$. Show that $f(x) = x^2$ is not uniformly continuous on $\mathbb{R}$.

16.5 Consider the function $f(x) = x\sqrt{x + e^x}$.

(a) Draw the graph of f and study the global invertibility of f.
(b) Find a neighborhood of $x_0 = 0$ in which f is invertible, call g the corresponding inverse function and compute, if existing, $g'(0)$.

16.6 Let a denote a real parameter and consider the function

$$f(x) = \begin{cases} \dfrac{x}{x+1} & x \le a \\[2ex] -\dfrac{x^3}{|x|^3 + 1} & x > a. \end{cases}$$

(a) Determine all the values of a for which f is invertible on its domain.
(b) Put $a = -3/4$. Find the explicit expression of the inverse of f, specifying its domain.

16.7 Consider the function

$$f(x) = \begin{cases} \dfrac{\sqrt{x+1}}{|x|+1} & x \ge -1 \\[2ex] 0 & x < -1. \end{cases}$$

(a) Where is f differentiable?
(b) Write, if possible, the McLaurin expansion of order 2 of f.

16.8 Take an integer $n \geq 2$, and consider the function $f(x) = x^n n^{-x} - 1$.

(a) Compute $f(0)$ and $\lim_{x \to +\infty} f(x)$.

(b) Use the inequality $(n/\log n) > e$, true for every integer $n \geq 2$, to find the number of zeroes of f in $[0, +\infty)$.

(c) How many positive solutions does the equation $n^x = x^n$ have if n is an integer greater than or equal to 2?

(d) Find explicitly the positive solutions of the equation $2^x = x^2$.

16.9 Consider the Cauchy problem

$$\begin{cases} y'(x) = |y(x) + \sin x| \\ y(0) = 0. \end{cases}$$

(a) Assuming that the solution exists, compute, if existing, $y''(0)$.

(b) Determine explicitly the solution in $(-\pi, \pi)$.

(c) Study the convergence of the improper integral $\int_0^1 \frac{1}{\sqrt[3]{y^2(x)}}\, dx$.

16.10 Consider the function: $f(x) = e^{\frac{4}{x}} - 4e^{\frac{2}{x}} + 4$.

(a) Find the domain I of f and compute the limits of f at the boundary of I.

(b) Study the sign of f and find its zeroes.

(c) Study the monotoneity of f.

(d) Put $A = \{f(x) : x \in I \cap (0, +\infty)\}$. Determine sup A, max A, inf A and min A.

(e) Draw the graph of f.

16.11 Consider the function $f(x) = x^2\sqrt{1 - x} + k \sin(x^2) + h$.

(a) For which $h, k \in \mathbb{R}$ does the function tend to 0 as $x \to 0^+$?

(b) For the values found in (a), determine the order with which $f(x) \to 0$ as $x \to 0^+$.

(c) Put $k = 0$. Determine for which $h \in \mathbb{R}$ the function tends to 0 as $x \to -3^+$ and in these cases compute the limit:

$$\lim_{x \to -3^+} \frac{f(x)}{(x^2 - 9)\log(x + 4)}.$$

16.12 Consider the function

$$g(x) = \begin{cases} \log\left(1 + \dfrac{4x}{x^2 + 1}\right) & x \geq 0 \\ \tan\left(\dfrac{\pi}{4}(5 - 2e^x)\right) & x < 0. \end{cases}$$

(a) Find the image of g.
(b) Denote by h the restriction of g to $I = (-\log 2, 1]$. Determine if h is invertible and, if yes, compute the inverse h^{-1}, specifying its domain.
(c) Compute $\lim\limits_{x \to 0^+} \dfrac{g(-x) + 1}{x^a}$ as a varies in $(0, +\infty)$.

16.13 Consider the function

$$f(x) = e^{\arctan\left(x^2/(1-x)\right)}.$$

(a) Determine the domain of f and compute the limits of f at the boundary of its domain.
(b) Study the monotoneity of f and find, if existing, the local or global extreme points.
(c) Draw the graph of f
(d) Find the maximal interval I containing $x = \pi$ on which f is invertible and find the explicit expression of the inverse function, specifying its domain.

16.14 Consider the differential equation: $y''(x) + 2y'(x) - 15y(x) = f(x)$.

(a) Determine all the solutions if $f(x) = e^x + e^{3x}$.
(b) Put $f(x) = 0$. Establish if the set of solutions that tend to 0 as $x \to -\infty$ is a vector space and, if yes, find a basis thereof.
(c) Put $f(x) = \int_0^x (1 - \cos t) t^{-2} \, dt$. Draw the graph in a neighborhood of $x_0 = 0$ of the solution $y(x)$ that satisfies $y(0) = y'(0) = 1$, specifying if it is three times differentiable at $x_0 = 0$.

16.15 Consider the function: $f(x) = \dfrac{2x - \sqrt{4x^2 - 1}}{|x| - x + 3}$.

(a) Draw the graph of f after an accurate study of: limits at the boundary of the domain, continuity and differentiability.
(b) Establish if f is invertible in its domain and, if so, compute the derivative of the inverse function at $(2 - \sqrt{3})/3$.
(c) Establish for which values of $k \in \mathbb{R}$ the series $\sum\limits_{n=1}^{+\infty} (f(n))^k$ converges.

16.16 Consider the function $f(x) = \left(\log\left(\dfrac{x^2}{x+1}\right)\right)^2$.

(a) Find $\mathrm{Dom}(f)$ and compute the limits of f at the boundary points of $\mathrm{Dom}(f)$.
(b) Determine where f is monotone.
(c) Draw a qualitative graph of f.
(d) Find the largest open interval containing $x_0 = 2$ on which f is invertible and compute the explicit expression of the inverse function specifying its domain.

16.17 Consider the function $f(x) = \log(e^x + x) - \log(e^x - ex)$.

(a) Find $\text{Dom}(f)$ and compute the limits of f at the boundary points of $\text{Dom}(f)$.
(b) Determine where f is monotone and find, if existing, local and global minima and maxima of f.
(c) Draw a qualitative graph of f.
(d) Find the largest open interval I containing $x_0 = 0$ on which f is invertible and, denoting g the inverse of $f|_I$, write the equation of the tangent line to the graph of g at $(f(0), 0)$.

16.18 Consider the series $\displaystyle\sum_{n=0}^{+\infty} \frac{e^{-kn}}{n^2 + 1}$.

(a) For which values of $k \in \mathbb{R}$ does the series converge?
(b) Put $k = 1$. How many terms of the series need to be summed in order to approximate the sum S with an error smaller than 10^{-3}?
(c) For which values of $k \in \mathbb{R}$ is $f(x) = \int_0^x e^{-kt}(t^2 + 1)^{-1}\, dt$ bounded?
(d) Put $k = 1$. Draw the graph of the function inverse to f.

16.19 Establish for which values of $x \in \mathbb{R}$ the series $\displaystyle\sum_{n=1}^{\infty} \frac{(n!)^2}{(2n)!} x^n$ converges.

16.20 Consider the function $f(x) = x - (x + 1)\log x$.

(a) Draw the graph of f after an accurate study of: limits at the boundary of the domain, monotoneity and convexity.
(b) Show that the function $g(x) = |f(x)|$ is invertible in a neighborhood I of $x_0 = e$ and compute, if existing, $(g|_I^{-1})'(-1)$.
(c) Establish if $h(x) = f(x)/(1 + |f(x)|)$ is bounded.

16.21 Consider the function $f(x) = \dfrac{1}{x^2 + 2kx + 1}$.

(a) Put $k = 1/2$. Compute, if existing, $\int_0^1 f(x)\, dx$.
(b) Put $k = 1/2$. Draw a qualitative graph of $F(x) = \int_0^x e^t f(t)\, dt$, after an accurate study of where the function is: monotone, convex, invertible.
(c) For which values of $k \in \mathbb{R}$ is f integrable in $[0, 1]$?

16.22 Consider the function $f(x) = \log\left(2x + \dfrac{1}{x}\right) - \dfrac{1}{6x}$.

(a) Find $\text{Dom}(f)$ and compute the limits of f at the boundary points of $\text{Dom}(f)$.
(b) Determine where f is monotone.
(c) Find, if existing, zeroes, local and global minima and maxima of f.
(d) Find a neighborhood J of $x_0 = 1$ on which f is invertible and, if existing, write the equation of the tangent line to the graph of g at the point $(f(1), 1)$, where g is the inverse of $f|_J$.

16.23 Consider the function $f(x) = \dfrac{1}{x} + \log x - (1 + \log x)^2$.

(a) Find $\mathrm{Dom}(f)$ and compute the limits of f at the boundary points of $\mathrm{Dom}(f)$.
(b) Determine where f is monotone and find, if existing, local and global minima and maxima of f.
(c) Compute, if existing, $\int_1^2 f(x)\,dx$.

16.24 Consider the Cauchy problem:

$$\begin{cases} y'(x) = (\sin x)y(x) + \sin|2x| \\ y(x_0) = -1. \end{cases}$$

(a) What type of equation is the differential equation?
(b) Discuss for which values of x_0, if any, the problem has a unique solution in a neighborhood of x_0.
(c) Determine the domain of the solutions as x_0 varies in $\mathbb{R}$.
(d) Determine, if existing, the solution (or the solutions) when $x_0 = 0$.
(e) Determine the regularity class of each of the solutions found in (d).

16.25 Consider the differential problem:

$$\begin{cases} y(x)y'(x) = e^{-y(x)}(x+1) \\ y(0) = y_0. \end{cases}$$

(a) Study local existence and uniqueness of solutions as y_0 varies in $\mathbb{R}$.
(b) Put $y_0 = 1$. Draw a qualitative graph of the solution in a neighborhood of $x_0 = 0$.
(c) Put $y_0 = 1$. Establish if there exists $\bar{x} > 0$ such that $y(\bar{x}) < 1$, where $y(x)$ is the solution of the problem.
(d) Put $y_0 = 1$. Prove that the solution is invertible in a neighborhood of $x_0 = 0$ and find an explicit expression of the inverse.

16.26 Establish when $\displaystyle\sum_{n=1}^{+\infty}(x-3)^{2n}\left(1+\frac{1}{n}\right)^{n^2}$ converges, as x varies in $\mathbb{R}$.

16.27 Consider the function

$$f(x) = \begin{cases} \dfrac{x}{\sqrt{x-1}} & x \geq 2 \\ ae^x & x < 2 \end{cases}$$

(a) For which values of the parameter a is the function continuous on $\mathbb{R}$?
(b) For the values of a found in (a), is the function bounded in $(-\infty, 5)$?
(c) Put $a = -1$. Is f invertible on $\mathbb{R}$? If yes, find an explicit expression of the inverse.

16.28 Consider the function $f(x) = \dfrac{x}{x + \sqrt{1+x}}$.

(a) Draw a qualitative graph of f.
(b) Determine if f is invertible and compute, if existing, $(f^{-1})'(0)$.
(c) Find an explicit analytic expression, if possible, of the inverse of the restriction of f to the interval $[0, +\infty)$.

16.29 Consider the function

$$f(x) = \begin{cases} \dfrac{e^{-x} - 1}{x} & x < 0 \\ k & x = 0 \\ \dfrac{x \log x}{1 - \cos x} - h & 0 < x \le 1. \end{cases}$$

(a) For which values of $h, k \in \mathbb{R}$ is the function continuous in its domain?
(b) For which values of $h, k \in \mathbb{R}$ is the function differentiable in its domain?
(c) Put $h = k = 0$. Draw a qualitative graph of f.

16.30 Consider the function $f(x) = \log(x + \sqrt{1 + x^2})$.

(a) Draw a qualitative graph of f.
(b) Determine if f is invertible and compute, if existing, $(f^{-1})'(1)$.
(c) Compute $\lim\limits_{x \to 0^+} (f(x))^2 / x \sin x$.

16.31 Suppose that $f : \mathbb{R} \to \mathbb{R}$ is three times differentiable and assume further that $f(0) = f'(0) = f''(0) = 0$ and $f'''(0) = -5$.

(a) Establish if $x_0 = 0$ is a local minimum or maximum.
(b) Establish if $g(x) - |x| f(x)$ is differentiable at $x_0 = 0$.

16.32 Consider the function $f(x) = x \log(2 - e^{-4x}) - 1$.

(a) Find $\mathrm{Dom}(f)$ and compute the limits of f at the boundary points of $\mathrm{Dom}(f)$.
(b) Determine where f is monotone and find, if existing, local and global minima and maxima of f.
(c) Study the equation $f(x) = 0$.

16.33 Consider the function

$$f(x) = \begin{cases} \dfrac{\arctan x^2}{x^2} & x > 0 \\ 1 & x = 0 \\ \dfrac{\sin(|x|^\alpha)}{|x|^\alpha} & x < 0. \end{cases}$$

(a) For which $\alpha > 0$ is f continuous in $\mathbb{R}$?
(b) For which $\alpha > 0$ is f differentiable in $\mathbb{R}$?

16.34 Consider the function

$$f(x) = \sqrt{x + \frac{1}{x + \sqrt{|x|}}}.$$

(a) Find $\mathrm{Dom}(f)$.
(b) Is f differentiable at $x_0 = 0$?
(c) Study the equation $f(x) = k$ as k varies in $\mathbb{R}$.

16.35 Consider the function

$$f(x) = \frac{x\sqrt{|x|}}{1 - \sqrt{2|x|}}.$$

(a) Find $\mathrm{Dom}(f)$.
(b) Determine where f is monotone.
(c) Prove that f is invertible in a neighborhood of $x_0 = 1$, and write, if existing, the
equation of the tangent line to the graph of the inverse at the point $(f(1), 1)$.

16.36 Consider the function

$$f(x) = \begin{cases} \dfrac{e^x - \cos\sqrt{2x}}{x} & x > 0 \\ 2 + \alpha x & x \le 0. \end{cases}$$

(a) For which $\alpha \in \mathbb{R}$ is f continuous in $\mathbb{R}$?
(b) For which $\alpha \in \mathbb{R}$ is f differentiable in $\mathbb{R}$?
(c) Find an interval $I \subset (0, +\infty)$ such that f is invertible on I.

16.37 Establish the order with which $f(x) = \cos x - e^{\sin(x^\alpha)}$ tends to 0 as $x \to 0^+$,
as α varies in $(0, +\infty)$.

16.38 Consider the function

$$f(x) = \frac{x}{\sqrt{x+1}} - \frac{\sqrt{x+1}}{x}.$$

(a) Find $\mathrm{Dom}(f)$.
(b) Determine where f is monotone.
(c) Find a neighborhood of $x_0 = 1$ on which f is invertible.

16.39 Prove that the series

$$\sum_{n=1}^{+\infty} \frac{\sqrt{n^2 + 2} - \sqrt{n^2 + 1}}{n + 1}$$

converges and find an upper bound for its sum.

16.40 Compute the following limits:

$$\lim_{x\to 0^+} \frac{\log(1+x)}{\sqrt{1-\cos x}} \qquad\qquad \lim_{x\to 0^-} \frac{\log(1+x)}{\sqrt{1-\cos x}}.$$

$$\lim_{x\to 0^+} \sin\left(\frac{1}{x^4}\right) - \log(4x) \qquad\qquad \lim_{x\to +\infty} \left(\cos\frac{1}{x}\right)^{\sqrt{1+x^4}}.$$

16.41 Consider the functions

$$f(t) = \frac{(e^t - e^{-t})\sqrt{|t|}}{\arctan t}, \qquad F(x) = \int_1^x f(t)\,dt.$$

(a) Compute the limits of f as $t \to 0^\pm$ and as $t \to \pm\infty$.
(b) Find $\mathrm{Dom}(F)$ and compute the limits of F at the boundary points of $\mathrm{Dom}(F)$.
(c) Determine where F is monotone.

16.42 Consider the function $f(x) = \log(2e^x - 3x - 2)$.

(a) Find $\mathrm{Dom}(f)$.
(b) Study the convexity of f and determine the equations of its asymptotes, if existing.

16.43 Consider the function $f(x) = x^{1/x} - x$.

(a) Study the convexity of f and determine its inflection points, if existing.
(b) Determine the equations of the asymptotes of f, if existing.
(c) Does there exist an extension of f to $(-\infty, 0]$ that is convex in a neighborhood of $x_0 = 0$?

16.44 Consider the functions $g(x) = \cos x - \arccos x$ and $h(x) = \cos x - x$.

(a) Prove that h has a unique zero $x_0 \in \mathbb{R}$.
(b) Prove that $g(x_0) = 0$.
(c) Compute, if existing, $\displaystyle\lim_{x\to x_0} \frac{g(x)}{h(x)}$.

16.45 Compute $\displaystyle\lim_{x\to \frac{\pi}{2}^-} \frac{e^{\cos x} - \sin x}{(\cos x)^\alpha}$ as α varies in $(0, +\infty)$.

16.46 Compute $\displaystyle\lim_{x\to +\infty} \left(1 + \frac{1}{x + \sin x}\right)^{(x^\alpha - x\log x)}$ as α varies in $(0, +\infty)$.

16.47 Suppose that $f \in C^1(\mathbb{R})$ is such that $f(0) = 0$ and $f'(0) = \ell \neq 0$.

(a) Prove that there exists a neighborhood I of $x_0 = 0$ such that the restriction of f to I is invertible.

(b) If g is the inverse found in (a), compute, if existing, $\lim\limits_{x \to 0} \dfrac{f(x)}{g(x)}$.

16.48 Compute the integral $\displaystyle\int_0^{\pi/3} \frac{1 + \tan x}{1 + \sin x}\, dx$.

16.49 Compute the integral $\displaystyle\int_{1/9}^{1/4} \frac{\log(\sqrt{x} - x^2)}{\sqrt{x}}\, dx$.

16.50 Draw the graph of $\displaystyle f(x) = \int_x^{1/x} \frac{t^2 - 2}{t\sqrt{t^2 + 1}}\, dt$.

16.51 Consider the function $\displaystyle f(x) = \int_x^{+\infty} \frac{\arctan |t|^\alpha}{\log(e^t - t)\sqrt{|t| + 1}}\, dt$, with $\alpha > 0$.

(a) Find $\mathrm{Dom}(f)$ as α varies in $(0, +\infty)$.
(b) Put $\alpha = 4/3$ and draw the graph of f.

16.52 Consider the function $\displaystyle f(x) = \int_{1/4}^x \frac{(9t^2 - 1)\tan(e^t)}{\sqrt{\arctan(\frac{1}{t}) + \frac{\pi}{2}}}\, dt$.

(a) Determine $\mathrm{Dom}(f)$.
(b) Study continuity and differentiability of f.
(c) Draw the graph of f.

Appendix A
Basic Facts and Notation

Here is a collection of basic notions and facts that are used throughout.

A.1 Set Theory

The standard set-theoretic notation is adopted. It is assumed that there exists one and only one set, called the *empty set*, and denoted $\emptyset$ which contains no elements. Normally, sets are denoted with capital letters and their elements with lower case letters. Thus, $a \in A$ stands for "a is an element of A". Take two sets A and B. Then:

- A is a *subset* of B if for every $a \in A$ it is also $a \in B$; in this case one writes $A \subset B$;
- the *union* of A and B is the set $A \cup B = \{x : \text{ either } x \in A \text{ or } x \in B\}$;
- the *intersection* of A and B is the set $A \cap B = \{x : x \in A \text{ and } x \in B\}$;
- the *difference* of A and B is the set $A \setminus B = \{x : x \in A \text{ but } x \notin B\}$;
- if $A \subset B$, then $B \setminus A$ is called the *complement* of A in B.

Apart from the real numbers, that are treated separately in the next Section, the basic numerical sets are:

- the set of *natural numbers* is $\mathbb{N} = \{0, 1, 2, \dots\}$;
- the set of *integers* is $\mathbb{Z} = \{\dots, -2, -1, 0, 1, 2, \dots\}$;
- the set of *rational numbers* is $\mathbb{Q} = \{p/q : p, q \in \mathbb{Z}, q \neq 0\}$.

A.2 Real Numbers

A.2.1 Axioms

The field of *real numbers* is introduced by means of a set of axioms. One defines first the abstract notions of *ordered field*, and then of *complete ordered field* as sets that

© Springer International Publishing Switzerland 2016
M. Baronti et al., *Calculus Problems*, UNITEXT - La Matematica per il 3+2 101,
DOI 10.1007/978-3-319-15428-2

satisfy certain axioms, and then proves that there is indeed a set that satisfies all these axioms. The existence of such a set is by no means trivial. Finally, it is possible to prove that such a set, the real numbers, is essentially unique. The axioms, in a sense, lay down the rules of the game.

Definition A.1 A *field* is a set F on which two operations are defined, called *addition (or sum)* and *multiplication (or product)*, that satisfy:

Addition axioms

- If $x \in F$ and $y \in F$, then $x + y \in F$;
- addition is commutative: $x + y = y + x$ for every $x, y \in F$;
- addition is associative: $x + (y + z) = (x + y) + z$ for every $x, y, z \in F$;
- there exists an element $0 \in F$ such that $x + 0 = x$ for every $x \in F$;
- for every $x \in F$ there exists an element, called the *opposite* and denoted $-x \in F$ such that $x + (-x) = 0$.

Multiplication axioms

- If $x \in F$ and $y \in F$, then $xy \in F$;
- multiplication is associative: $xy = yx$ for every $x, y \in F$;
- multiplication is associative: $x(yz) = (xy)z$ for every $x, y, z \in F$;
- there exists an element $1 \in F$, $1 \neq 0$, such that $1x = x$ for every $x \in F$;
- for every $x \in F$, $x \neq 0$, there exists an element in F, called the *reciprocal* and denoted $1/x$ or x^{-1}, such that $x(1/x) = 1$.

Distributive property

For every $x, y, z \in F$ it holds $x(y + z) = xy + xz$.

Definition A.2 An *ordered field* is a field F in which there is an *order relation*, denoted $x < y$, that satisfies:

Order axioms

- If $x, y, z \in F$ and $y < z$, then $x + y < x + z$;
- if $x, y \in F$ and $x > 0$, $y > 0$, then $xy > 0$.

The notation $x \leq y$ means that either $x < y$ or $x = y$. The set of rational numbers $\mathbb{Q}$ is an ordered field, but it is not complete, in the sense of the next definition.

Definition A.3 A set F, endowed with the order relation $<$ is called (ordinally) *complete* if whenever A and B are non-empty subsets of F such that $a \leq b$ for every $a \in A$ and every $b \in B$ (sets with this property are called *separated*), then there exists an element $s \in F$, called a *separating element*, such that $a \leq s \leq b$ for every $a \in A$ and every $b \in B$.

Theorem A.1 *There exists a complete ordered field, denoted* $\mathbb{R}$.

Finally, it is possible to prove that the complete ordered field is essentially unique, in the sense that any two complete ordered fields can be related by a bijection that respects all the operations.

A.2.2 Properties

Definition A.4 (*Absolute value and distance*) Given $x \in \mathbb{R}$, its *absolute value* is

$$|x| = \begin{cases} x & x \geq 0 \\ -x & x < 0. \end{cases}$$

If $x, y \in \mathbb{R}$, the quantity $d(x, y) = |x - y|$ is called the *distance* between x and y.

Proposition A.1 *The absolute value satisfies the following properties:*

(i) $|-x| = |x|$ *for every* $x \in \mathbb{R}$;
(ii) *if* $a > 0$, *then the inequality* $|x| < a$ *is equivalent to* $-a < x < a$ *and, similarly,* $|x| \leq a$ *is equivalent to* $-a \leq x \leq a$;
(iii) *for every* $x, y \in \mathbb{R}$ *the* triangle inequality *holds, namely* $|x + y| \leq |x| + |y|$;
(iv) $|xy| = |x| |y|$ *for every* $x, y \in \mathbb{R}$;
(v) $||x| - |y|| \leq |x - y|$ *for every* $x, y \in \mathbb{R}$.

Proposition A.2 *The distance function satisfies the following properties:*

(i) $d(x, y) \geq 0$ *for every* $x, y \in \mathbb{R}$ *and* $d(x, y) = 0$ *if and only if* $x = y$;
(ii) $d(x, y) = d(y, x)$ *for every* $x, y \in \mathbb{R}$;
(iii) $d(x, z) \leq d(x, y) + d(y, z)$ *for every* $x, y, z \in \mathbb{R}$.

Theorem A.2 (The field $\mathbb{R}$ is Archimedean) *For every* $y \in \mathbb{R}$ *and every* $x \in \mathbb{R}$, $x > 0$, *there exists a positive integer* n *such that* $nx > y$.

Corollary A.1 *If* $x \in \mathbb{R}$ *and for every positive integer* n *it is* $|x| \leq 1/n$, *then* $x = 0$.

Theorem A.3 (Rationals are dense) *If* $x, y \in \mathbb{R}$ *and* $x < y$, *there exists* $r \in \mathbb{Q}$ *such that* $x < r < y$.

A.3 Topology

A basic type of subset of $\mathbb{R}$ is an *interval*, which consists of a set that is defined in terms of one or two order relations. Table A.1 below contains the complete list of possible intervals, where $a, b \in \mathbb{R}$ are arbitrary real numbers satisfiying $a < b$.

The intervals (a, b), $(-\infty, a)$, $(a, +\infty)$, are the *open intervals*, and the intervals $[a, b]$, $(-\infty, a]$, $[a, +\infty)$ are the *closed intervals*, as well as the degenerate interval $[a, a]$ which consists of the *singleton* $\{a\}$. The interval $\mathbb{R}$ is both open and closed. It should be noticed that a round bracket indicates that the adjacent element is not in the interval, whereas a square bracket indicates that the adjacent element is indeed in the interval.

Table A.1 Intervals

Bounded intervals	Unbounded intervals
$(a, b) = \{x \in \mathbb{R} : a < x < b\}$	$(-\infty, a) = \{x \in \mathbb{R} : x < a\}$
$[a, b] = \{x \in \mathbb{R} : a \leq x \leq b\}$	$(-\infty, a] = \{x \in \mathbb{R} : x \leq a\}$
$(a, b] = \{x \in \mathbb{R} : a < x \leq b\}$	$(a, +\infty) = \{x \in \mathbb{R} : a < x\}$
$[a, b) = \{x \in \mathbb{R} : a \leq x < b\}$	$[a, +\infty) = \{x \in \mathbb{R} : a \leq x\}$
$[a, a] = \{a\}$ (degenerate interval)	$(-\infty, \infty) = \mathbb{R}$

The most basic (open, see *infra*) sets are the *open balls*: given any $x_0 \in \mathbb{R}$ and any $R > 0$, the open ball centered at x_0 with radius R is

$$B(x_0, R) = \left\{x \in \mathbb{R} : |x - x_0| < R\right\} = \left\{x \in \mathbb{R} : d(x, x_0) < R\right\}.$$

Definition A.5 A set $\mathscr{U}$ is a *neighborhood* of the point $x_0 \in \mathbb{R}$ if there exists $\delta > 0$ such that $B(x_0, \delta) \subset \mathscr{U}$. A set $\mathscr{V}$ is a *punctured neighborhood* of x_0 if it is of the form $\mathscr{V} = \mathscr{U} \setminus \{x_0\}$ where $\mathscr{U}$ is a neighborhood of x_0. The punctured neighborhoods of $+\infty$ are the sets that contain an interval of the form $(a, +\infty)$ and the punctured neighborhoods of $-\infty$ are the sets that contain an interval of the form $(-\infty, a)$.

Definition A.6 A subset G of $\mathbb{R}$ is called *open* if for every $x \in G$ there exists an open interval containing x and contained in G. Thus, G is open if and only if it is a neighborhood of each of its points. A subset F of $\mathbb{R}$ is called *closed* if its complement $\mathbb{R} \setminus F$ is open.

Definition A.7 A pont x_0 in the subset A of $\mathbb{R}$ is an *interior point* of A if there exists a neighborhood $\mathscr{U}$ of x_0 such that $\mathscr{U} \subseteq A$. Thus, a set is open if and only if every point is an interior point.

Definition A.8 A subset K of $\mathbb{R}$ is called *bounded* if there exists $M > 0$ such that $|x| \leq M$ for every $x \in K$, that is, if it is bounded above and below (see Definition 3.1 in Chap. 3). It is called *compact* if it is bounded and closed.

A.4 Functions

In this Section, we list the main properties of the elementary functions. All the listed formulae are valid when both sides of the equality are defined.

Trigonometric Functions

Basic trigonometric relations

$$\sin^2 x + \cos^2 x = 1$$

$$\sin^2 x = \frac{\tan^2 x}{1 + \tan^2 x} \qquad\qquad \cos^2 x = \frac{1}{1 + \tan^2 x}$$

$$\sin(\tfrac{\pi}{2} - x) = \sin(\tfrac{\pi}{2} + x) = \cos x \qquad \cos(\tfrac{\pi}{2} - x) = -\cos(\tfrac{\pi}{2} + x) = \sin x$$

$$\sin(\pi - x) = -\sin(\pi + x) = \sin x \qquad \cos(\pi - x) = \cos(\pi + x) = -\cos x.$$

Addition formulae

$$\sin(\alpha + \beta) = \sin\alpha\cos\beta + \cos\alpha\sin\beta \qquad \sin(\alpha - \beta) = \sin\alpha\cos\beta - \cos\alpha\sin\beta$$

$$\cos(\alpha + \beta) = \cos\alpha\cos\beta - \sin\alpha\sin\beta \qquad \cos(\alpha - \beta) = \cos\alpha\cos\beta + \sin\alpha\sin\beta$$

$$\tan(\alpha + \beta) = \frac{\tan\alpha + \tan\beta}{1 - \tan\alpha\tan\beta} \qquad \tan(\alpha - \beta) = \frac{\tan\alpha - \tan\beta}{1 + \tan\alpha\tan\beta}.$$

Duplication and bisection formulae for $x \in (0, \pi)$

$$\sin(2x) = 2\sin x \cos x \qquad\qquad\qquad \cos(2x) = \cos^2 x - \sin^2 x$$

$$\sin\left(\frac{x}{2}\right) = \sqrt{\frac{1 - \cos x}{2}} \qquad\qquad \cos\left(\frac{x}{2}\right) = \sqrt{\frac{1 + \cos x}{2}}$$

$$\tan\left(\frac{x}{2}\right) = \sqrt{\frac{1 - \cos x}{1 + \cos x}} = \frac{1 - \cos x}{\sin x} = \frac{\sin x}{1 + \cos x}.$$

Parametric formulae: $t = \tan(x/2)$

$$\sin x = \frac{2t}{1 + t^2} \qquad\qquad \cos x = \frac{1 - t^2}{1 + t^2} \qquad\qquad \tan x = \frac{2t}{1 - t^2}.$$

Sum-to-product formulae

$$\sin\alpha + \sin\beta = 2\sin(\frac{\alpha + \beta}{2})\cos(\frac{\alpha - \beta}{2}) \qquad \sin\alpha - \sin\beta = 2\cos(\frac{\alpha + \beta}{2})\sin(\frac{\alpha - \beta}{2})$$

$$\cos\alpha + \cos\beta = 2\cos(\frac{\alpha + \beta}{2})\cos(\frac{\alpha - \beta}{2}) \qquad \cos\alpha - \cos\beta = -2\sin(\frac{\alpha + \beta}{2})\sin(\frac{\alpha - \beta}{2})$$

$$\tan\alpha \pm \tan\beta = \frac{\sin(\alpha \pm \beta)}{\cos\alpha\cos\beta}.$$

Powers and Roots

For any $a \neq 0$ and any $n \in \mathbb{Z}$, the nth *power* of a is

$$a^n = \begin{cases} \underbrace{a \cdot a \cdots a}_{n \text{ times}} & n > 0 \\ 1 & n = 0 \\ \frac{1}{a^{-n}} & n < 0. \end{cases}$$

Theorem A.4 (Existence of the nth positive root) *Given any positive real number y and any positive integer n there exists one and only one nth positive root of y, that is, a unique positive real number x such that $x^n = y$, that is denoted $x = y^{1/n}$.*

For $a > 0$ and $r = p/q \in \mathbb{Q}$ with $q > 0$, the power a^r is defined by

$$a^r = (a^p)^{1/q} = (a^{1/q})^p.$$

It turns out that, given a real number $a > 1$ and any $x \in \mathbb{R}$, the sets $\{a^r : r \in \mathbb{Q}, \ r < x\}$ and $\{a^s : s \in \mathbb{Q}, \ s > x\}$ are separated and have a unique separating element. Consequently, for $a > 1$ and any $x \in \mathbb{R}$ the power a^x is defined by

$$a^x = \sup\{a^r : r \in \mathbb{Q}, \ r < x\} = \inf\{a^s : s \in \mathbb{Q}, \ s > x\} \qquad a > 1.$$

Finally, if $0 < a < 1$ and any $x \in \mathbb{R}$ the power a^x is defined by

$$a^x = (1/a)^{-x} \qquad a < 1.$$

Exponentials and Logarithms

For any $a > 0$ and any $x \in \mathbb{R}$ the quantity $\exp_a(x) = a^x$ is well defined and consequently is defined the *exponential*, denoted by

$$\exp_a : \mathbb{R} \to (0, +\infty), \qquad x \mapsto a^x.$$

The exponential functions satisfy the following properties:

- $a^0 = 1$ and $a^1 = a$;
- $a^{x+y} = a^x a^y$ for every $x, y \in \mathbb{R}$;
- the image of $\exp_a$ is $(0, +\infty)$ if $a \neq 1$;
- $x \mapsto a^x$ is strictly increasing if $a > 1$ and strictly decreasing if $a < 1$;
- $(a^x)^y = a^{xy}$ for every $x, y \in \mathbb{R}$;
- $1^x = 1$ for every $x \in \mathbb{R}$.

For any fixed $a > 0$ and $a \neq 1$ the exponential function $\exp_a : \mathbb{R} \to (0, +\infty)$ is bijective hence invertible, and its inverse function

$$\log_a : (0, +\infty) \to \mathbb{R}, \qquad x \mapsto \log_a x$$

is called the *logarithm with basis a*. The fact that exponential and logarithm are inverse functions of eachother is expressed by:

$$a^{\log_a x} = x, \qquad x > 0$$
$$\log_a(a^x) = x, \qquad x \in \mathbb{R}.$$

The logarithm, defined for $a > 0$ and $a \neq 1$, satisfies the following properties:

- $\log_a(1) = 0$ and $\log_a(a) = 1$;
- $\log_a(xy) = \log_a(x) + \log_a(y)$ for every $x, y \in (0, +\infty)$;
- the image of $\log_a$ is $\mathbb{R}$;
- $\log_a$ is strictly increasing $a > 1$ and strictly decreasing if $a < 1$;
- $\log_a(x^y) = y \log_a(x)$ for every $x \in (0, +\infty)$ and every $y \in \mathbb{R}$;
- $\log_b(x) = (\log_b a)(\log_a x)$ for every $x \in (0, +\infty)$ and every $b > 0, b \neq 1$.

Of particular importance are the exponential function and logarithm associated with Neper's number e (see Sect. B.1 below). The latter is called the *natural logarithm* and denoted log. All exponentials and logarithms are expressible in terms of the natural ones by:

$$a^x = e^{x \log a}$$
$$\log_a(x) = \log_a(e) \log(x).$$

A.4.1 Hyperbolic Functions

The *hyperbolic* functions are the *hyperbolic sine* and the *hyperbolic cosine*, respectively, and are defined as the odd and even parts of the exponential:

$$\sinh x = \frac{e^x - e^{-x}}{2}$$
$$\cosh x = \frac{e^x + e^{-x}}{2}.$$

They satisfy the following properties:

$$\cosh^2 x - \sinh^2 x = 1$$
$$\cosh x + \sinh x = e^x$$

as well as many others that are similar to those that arise in (circular) trigonometry.

A.5 Graphs

Below are the graphs of the most common elementary functions (Figs A.1, A.2, A.3, A.4, A.5, A.6, A.7, A.8, A.9, A.10, A.11 and A.12).

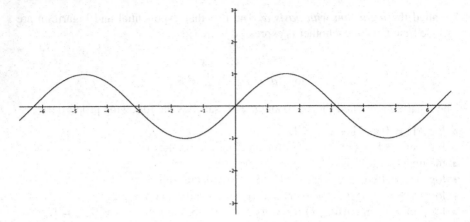

Fig. A.1 The sine function

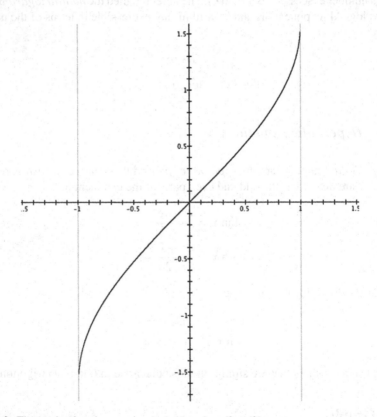

Fig. A.2 The arcsine function, arcsin : $[-1, 1] \rightarrow [-\pi/2, \pi/2]$

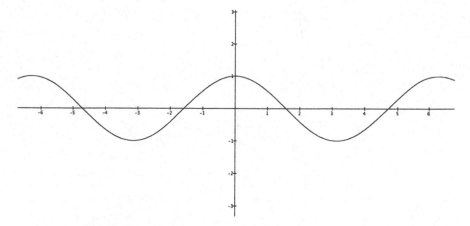

Fig. A.3 The cosine function

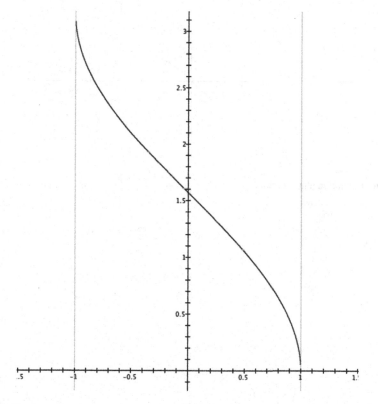

Fig. A.4 The arccosine function arccos : $[-1, 1] \rightarrow [0, \pi]$

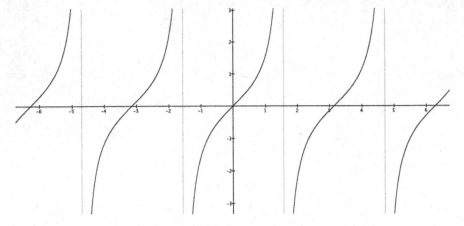

Fig. A.5 The tangent function

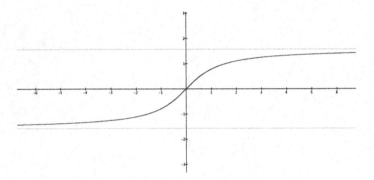

Fig. A.6 The arctangent function

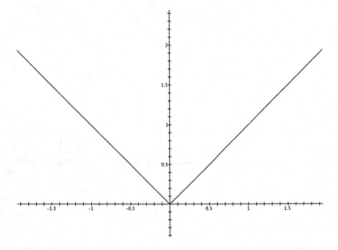

Fig. A.7 The absolute value

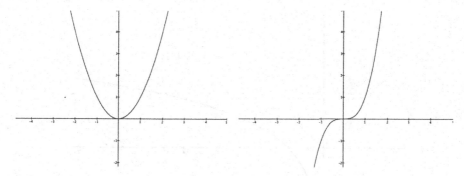

Fig. A.8 The powers x^2 and x^3

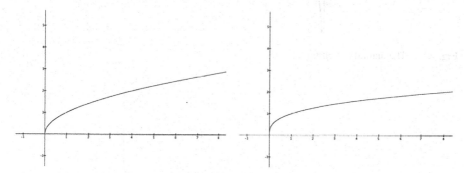

Fig. A.9 The roots $\sqrt{x}$ and $x^{1/3}$

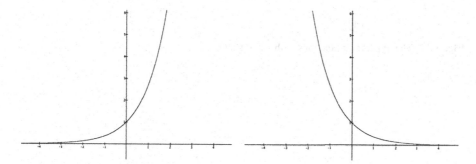

Fig. A.10 The exponential functions e^x and e^{-x}

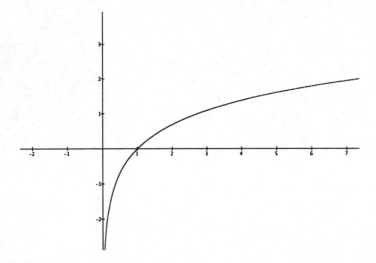

Fig. A.11 The natural logarithm

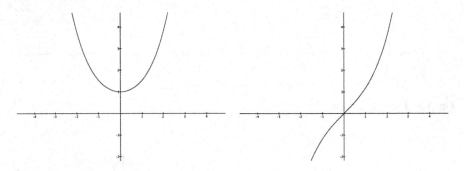

Fig. A.12 The hyperbolic functions $\cosh x$ and $\sinh x$

Appendix B
Calculus

In this Appendix, the most common Calculus formulae are collected. The meaning of the various symbols is to be found in the main body of the book.

B.1 Sequences and their Limits

Powers and Roots

Below, a is a positive number and $0 < b < 1$.

$$\lim_n n^a = +\infty \qquad\qquad \lim_n n^{-a} = 0 \qquad\qquad \lim_n a^{1/n} = \lim_n \sqrt[n]{a} = 1$$

$$\lim_n b^n = 0 \qquad\qquad \lim_n b^{-n} = +\infty \qquad\qquad \lim_n nb^n = 0$$

$$\lim_n n^{1/n} = \lim_n \sqrt[n]{n} = 1.$$

Neper's Number

Proposition B.1 *The sequence* $(e_n)_{n \geq 1}$ *defined by*

$$e_n = \left(1 + \frac{1}{n}\right)^n$$

is increasing and bounded. Therefore it converges to a real number, which is called Neper's number and is denoted e. *It is not a rational number and satisfies the inequality* $2 <$ e < 3. *In fact,* e $= 2.71828182845904523536...$, *in the sense that the first 20 exact decimals of* e *are the indicated ones.*

© Springer International Publishing Switzerland 2016
M. Baronti et al., *Calculus Problems*, UNITEXT - La Matematica per il 3+2 101,
DOI 10.1007/978-3-319-15428-2

Proposition B.2 *For every* $x \in \mathbb{R}$ *the sequence*

$$e_n(x) = \left(1 + \frac{x}{n}\right)^n$$

is bounded and eventually increasing, and

$$\lim_n \left(1 + \frac{x}{n}\right)^n = \sup\left\{\left(1 + \frac{x}{n}\right)^n : n = 1, 2, \ldots\right\} = e^x.$$

Exponentials and Logarithms

Below, a and α are positive numbers.

$$\lim_n \frac{n^n}{n!} = +\infty \qquad \lim_n \frac{n!}{a^n} = +\infty \qquad \lim_n n^\alpha e^{-n} = 0.$$

B.2   Limits of Functions

Rational Functions

Below p and q are non-negative integers, and $a_0 \neq 0 \neq b_0$.

$$\lim_{x \to \pm\infty} \left(\frac{a_0 x^p + a_1 x^{p-1} + \cdots + a_{p-1} x + a_p}{b_0 x^q + b_1 x^{q-1} + \cdots + b_{q-1} x + b_q}\right) = \begin{cases} 0 & p < q \\ a_0/b_0 & p = q \\ \pm\infty & p > q. \end{cases}$$

In the case $p > q$ the limit is $+\infty$ if $a_0/b_0 > 0$ and is $-\infty$ if $a_0/b_0 < 0$.
Below $p \geq h$ and $q \geq k$ are non-negative integers, and $a_h \neq 0 \neq b_k$.

$$\lim_{x \to 0} \left(\frac{a_0 x^p + a_1 x^{p-1} + \cdots + a_h x^h}{b_0 x^q + b_1 x^{q-1} + \cdots + b_k x^k}\right) = \begin{cases} 0 & h > k \\ a_k/b_k & h = k \\ \pm\infty & h < k. \end{cases}$$

In the case $h < k$, the limit is $+\infty$ if $a_h/b_k > 0$ and is $-\infty$ if $a_h/b_k < 0$.

Trigonometric Functions

$$\lim_{x \to 0} \frac{\sin x}{x} = 1 \qquad\qquad \lim_{x \to 0} \frac{1 - \cos x}{x^2} = \frac{1}{2}$$

$$\lim_{x \to 0} \frac{\tan x}{x} = 1 \qquad\qquad \lim_{x \to 0} \frac{\arctan x}{x} = 1.$$

Exponentials and Logarithms

Below, a is a positive number, $a \neq 1$ and α is a positive number.

$$\lim_{x \to \pm\infty} \left(1 + \frac{1}{x}\right)^x = e \qquad \lim_{x \to 0} \frac{e^x - 1}{x} = 1 \qquad \lim_{x \to 0} \frac{a^x - 1}{x} = \log a$$

$$\lim_{x \to 0} \frac{\log(1 + x)}{x} = 1 \qquad \lim_{x \to 0} \frac{\log_a(1 + x)}{x} = \frac{1}{\log a} \qquad \lim_{x \to 0^+} x^\alpha \log x = 0$$

$$\lim_{x \to +\infty} e^x x^{-\alpha} = +\infty \qquad \lim_{x \to -\infty} e^x x^\alpha = 0 \qquad \lim_{x \to +\infty} x^{-\alpha} \log x = 0.$$

B.3 Inequalities

$$|\sin x| \leq |x| \qquad\qquad\qquad x \in \mathbb{R}$$

$$\cos x \leq \frac{\sin x}{x} \leq 1 \qquad\qquad -\pi < x < \pi, \ x \neq 0$$

$$1 + x \leq e^x$$

$$\frac{x}{x + 1} \leq \log(1 + x) \leq x \qquad\qquad x > -1.$$

B.4 Derivatives

Below, $\alpha \in \mathbb{R}$ and $a > 0$, $a \neq 1$. Formulae are true when both sides are defined.

$$\frac{d}{dx} \sin x = \cos x \qquad\qquad \frac{d}{dx} \arcsin x = \frac{1}{\sqrt{1 - x^2}}$$

$$\frac{d}{dx} \cos x = -\sin x \qquad\qquad \frac{d}{dx} \arccos x = -\frac{1}{\sqrt{1 - x^2}}$$

$$\frac{d}{dx} \tan x = 1 + \tan^2 x = \frac{1}{\cos^2 x} \qquad\qquad \frac{d}{dx} \arctan x = \frac{1}{1 + x^2}$$

$$\frac{d}{dx} x^\alpha = \alpha x^{\alpha - 1} \qquad\qquad \frac{d}{dx} f(x)^{g(x)} = f(x)^{g(x)} \left[g'(x) \log f(x) + \frac{g(x)}{f(x)} f'(x) \right]$$

$$\frac{d}{dx} e^x = e^x \qquad\qquad \frac{d}{dx} a^x = a^x \log a$$

$$\frac{d}{dx} \log x = \frac{1}{x} \qquad\qquad \frac{d}{dx} \log_a x = \frac{1}{x \log a}$$

$$\frac{d}{dx} \sinh x = \cosh x \qquad\qquad \frac{d}{dx} \cosh x = \sinh x$$

B.5 McLaurin Expansions

$$e^x = 1 + x + \frac{1}{2!}x^2 + \frac{1}{3!}x^3 + \frac{1}{4!}x^4 + \cdots + \frac{1}{n!}x^n + o\left(x^n\right)$$

$$\log\left(1+x\right) = x - \frac{1}{2}x^2 + \frac{1}{3}x^3 - \frac{1}{4}x^4 + \cdots + (-1)^{n+1}\frac{1}{n}x^n + o\left(x^n\right)$$

$$\sin x = x - \frac{1}{3!}x^3 + \frac{1}{5!}x^5 - \frac{1}{7!}x^7 + \cdots + (-1)^n\frac{1}{(2n+1)!}x^{2n+1} + o\left(x^{2n+2}\right)$$

$$\cos x = 1 - \frac{1}{2!}x^2 + \frac{1}{4!}x^4 - \frac{1}{6!}x^6 + \cdots + (-1)^n\frac{1}{(2n)!}x^{2n} + o\left(x^{2n+1}\right)$$

$$\arctan x = x - \frac{1}{3}x^3 + \frac{1}{5}x^5 - \cdots + (-1)^n\frac{1}{2n+1}x^{2n+1} + o\left(x^{2n+2}\right)$$

$$(1-x)^{-1} = 1 + x + x^2 + \cdots + x^n + o\left(x^n\right)$$

$$\tan x = x + \frac{1}{3}x^3 + \frac{2}{15}x^5 + o\left(x^6\right)$$

$$\sqrt{1+x} = 1 + \frac{1}{2}x - \frac{1}{8}x^2 + \frac{1}{16}x^3 + o\left(x^3\right).$$

B.6 Primitive Functions

All formulae are true on each interval on which the integrands are defined. Below $\alpha \in \mathbb{R} \setminus \{-1\}$.

$$\int x^\alpha \, dx = \frac{x^{\alpha+1}}{\alpha+1} + c \qquad\qquad \int \frac{1}{x} \, dx = \log|x| + c$$

$$\int \sin x \, dx = -\cos x + c \qquad\qquad \int \cos x \, dx = \sin x + c$$

$$\int \sin^2 x \, dx = \frac{x - \sin x \cos x}{2} + c \qquad\qquad \int \cos^2 x \, dx = \frac{x + \sin x \cos x}{2} + c$$

$$\int \frac{1}{1+x^2} \, dx = \arctan x + c \qquad\qquad \int \tan x \, dx = -\log|\cos x| + c$$

$$\int e^x \, dx = e^x + c \qquad\qquad \int \log x \, dx = x \log x - x + c$$

$$\int \cosh x \, dx = \sinh x + c \qquad\qquad \int \sinh x \, dx = \cosh x + c.$$

B.7 Sums and Series

The *binomial coefficient* is defined by the formula

$$\binom{n}{k} = \frac{n!}{k!(n-k)!}, \qquad n! = n(n-1)\cdots 3\cdot 2\cdot 1,$$

and, by definition, $0! = 1$. The *binomial theorem* asserts that for every $a, b \in \mathbb{R}$:

$$\sum_{k=0}^{n}\binom{n}{k}a^k b^{n-k} = (a+b)^n.$$

Finite sums

$$\sum_{k=0}^{n} k = \frac{n(n+1)}{2} \qquad\qquad \sum_{k=0}^{n} k^2 = \frac{n(n+1)(2n+1)}{6}$$

$$\sum_{k=0}^{n} k^3 = \left(\frac{n(n+1)}{2}\right)^2 \qquad\qquad \sum_{k=0}^{n} x^k = \frac{1-x^{n+1}}{1-x}.$$

Sums of Series

$$\sum_{n=N}^{+\infty} x^n = \frac{x^N}{1-x} \quad (|x|<1) \qquad\qquad \sum_{n=0}^{+\infty} \frac{x^n}{n!} = e^x$$

$$\sum_{n=0}^{+\infty} \frac{x^{2n+1}}{(2n+1)!} = \sinh x \qquad\qquad \sum_{n=0}^{+\infty} \frac{x^{2n}}{(2n)!} = \cosh x$$

$$\sum_{n=0}^{+\infty} (-1)^n \frac{x^{2n+1}}{(2n+1)!} = \sin x \qquad\qquad \sum_{n=0}^{+\infty} (-1)^n \frac{x^{2n}}{(2n)!} - \cos x$$

$$\sum_{n=1}^{+\infty} (-1)^{n+1} \frac{x^n}{n} = \log(1+x) \quad (-1<x\le 1) \qquad\qquad \sum_{n=0}^{+\infty} (-1)^n \frac{x^{2n+1}}{2n+1} = \arctan x.$$

Converging and Diverging Series
   Below $p > 0$.

$$\sum_{n=1}^{+\infty} \frac{1}{n^p} < +\infty \qquad\qquad \Longleftrightarrow \qquad p > 1$$

$$\sum_{n=2}^{+\infty} \frac{1}{n \log^p n} < +\infty \qquad\qquad \Longleftrightarrow \qquad p > 1$$

$$\sum_{n=0}^{+\infty} p^n n^\alpha < +\infty \qquad\qquad \Longleftrightarrow \qquad \begin{cases} 0 < p < 1 \\ \text{or} \\ p = 1 \text{ and } \alpha < -1. \end{cases}$$

Solutions

Problems of Chapter 1

1.7 The required graphs are in Figs. B.2 and B.3.

1.8 The required graph is in Fig. B.1.

1.9 The required graphs is are in Figs. B.4, B.5, B.6, B.7, B.8 and B.9,

1.10

$$f(-x) = \begin{cases} -x & x < 0 \\ 1 & x = 0 \\ -1 & x > 0 \end{cases}, \quad f(x+1) = \begin{cases} -1 & x < -1 \\ 1 & x = -1 \\ x+1 & x > -1 \end{cases}, \quad f(|x|) = \begin{cases} |x| & x \neq 0 \\ 1 & x = 0. \end{cases}$$

The required graphs are as in Figs. B.10 and B.11

1.11 The required graphs are as in Figs. B.12 and B.13

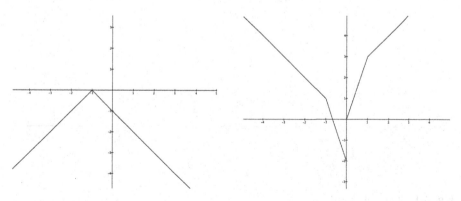

Fig. B.1 Graph of $x/(x-2)$

© Springer International Publishing Switzerland 2016
M. Baronti et al., *Calculus Problems*, UNITEXT - La Matematica per il 3+2 101,
DOI 10.1007/978-3-319-15428-2

326   Solutions

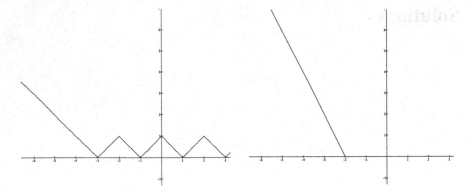

Fig. B.2 Graphs of f_1 and f_2

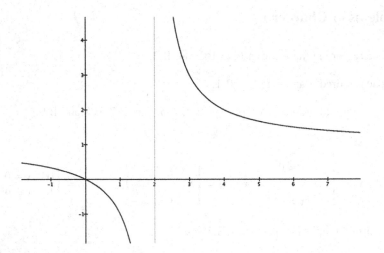

Fig. B.3 Graphs of f_3 and f_4

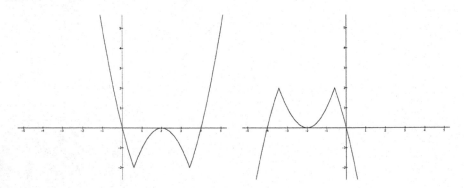

Fig. B.4 Graphs of f_1 and f_2

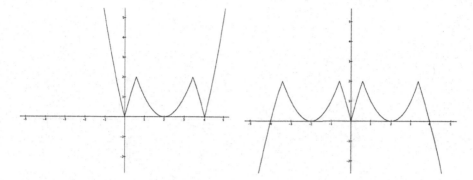

Fig. B.5 Graphs of f_3 and f_4

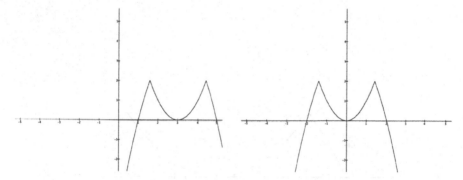

Fig. B.6 Graphs of f_5 and f_6

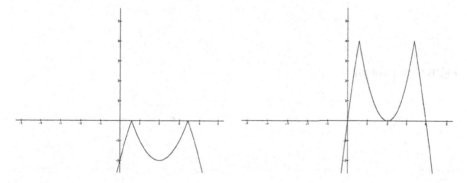

Fig. B.7 Graphs of f_7 and f_8

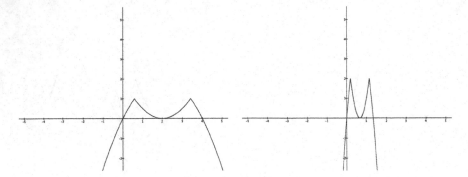

Fig. B.8 Graphs of f_9 and f_{10}

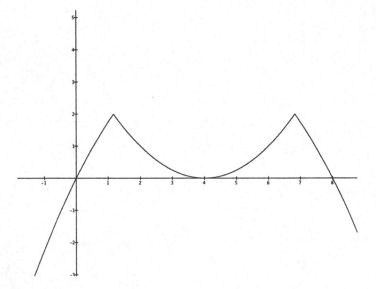

Fig. B.9 Graph of f_{11}

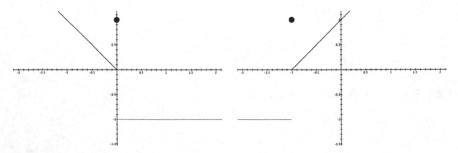

Fig. B.10 Graphs of $f(-x)$ and $f(x+1)$

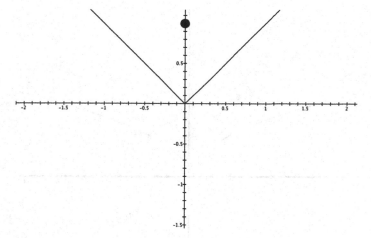

Fig. B.11 Graphs of $f(|x|)$

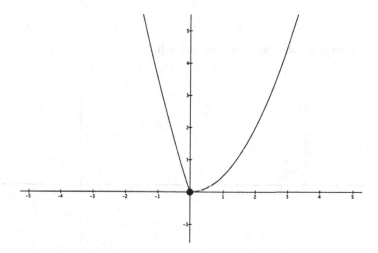

Fig. B.12 Graph of f

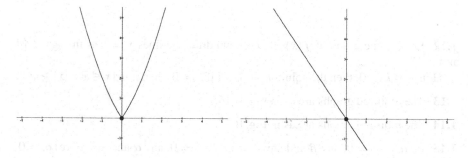

Fig. B.13 Graphs of f_e and f_o

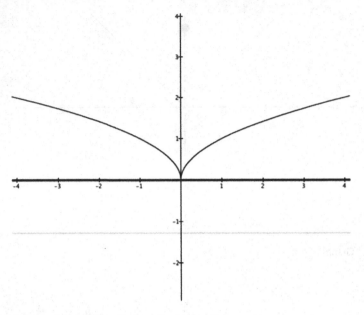

Fig. B.14 Graphs of f and of the line $y = k$ with $k < 0$

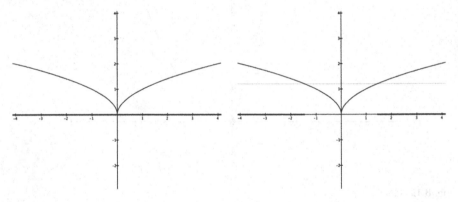

Fig. B.15 Graphs of f and of the line $y = k$ with $k = 0$ (*left*) and $k > 0$ (*right*)

1.12 Look at the graph of $f(x) = \sqrt{|x|}$ and draw the lines $y = k$ as in Figs. B.14 and B.15.

Hence, if $k \leq 0$, then the solution set is $\mathbb{R}$; if $k > 0$, then it is $\{x \in \mathbb{R} : |x| \geq k^2\}$.

1.13 The required graphs are as in Fig. B.16.

1.14 The required graphs are as in Fig. B.17.

1.15 (a) $\alpha = \gamma = 0$, any β and any δ. (b) $\beta = \delta = 0$, any α and any γ. (c) $\alpha = 0$, $\beta = 1/3$, $\gamma = 0$ and $\delta = -1/3$.

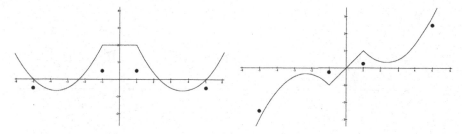

Fig. B.16 Graphs of f_e and f_o

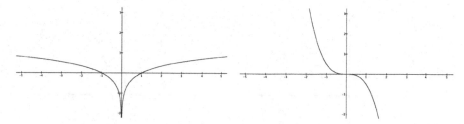

Fig. B.17 Graphs of $f_e = (\log |x|)/2$ (for $x \neq 0$) and of $f_o = -x^3/2$

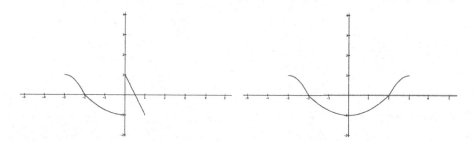

Fig. B.18 Graphs of $f(-x)$ and of $f(|x|)$

1.16 The required graphs are as in Figs. B.18, B.20, B.21 and B.22.

1.17 The required graphs are as in Fig. B.23.

1.18 The required graphs are as in Fig. B.19.

Problems of Chapter 2

2.7 (a) $\mathbb{R} \setminus \{0\}$; $(-\infty, 0) \cup (1, +\infty)$.
(b) It is invertible but not monotone.
(c) The inverse function $f^{-1} : (-\infty, 0) \cup (1, +\infty) \to (-\infty, 0) \cup (0, +\infty)$, is given by $f^{-1}(x) = \log_3((x - 1)/x)$.

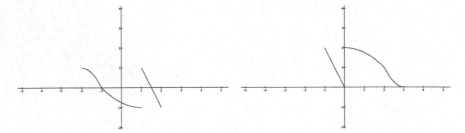

Fig. B.19 Graphs of $|q(-x)|$ and of $|3 - q(x + 2)|$

Fig. B.20 Graphs of $f(1 - x)$ and of $1 - f(x)$

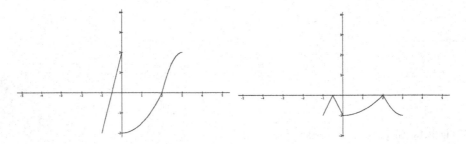

Fig. B.21 Graphs of $3 + f(x + 3)$ and of $f(3x)$

2.8 (a) $\mathrm{Dom}(f) = \mathbb{R} \setminus \{-1/2, 0, 1/2\}$, $\mathrm{Im}(f) = \mathbb{R} \setminus \{0\}$.
(b) It is decreasing in $(-\infty, -1/2)$ and in $(-1/2, 0)$ and increasing in $(0, 1/2)$ and in $(1/2, +\infty)$; $f\,is\,not\,invertible$.
(c) g is invertible; $g^{-1} : (-\infty, 0) \to (1/2, +\infty)$ and $g^{-1}(x) = 2^{-(1/x)-1}$.

2.9 (a) $(-\infty, (3 - \sqrt{5})/2) \cup ((3 + \sqrt{5})/2, +\infty)$.
(b) $((3 + \sqrt{5})/2, +\infty)$.
(c) $J = \mathbb{R}$ and $f^{-1}(x) = (3 + \sqrt{5 + 4e^x})/2$.

2.10 $g^{-1} : (-3/11, 1/11] \to [1/2, +\infty)$ is given by $g^{-1}(x) = \dfrac{1}{2} + \dfrac{1}{2}\sqrt{\dfrac{11 - 121x}{11x + 3}}$.

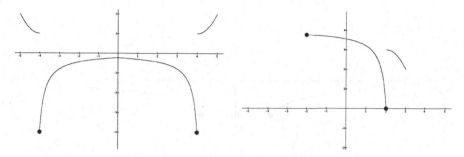

Fig. B.22 Graphs of $2f(x)$ and of $-|f(x)|$

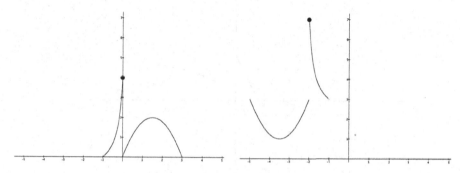

Fig. B.23 Graphs of $p(4 - |x|)$ and of $4 - |p(2 - x)|$

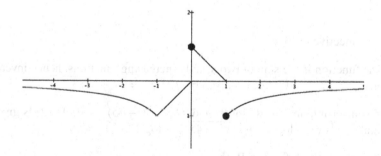

Fig. B.24 Graph of the function f of Problem 2.11

2.11 (a) The graph of f is in Fig. B.24.
(b) f is neither surjective nor injective.
(c) It is increasing in $[-1, 0]$ and $[1, +\infty)$.
(d) The graph of $f(|x|)$ is in Fig. B.25

2.12 (a) $\mathbb{R}$.
(b) f is invertible and $g^{-1}(x) = 2 + ((1 - x)/x)^2$.
(c) $f_e = f$ and $f_o = 0$.

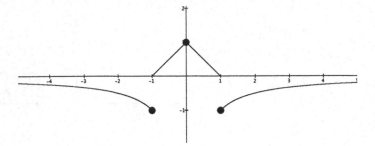

Fig. B.25 Graph of $f(|x|)$, Problem 2.11

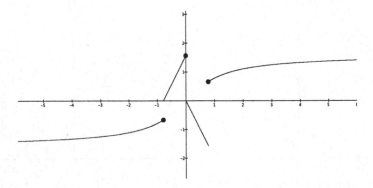

Fig. B.26 Graph of f, Problem 2.16

2.13 f is injective in $(0, e)$.

2.14 The function is the sum of two strictly increasing functions, hence inveritble. Its inverse $f^{-1} : \mathbb{R} \to (0, +\infty)$ is given by $f^{-1}(x) = (x + \sqrt{x^2 + 4})/2$.

2.15 f is invertible in $I = (0, \log 2)$ and $f^{-1} : (2, +\infty) \to (0, \log 2)$ is given by the formula $f^{-1}(x) = \log(x^2 + \sqrt{x^4 - 4x^2}) - \log 2$.

2.16 (a) The graph of f is Fig. B.26.
(b) It is neither surjective nor injective.
(c) It is increasing in $(-\infty, 0]$ and in $[\pi/4, +\infty)$.

2.17 f is injective in $(e^5, +\infty)$.

2.18 (a) $k \geq 0$. (b) $I = (-\infty, 0)$ and $g^{-1}(x) = \log((1 - x - \sqrt{1 - 2x - 3x^2})/2x)$.

2.19 f is not invertible because it is not injective: $f(0) = 1 = f(-2)$.

2.20 (a) $[1/2, 1) \cup (1, +\infty)$. (b) g is not invertible.

2.21 (a) $[-1/5, 1/4]$. (b) g is invertible and

$$g^{-1}(x) = \begin{cases} \tan\left(\dfrac{\arccos(-5x)}{2}\right) & -1/5 \le x < 0 \\[3mm] \dfrac{\log_2 x + \sqrt{\log_2^2 x - 4}}{2} & 0 < x < 1/4. \end{cases}$$

(c) $(-\infty, -1)$.

Problems of Chapter 3

3.7 (a) $S = (-\infty, 1] \cup \{3\}$. (b) $\max S = \sup S = 3$, $\inf S = -\infty$, $\min S$ does not exist.

3.8 Since $A = (-\infty; 2]$, there is no minimum and $\max A = 2$.

3.9 $\sup A = 2$.

3.10 $\max A = \sup A = 1/2$ and $\inf A = 0$. There is no minimum.

3.11 $\max A = \sup A = 2/5$ and $\min A = \inf A = -3/10$.

3.12 $\max A = \sup A = (\sqrt{2}+1)/2$ and $\min A = \inf A = (1 - \sqrt{2})/2$; $\sup B = +\infty$ and $\min B = \inf B = 0$; $\max C = \sup C = 5$ and $\min C = \inf C = -2$.

3.13 $\max A = \sup A = 2 + 1/6$ and $\min A = \inf A = -(2 + 1/5)$.

3.14 Just use the definition of "sup".

Problems of Chapter 4

4.9 The limit is 1.

4.10 The limit is 0.

4.11 The limit is $-\infty$.

4.12 The limit is 1.

4.13 The limit is: $+\infty$ for $k > 1$, 1 for $k = 1$, 0 for $k < 1$.

4.14 The limit is: $+\infty$ for $k > 0$, 0 for $k < 0$.

4.15 The limit is 1.

4.16 The limit is 0.

4.17 The limit is: 0 for $\alpha > 1$, 1 for $\alpha = 1$ and $+\infty$ for $\alpha < 1$.

4.18 The limit is 1.

4.19 (a) It is bounded. (b) It is strictly increasing. (c) The limit is $1 + \sqrt{1 + k}$.

4.20 (a) It is strictly increasing for $x_0 < 0$; it is a constant sequence for $x_0 = 0$; it is strictly decreasing for $x_0 > 0$. (b) $a_n = a/n$, for every $n \in \mathbb{N}$, $n \geq 1$.

4.21 (a) It is strictly increasing. (b) The limit is 1.

4.22 (a) It is bounded. (b) No.

4.23 The limits are $+\infty$ and 0.

Problems of Chapter 5

5.15 (a) $+\infty$. (b) $-\infty$. (c) $+\infty$. (d) 0. (e) 0.

5.16 (a) $-\infty$. (b) -1. (c) $\sqrt{2}/2$. (d) 1. (e) 0. (f) $+\infty$.

5.17 (a) $+\infty$ if $k > -1$; $-\infty$ if $k < -1$; 2 if $k = -1$. (b) 0 if $k > 0$; 1 if $k = 0$; $+\infty$ if $k < 0$. (c) $1 - k$ for all $k \in \mathbb{R}$.

5.18 (a) It doesn't exist for $a > 0$, it is 0 for $a = 0$ and for all $b > 0$. (b) $+\infty$ for $a > 0$; 0 for $a = 0$ and for all $b > 0$. (c) $+\infty$ for $a > 1$ and $b \geq 1$; it doesn't exists for $a \geq 1$ and $0 < b < 1$; 1 for $a = 1$ and $b > 1$; $\cos 1$ for $a = b = 1$; 0 for $a \in (0, 1)$ and for all $b > 0$.

5.19 $e^{\pi/2}$ and $-e^{\pi/2}$.

5.20 If $x \to +\infty$, then the limit is $+\infty$ for $\lambda > 1$; it is 0 for $\lambda = 1$ and $-\infty$ for $\lambda < 1$. If $x \to -\infty$, then the limit is $+\infty$ for all $\lambda \in \mathbb{R}$.

5.21 (a) 1. (b) 1/3. (c) 6. (d) 1/3.

5.22 f tends to $+\infty$ for $a \in (0, 1)$ and its order is greater than a but it is less than b for all $b > a$; further, f tends to $-\infty$ for $a \geq 1$ and its order is less than 1 but greater than b for every $b > 1$.

5.23 The limit is -1.

5.24 The limit is $+\infty$. The order is greater than α for every $\alpha < 1$ but smaller than β for every $\beta > 1$.

5.25 (a) The limit is: $-\infty$ with order 1/2 if $k < 1$; $+\infty$ with order 1/2 if $k > 1$; 0 with order 1/2 if $k = 1$. (b) For $k > 1$.

5.26 The limit is: 1 if $\alpha > -1$; 1/4 if $\alpha = -1$ and 0 if $\alpha < -1$.

Problems of Chapter 6

6.13 One zero in $[-1, -1/2]$.

6.14 All $\alpha, \beta \in \mathbb{R}$.

6.15 All $a \in \mathbb{R}$.

6.16 All $\alpha, \beta \in \mathbb{R}$.

6.17 For every $k \in \mathbb{R}$ the only continuous extension is $\tilde{f}(x) = \begin{cases} f(x) & x \neq 0 \\ k - 1 & x = 0. \end{cases}$

6.18 (a) For $a = b = 0$. (b) Yes.

6.19 At 0 and 1.

6.20 Continuity of f: for $a = c = 0$, all $b \in \mathbb{R}$. Continuity of g: for $a = k\pi, k \in \mathbb{Z}$.

6.21 (a) f has one zero if $k > 0$. (b) It belongs to $[-1, -1/2]$.

6.22 If $a < 2$ and $b = 0$ or if $a = 2$ and $b = 1$.

6.23 For $k \geq 0$.

6.24 Apply the intermediate zero theorem to $f(x) - x$.

6.25 Yes.

6.26 The approximate value is $x_0 = (3\sqrt{3} + 4)/(8\sqrt{3})$.

6.27 Put $f(x) = \log_2(1+x) - \sqrt{x}$. Since $f(0) = 0$ and $f(x) \to -\infty$ as $x \to +\infty$, the intermediate value theorem applies. A possible choice is $a = f(7) = 3 - \sqrt{7}$.

6.28 (a) $\text{Im}(f) - (-\pi/2, \pi/2)$. (b) f has neither global maxima nor global minima.

6.29 (a) $\text{Dom}(f) = (0, x_0) \cup (x_0, +\infty)$, where x_0 is the positive root of $x^4 + x^3 - 1 = 0$. (b) No.

6.30 (a) $\text{Dom}(f) = (1, +\infty)$. (b) Yes.

6.31 Use the definition of limit and Weierstrass' theorem.

6.32 The function admits a global maximum if $b > 0$ and $0 < a \leq \sqrt{2}$, and it does not have a global minimum for any value of a and b.

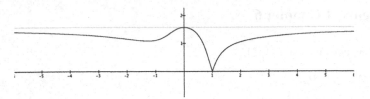

Fig. B.27 Graph of f, Problem 7.23

Problems of Chapter 7

7.14 $p(y) = 1 + \frac{1}{e+5}(y - 1 - e)$.

7.15 (a) $p(x) = \frac{\pi}{6} + (\frac{1}{\sqrt{3}} - \frac{\pi}{3})(x - \frac{3}{4})$. (b) $\alpha = \frac{\pi}{3} - \frac{1}{\sqrt{3}}, \beta = \frac{\sqrt{3}}{4} - \frac{5\pi}{12}$.

7.16 (a) 0 if $0 < a < 1$; it does not exist if $a \geq 1$.
(b) $p(x) = \sin(\frac{\sqrt{2}\pi}{4}) + (\frac{\pi}{4} - 1)\cos\frac{\sqrt{2}\pi}{4}(x - \sqrt{2})$.

7.17 (a) $p(x) = \log(e + 2) - e^2 + 2(\frac{1}{e+2} - e^2)(x - \log 2)$.
(b) 0 if $0 < a < 1$; it does not exist if $a \geq 1$.
(c) $g'(x) = \frac{1}{e+x} - e^x < 0$ in $[0, +\infty)$.

7.18 (a) $(-\infty, -1] \cup [1, +\infty)$. (b) $(\frac{3}{2}, 2)$. (c) 0 if $0 < a < 1$; it does not exist if $a \geq 1$.

7.19 (a) $[0, +\infty)$. (b) 0 if $a < 1$; $(\frac{1}{3\sqrt[3]{\pi^2}} + 1)/(2\sqrt[6]{\pi})$ if $a = 1$; $+\infty$ if $a > 1$.

7.20 (a) 4. (b) $p(x) = -\log 2 - \frac{1}{2}\left(x - \frac{3\pi}{4}\right)$.

7.21 (a) There is a unique zero; its approximate value is $(\pi + 3)/12$. (b) 0 if $b < 3$, it does not exist if $b \geq 3$.

7.22 (a) $\bigcup_{k\in\mathbb{Z}}[k\pi, k\pi + \frac{\pi}{2}]$. (b) No.

7.23 (a) $\text{Dom}(f) = \mathbb{R} \setminus \{0\}$, $\lim_{x\to-\infty} f(x) = \frac{\pi}{2} = \lim_{x\to+\infty} f(x) = \lim_{x\to0} f(x)$.
(b) f is strictly decreasing in $(-\infty, \sqrt[3]{-2}]$ and in $(0, 1)$, and is strictly increasing in $[-\sqrt[3]{2}, 0)$ and in $[1, +\infty)$.
(c) There are no local maxima; 1 is the global minimum and $\sqrt[3]{-2}$ is a local minimum.
(d) The qualitative graph is in Fig. B.27.
(e) $y = 2 + \frac{13}{4}(x - \arctan(\frac{7}{4}))$.

7.24 (a) $\text{Dom}(f) = (1, +\infty)$, $\lim_{x\to1+} f(x) = 0$, $\lim_{x\to+\infty} f(x) = -\infty$.
(b) f is strictly increasing in $(1, \bar{x})$ and strictly decreasing in $(\bar{x}, +\infty)$, where $\bar{x} \in (1, 2)$.
(c) $\bar{x}$ is the global maximum, there are no local minima.
(d) The qualitative graph is in Fig. B.28.
(e) $y = 2 - 3x$.

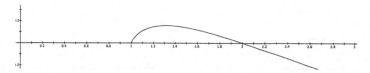

Fig. B.28 Graph of f, Problem 7.24

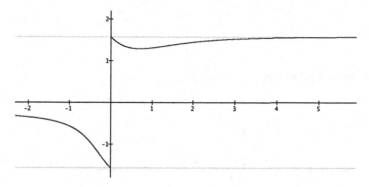

Fig. B.29 Graph of f, Problem 7.25

7.25 (a) $\lim_{x \to -\infty} f(x) = 0$, $\lim_{x \to 0^-} f(x) = -\frac{\pi}{2}$, $\lim_{x \to 0^+} f(x) = \frac{\pi}{2} = \lim_{x \to +\infty} f(x)$.

(b) f is strictly decreasing in $(-\infty, 0)$ and in $(0, \bar{x})$ and strictly increasing in $(\bar{x}, +\infty)$, where $\bar{x} \in (0, 1)$.

(c) f has a unique local minimum $\bar{x} \in (\frac{1}{2}, 1)$.

(d) The qualitative graph is in Fig. B.29.

(e) No solutions for $|k| \geq \frac{\pi}{2}$ and $0 \leq k < f(\bar{x})$; one solution if $-\frac{\pi}{2} < k < 0$ and $k = f(\bar{x})$; two solutions if $f(\bar{x}) < k < \frac{\pi}{2}$.

7.26 (a)

$$\lim_{x \to 0^+} f(x) = \begin{cases} 0 & \alpha < 1 \\ -1 & \alpha = 1 \\ -\infty & \alpha > 1, \end{cases} \qquad \lim_{x \to +\infty} f(x) = \begin{cases} 0 & \alpha > 0 \\ -\frac{\pi}{2} & \alpha = 0 \\ -\infty & \alpha < 0. \end{cases}$$

(b) $f'(1) = -\frac{e}{1+(1-e)^2} - 2\arctan(1 - e)$.

(c) g is strictly decreasing in $\mathbb{R}$. The qualitative graph is in Fig. B.30.

7.27 (a) $\text{Dom}(f) = (\frac{-1-\sqrt{5}}{2}, \frac{-1+\sqrt{5}}{2}) \cup (1, +\infty)$.

(b) $\lim_{x \to (\frac{-1 \mp \sqrt{5}}{2})^{\pm}} f(x) = -\infty$, $\lim_{x \to +\infty} f(x) = 0$, $\lim_{x \to 1^+} f(x) = -\infty$.

(c) f is strictly increasing in $((-1 - \sqrt{5})/2, -\sqrt{2})$, in $(-\sqrt{2/3}, 0)$ and in $(1, \sqrt{2})$. It is strictly decreasing in $(-\sqrt{2}, -\sqrt{2/3})$, in $(0, (-1 + \sqrt{5})/2)$ and in $(\sqrt{2}, +\infty)$. The points $0, -\sqrt{2}, \sqrt{2}$ are the global maxima, $-\sqrt{2/3}$ is a local minimum; there are no global minima.

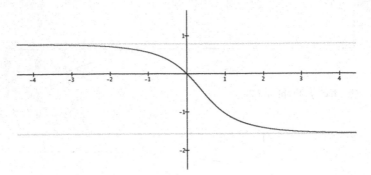

Fig. B.30 Graph of g, Problem 7.26

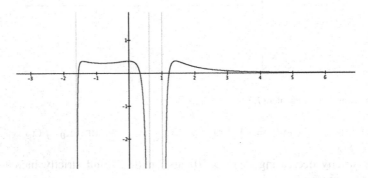

Fig. B.31 Graph of f, Problem 7.27

(d) There are 3 zeroes.
(e) The qualitative graph is in Fig. B.31.

7.28 (a) $\mathrm{Dom}(f) = \mathbb{R}$, $\lim_{x \to -\infty} f(x) = -\frac{\pi}{2} e^{-\frac{\pi}{2}}$, $\lim_{x \to +\infty} f(x) = \frac{\pi}{2} e^{\frac{\pi}{2}}$.
(b) $f'(x) = e^{\arctan x}(1 + \arctan x)/(1 + x^2)$.
(c) f is strictly decreasing in $(-\infty, -\tan 1)$ and strictly increasing in $(-\tan 1, +\infty)$.
(d) $x = -\tan 1$ is the global minimum; there are no local maxima. The qualitative graph is in Fig. B.32.

7.29 (a) $\mathrm{Dom}(f) = \mathbb{R} \setminus \{0\}$; $\lim_{x \to -\infty} f(x) = \lim_{x \to 0^-} f(x) = +\infty$,
$\lim_{x \to 0^+} f(x) = -\infty$, $\lim_{x \to +\infty} f(x) = 0$.
(b) f is strictly increasing in $(x_1, 0)$ and $(0, x_2)$, where $x_1 < 0$, $x_2 > 0$; f is strictly decreasing in $(-\infty, x_1)$ and $(x_2, +\infty)$.
(c) x_1 is a local minimum, x_2 a local maximum; there are no global extreme points.
(d) The qualitative graph is in Fig. B.33.
(e) $y = \log 2 + x \log 2$.

7.30 (a) $\lim_{x \to 0^+} f(x) = -\sqrt[3]{e}$, $\lim_{x \to +\infty} f(x) = +\infty$.
(b) f is strictly increasing in $(0, 1/e)$ and in $(1, +\infty)$, strictly decreasing in $(1/e, 1)$.

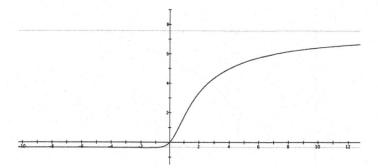

Fig. B.32 Graph of f, Problem 7.28

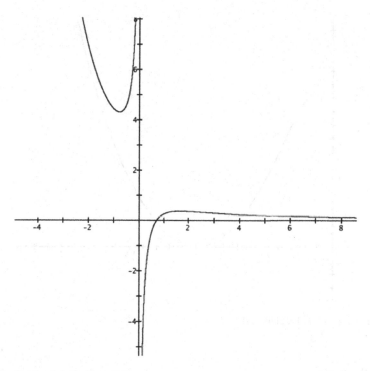

Fig. B.33 Graph of f, Problem 7.29

(c) 1 is the global minimum, $1/e$ a local maximum; there are no global maxima.
(d) f has one zero. (e) The qualitative graph is in Fig. B.34.

7.31 (a) $\lim_{x \to 0^+} f(x) = 0$, $\lim_{x \to +\infty} f(x) = +\infty$.
(b) f is strictly increasing in $(0, 1/e^2)$ and in $(1, +\infty)$, strictly decreasing in $(1/e^2, 1)$.
(c) $x = 1$ is a global minimum, $x = 1/e^2$ is a local maximum; there are no global maxima. (d) The qualitative graph is in Fig. B.35.

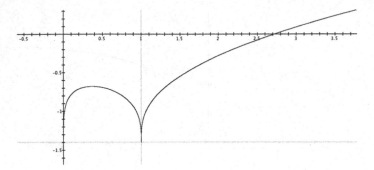

Fig. B.34 Graph of f, Problem 7.30

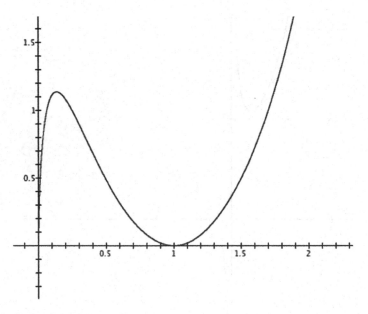

Fig. B.35 Graph of f, Problem 7.31

7.32 (a) $a = b - 1$.
(b) For no values of a and b the function f is differentiable in $\mathbb{R}$; the derivative is

$$f'(x) = \begin{cases} \dfrac{1}{2\sqrt{\arctan x}\,(1 + x^2)} & x > 0 \\ be^x + e^{-x} & x < 0. \end{cases}$$

(c) f is strictly increasing in $(-\infty, -1/2]$ and in $(0, +\infty)$, strictly decreasing in $(-1/2, 0]$. The qualitative graph is in Fig. B.36.

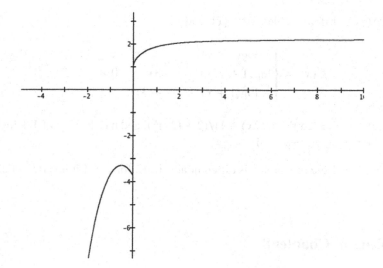

Fig. B.36 Graph of f, Problem 7.32

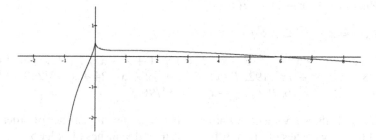

Fig. B.37 Graph of f, Problem 7.33

7.33 (a) $\lim_{x \to -1^+} f(x) = -\infty$, $\lim_{x \to +\infty} f(x) = -\infty$.
(b) f is differentiable in $(-1, 0) \cup (0, +\infty)$ and its derivative is

$$\begin{cases} \frac{1}{1+x} - \frac{1}{2\sqrt{x}} & x > 0 \\ \frac{1}{1+x} + \frac{1}{2\sqrt{-x}} & x < 0. \end{cases}$$

(c) f is increasing in $(-1, 0]$ and decreasing elsewhere.
(d) No global minima exist; 0 is the global maximum, whose value is $1/2$.
(e) f has two zeroes, one in $(-1, 0)$, one in $(1, +\infty)$.
(f) The qualitative graph is in Fig. B.37.

7.34 (a) g has two zeroes, at 0 and at a point $\bar{x} > 0$; $g(x) < 0$ if $x \in (-1, 0) \cup (\bar{x}, +\infty)$ and $g(x) > 0$ if $x \in (0, \bar{x})$.

(b) $f(x)$ is differentiable in $\mathbb{R} \setminus \{\overline{x}\}$, and

$$
f'(x) = \begin{cases} -\log(1-x) + \frac{x}{1-x} - 3x^2 & x \le 0 \\ \log(1+x) + \frac{x}{1+x} - 3x^2 & 0 < x < \overline{x} \\ -\log(1+x) - \frac{x}{1+x} + 3x^2 & x > \overline{x}. \end{cases}
$$

7.35 (a) $f'(x) = (2x)^x (\log(2x) + 1)/(2 + (2x)^x)$, and has the sign of $1 + \log(2x)$;
(b) $y - 1 = \frac{2}{1+\log 2}(x - \log 4)$.

7.36 For $k < 1$ the function f is differentiable in $\mathbb{R}$; for $k \ge 1$ it is differentiable in $\mathbb{R} \setminus \{0\}$.

Problems of Chapter 8

8.12 $P(x) = x - x^2/2$.

8.13 $P(x) = 1 + (x-1)^2 - 2(x-1)^3/3$.

8.14 $P(x) = x - x^2/2 + x^3/6 - x^4$.

8.15 $P_f(x) = x + x^2 - x^3/6 - x^4$, $P_g(x) = 1 + x - x^2/2 + 11x^3/6$, $P_h(x) = 1 - x^2/8 + x^3/16 - 7x^4/192$, $P_k(x) = -3x^2/2 + x^4/24 + 17x^6/240$, $P_\ell(x) = 1 + x + 3x^2/2 + 7x^3/6$, $P_m(x) = -x^2/4 - x^4/96$.

8.16 If $\lambda \ne 1$, then f vanishes of order 1; If $\lambda = 1$, then it vanishes of order 2.
If $\lambda \ne 1$ then g vanishes of order 3; If $\lambda = 1$, then it vanishes of order 5.
If $\lambda \ne 1/2$, then h vanishes of order 2; If $\lambda = 1/2$, then it vanishes of order 4.
If $\lambda \ne \pm 1$ k vanishes of order 2; If $\lambda = \pm 1$ k vanishes of order 4.
If $\lambda \ne \pm\sqrt{2}$ ℓ vanishes of order 2; If $\lambda = \pm\sqrt{2}$ ℓ vanishes of order 4.
$\forall \lambda \in \mathbb{R}$ m vanishes of order 2.
If $\lambda \ne 1/2$ n vanishes of order 3; If $\lambda = 1/2$ n vanishes of order 4.

8.17 f vanishes at $x_0 = 0$ of order 2 and $\lim_{x \to 0} f(x)/(1 - \cos x) = 3/4$.

8.18 If $\lambda \ne \pm\sqrt{2}$, then f vanishes at $x_0 = 0$ of order 2; If $\lambda = \pm\sqrt{2}$, then f vanishes at $x_0 = 0$ of order 4.
$\lim_{x \to 0} f(x)/x^2 = 1/2$ for $\lambda = 1$.

8.19 If $\lambda \ne -1/3$, then f vanishes at $x_0 = 0$ of order 3; If $\lambda = -1/3$, then f vanishes at $x_0 = 0$ of order 5.
$\lim_{x \to 0} f(x)/x^3 = 7/3$ for $\lambda = 2$.

8.20 The requested limits are

$$\lim_{x \to 0} \frac{x - \tan x + e^{\lambda x^2/2} - \cos \lambda x}{x^3 - \lambda x^2} = \begin{cases} -1/2 - \lambda/2 & \lambda \neq 0 \\ -1/3 & \text{otherwise} \end{cases}$$

$$\lim_{x \to 0^+} \frac{e^x - \cos\left(\lambda \sqrt{x}\right)}{\sqrt{1 + \sin(\lambda x)} - 1} = 1/\lambda + \lambda/2$$

$$\lim_{x \to 0} \frac{e^{1 - \cos(\lambda x)} - 1}{x^2} = \lambda^2/2$$

$$\lim_{x \to 0^+} \frac{\log(1 + x^2) - \cos x + e^{x^2} + \lambda x^2}{x^4} = \begin{cases} +\infty & \lambda > -5/2 \\ -\infty & \lambda < -5/2 \\ -1/24 & \lambda = -5/2. \end{cases}$$

8.21 $\displaystyle \lim_{x \to 0^+} \left(\frac{\sin x}{x}\right)^{1/(\cos x - 1)} = e^{1/3}.$

8.22 f admits a finite limit at $x_0 = 0$ if $\alpha > 1$. The continuous function, obtained by extending f with the value of the limit at $x_0 = 0$, is differentiable if $\alpha \geq 2$.

8.23 $P(x) = 1 - x^2/2$ approximates f in $[-1/10, 1/10]$ with an error less than 10^{-3}.

8.24 $P(x) = x^2 - 1/3.$

8.25 The function f is continuous at $x_0 = 0$ for $\alpha > 1/3$ and $\beta = 0$. It is also differentiable at $x_0 = 0$ for $\alpha \geq 2/3$ and $\beta = 0$.

8.26 The function f is differentiable at $x_0 = 0$ and its first order McLaurin polynomial is $P(x) = -x$.

8.27 There is no continuos extension of f at $x_0 = 0$ wich is also differentiable at that point.

8.28 0.45.

8.29 3.31.

8.30 $P(x) = 1 - x^2/2 + x^4/24$ and $\delta = 0.101$.

8.31 $P(x) = 1 - x^6/72.$

8.32 The function f is differentiable on $\mathbb{R} \setminus \{0\}$.

8.33 $a = 1/6; b = -1/2; c = 1/2; d = -1/6.$

8.34 $P(x) = e(1 + \frac{5}{2}(x - 1) + \frac{19}{6}(x - 1)^2)$ with $k = e$.

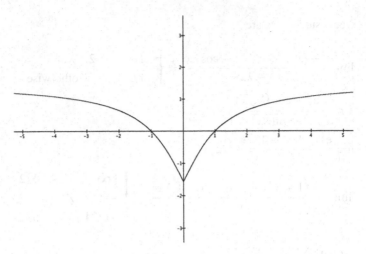

Fig. B.38 Graph of f, Exercise 9.11

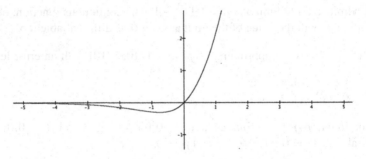

Fig. B.39 Graph of f, Exercise 9.12

Problems of Chapter 9

9.11 f is concave in $(-\infty, 0)$ and in $(0, +\infty)$. The graph is as in Fig. B.38

9.12 The graph is as in Fig. B.39

9.13 (a) f is convex in $(0, +\infty)$ for $k \geq 0$, in $(0, -1/2k)$ for $k < 0$. (b) $k \leq -e/2$. (c) The graph is as in Fig. B.40

9.14 f is concave in its domain. The graph is as in Fig. B.41

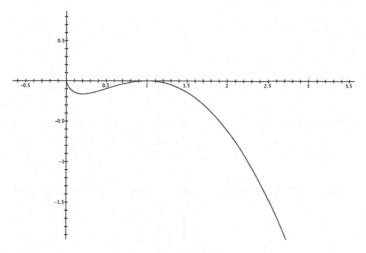

Fig. B.40 Graph of f, Exercise 9.13

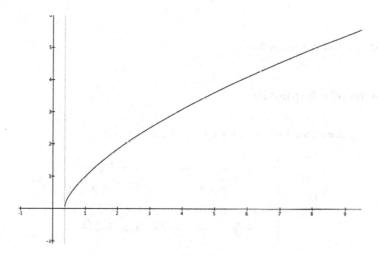

Fig. B.41 Graph of f, Exercise 9.14

9.15 f is convex in $(-\infty, -\sqrt{3})$ and in $(\sqrt{3}, x_1)$ and it is concave in $(-\sqrt{3}, \sqrt{3})$ and in $(x_1, +\infty)$ where $x_1 > 3$ is the unique inflection point. The graph is as in Fig. B.42 .

9.16 It is sufficient to apply the definitions of a convex and monotonic function.

9.17 The function is convex in $((5 - \sqrt{13})/2, (5 + \sqrt{13})/2)$ and concave in $(-\infty, (5 - \sqrt{13})/2)$ and in $((5 + \sqrt{13})/2, +\infty)$. Inflection points: $(5 \pm \sqrt{13})/2$.

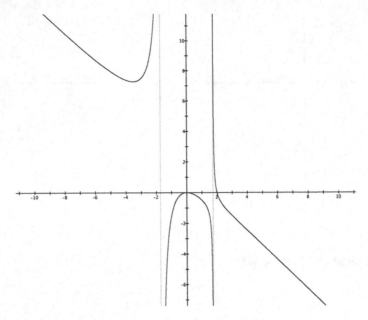

Fig. B.42 Graph of f, Exercise 9.15

Problems of Chapter 10

10.12 They are of the form $F(x) + c$ for some $c \in \mathbb{R}$, where

$$F(x) = \begin{cases} \frac{x^3}{3} - \frac{x^2}{2} - x + \frac{7-5\sqrt{5}}{6} & x < \frac{1-\sqrt{5}}{2} \\ -\frac{x^3}{3} + \frac{x^2}{2} + x & \frac{1-\sqrt{5}}{2} \le x \le 0 \\ -\frac{x^3}{3} - \frac{x^2}{2} + x & 0 < x \le \frac{-1+\sqrt{5}}{2} \\ \frac{x^3}{3} + \frac{x^2}{2} - x + \frac{-7+5\sqrt{5}}{6} & x > \frac{-1+\sqrt{5}}{2}. \end{cases}$$

10.13 If either $a < b < -3$ or $-3 < a < b < 1$ or $1 < a < b$, then f has primitive functions in $[a, b]$.

10.14 They are all of the form

$$F(x) = \begin{cases} \frac{1}{2}\log(1-x) - \frac{1}{2}\log(-1-x) + c_1 & x < -1 \\ \frac{1}{2}\log(1+x) - \frac{1}{2}\log(1-x) + c_2 & -1 < x < 1 \\ \frac{1}{2}\log(x-1) - \frac{1}{2}\log(1+x) + c_3 & x > 1 \end{cases}$$

for some $c_1, c_2, c_3 \in \mathbb{R}$.

10.15 (a) $\log 4 + \dfrac{5}{4}\log 5 - 2$.

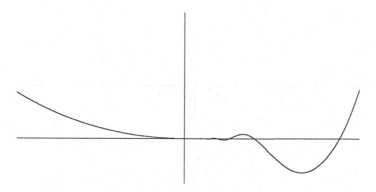

Fig. B.43 Graph of F, Problem 10.18

(b) $\dfrac{2}{3} + \pi + \log\left(\dfrac{e^{\pi}}{2\sqrt[3]{e^{\pi}-1}}\right) + \cos(\log 2) - \dfrac{1}{3}\cos^3(\log 2)$.

(c) $2\log\left(\dfrac{1-\sqrt{2}+\sqrt{3}}{-1+\sqrt{2}+\sqrt{3}}\right)$.

(d) $\dfrac{\sqrt{3}}{3}\pi - \dfrac{\pi}{2} + \log 2$.

(e) $-\log(\tan(\pi/8))$.

(f) $2\left(\log\dfrac{\sqrt{3}}{3} - \log\tan\dfrac{\pi}{8}\right)$.

(g) $-\dfrac{1}{2}\log\dfrac{\sqrt{2}}{2} - \dfrac{1}{4}$.

(h) $\pi/4$.

(i) $\log(8/9)$.

(l) $\dfrac{\pi}{12} + \dfrac{\sqrt{3}}{2}$.

10.16 If either $a < b < 0$ or $0 \le a \le b \le 1$ or $1 < a < b \le 2$, then f has primitive functions in $[a, b]$.

10.17 (a) For all $k \in \mathbb{R}$. (b) $k = -\frac{1}{2}$.

10.18 (a) For all $\alpha, \beta \in \mathbb{R}$. (b) A qualitative graph is in Fig. B.43.

10.19 $F(x) = (\frac{1}{2} - e^{-x})\arctan(1 - e^x) + \frac{1}{4}\log((1 - e^x)^2 + 1) - \frac{1}{2}x + c$, with $c \in \mathbb{R}$. The graph is in Fig. B.44.

10.20 $\mathrm{Dom}(f) = \mathbb{R}$; the second order McLaurin polynomial does not exist because f is not differentiable at 0.

10.21 $F(x) = 16\left(\dfrac{1}{20}x^{\frac{5}{4}}\log x - \dfrac{1}{25}x^{\frac{5}{4}} + \dfrac{1}{25}\right)$. The graph is in Fig. B.45.

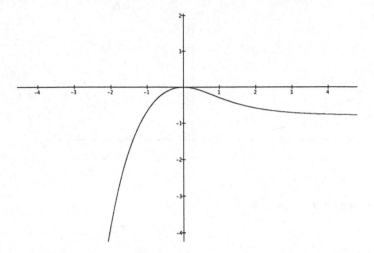

Fig. B.44 Graph of F such that $F(0) = 0$, Problem 10.19

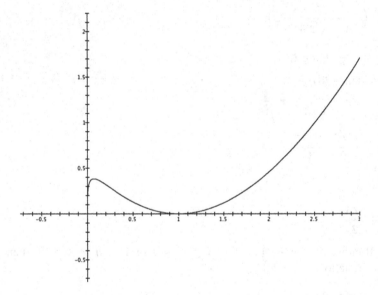

Fig. B.45 Graph of F such that $F(1) = 0$, Problem 10.21

10.22 $F(x) = x \log(\frac{x^2-1}{x}) + \log \left(\frac{x+1}{x-1}\right) + \log \left(\frac{e-1}{e+1}\right) - \frac{x^3}{3} - x - e \log(\frac{e^2-1}{e}) + \frac{e^3}{3} + e$,

defined for $x > 1$; it is convex in $(1, \frac{\sqrt{3+\sqrt{17}}}{2}]$ and concave in $[\frac{\sqrt{3+\sqrt{17}}}{2}, +\infty)$.

10.23 F is defined in $(\alpha, 1)$ with $\alpha < 0$ such that $e^\alpha = -\alpha$, it is decreasing in $(\alpha, 0)$ and increasing in $(0, 1)$, and it is convex. Hence the graph is in Fig. B.46.

10.24 $p(x) = \frac{1}{2e}(x - e + 1)^2$.

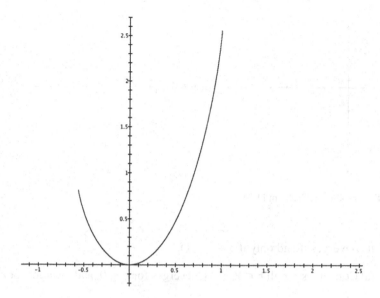

Fig. B.46 Graph of F such that $F(0) = 0$, Problem 10.23

10.25 0.

10.26 $\pi^2/72$.

10.27 $\log|\log(\frac{\sqrt{3}}{2})| - \log|\log(\frac{1}{2})|$.

10.28 (a) $r = 0$. (b) For all $r \in \mathbb{R}$. (c) $\frac{\sqrt{2}}{2}(\log\frac{\sqrt{2}}{2} - 1) - \log(|\tan\frac{\pi}{8}|)$.

10.29 $-2/\pi^2$.

Problems of Chapter 11

11.16 The integrand function has a continuous extension at $t_0 = 0$ and diverges at $t_1 = 1$ with order $1/2$.

11.17 The integral is convergent to a number $l \leq \frac{8}{3}$.

11.18 The integral is divergent for $\alpha = 3$. For $\alpha = 1$ it converges to $\frac{1}{4}\log(\frac{5}{3})$.

11.19 The integrand vanishes at $+\infty$ of order greater than any $\alpha > 0$.

11.20 It converges to $\frac{\pi}{8}$.

11.21 (a) For all $[a, b] \subset [0, +\infty)$. (b) No.

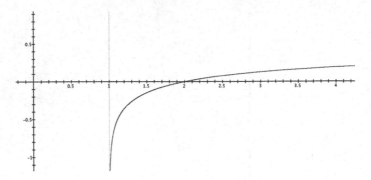

Fig. B.47 Graph of f, Problem 11.25

11.22 It converges if and only if $\alpha \in (\frac{1}{4}, 1)$.

11.23 (a) Converges for all $r \in \mathbb{R}$. (b) Converges for $r \le 0$, and diverges for $r > 0$.

11.24 (a) $[-1, +\infty)$; $\lim_{x \to -1^+} f(x) = \int_0^{-1} \frac{e^t - 1}{t\sqrt{t+1}} \, dt$; $\lim_{x \to +\infty} f(x) = +\infty$.
(b) It is invertible.

11.25 (a) The graph is in Fig. B.47.
(b) $f(x) = \frac{1}{4} \log(\frac{5x-5}{x+3})$ for $x > 1$.

11.26 (a) $[-4, 1)$; $[-4, 1)$; $(-4, 1)$.
(b) $f(1/2) = \frac{1}{\sqrt{5}}(\log(\sqrt{5} - \frac{3}{\sqrt{2}}) - \log(\frac{3}{\sqrt{2}} + \sqrt{5}) - \log(\sqrt{5} - 2) + \log(2 + \sqrt{5}))$.

11.27 (a) $(1, +\infty)$. (b) f is strictly increasing in its whole domain.

11.28 $[-1, +\infty)$.

11.29 (a) $\mathrm{Dom}(f) = (-\infty, 0)$ for $a < 0$; $\mathrm{Dom}(f) = [0, +\infty)$ for $a \ge 0$.
(b) The graph is in Fig. B.48.

11.30 (a) $(-\infty, \frac{1}{4})$. (b) $(0, \frac{1}{4})$.

11.31 (a) $(0, +\infty)$. (b) f is strictly increasing in $(2, +\infty)$. (c) The limit is 0 if $a \le 3$ and it does not exist if $a > 3$.

11.32 (a) $[\frac{3}{4}, +\infty]$ for $r \ge \frac{3}{4}$; $(-\infty, -1)$ for $r < -1$; $\emptyset$ for $-1 \le r < \frac{3}{4}$.
(b) $\lim_{x \to +\infty} f(x) = +\infty$; $[\frac{3}{4}, +\infty)$.
(c) f is strictly increasing and its graph is in Fig. B.49.
(d) $+\infty$.

11.33 (a) It is empty when $r \ne \log(2/\pi)$; $(-\infty, x_1)$ with $r = \log(2/\pi)$ and $x_1 \in (0, 1)$.
(b) The function is differentiable in $(-\infty, 0) \cup (0, x_1)$. (c) The limit does not exist.

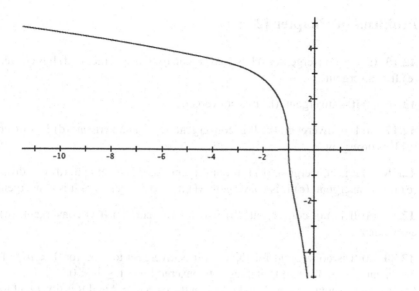

Fig. B.48 Graph of f for $a = -1$, Problem 11.29

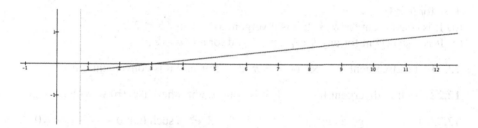

Fig. B.49 Graph of f for $r = 3$, Problem 11.32

11.34 (a) $(-\infty, \ 1) \cup (0, +\infty)$; $(-\infty, -1) \cup (0, 1) \cup (1, 2) \cup (2, +\infty)$. (b) $I = (1, 2)$.

11.35 (a) $(-\infty, -1) \cup (2, +\infty)$.
(b) $\lim_{x \to +\infty} f(x) = \lim_{x \to -\infty} f(x) = 0, \lim_{x \to 2^+} f(x) = +\infty, \lim_{x \to -1^-} f(x) = +\infty$.
(c) $n \geq 9$. (d) The integral is convergent.

11.36 (a) In $[a, b]$ for $a < b < -1$ or $-1 \leq a < b$. (b) $[-1, +\infty)$; $[-1, +\infty)$; $(-1, +\infty)$.

354 Solutions

Problems of Chapter 12

12.15 (a) It is convergent. (b) It is convergent. (c) It is divergent. (d) It is convergent. (e) It is divergent.

12.16 (a) It is divergent. (b) It is convergent.

12.17 (a) It is divergent. (b) It is convergent. (c) It is divergent. (d) It is divergent. (e) It is divergent.

12.18 (a) It is convergent. (b) It is convergent. (c) It is divergent. (d) It is divergent. (e) It is convergent. (e) It is convergent. (f) It is convergent. (g) It is convergent.

12.19 (a) It is not convergent. (b) It is convergent. (c) It is convergent. (d) It is convergent.

12.20 (a) It is convergent for $|b| > 1$; it is divergent to $+\infty$ for $0 < b \leq 1$; it is divergent to $-\infty$ for $b = 0$; it does not converge for $-1 \leq b < 0$.
(b) It is convergent for $c > k$ and $c = k$ in the case $k \geq 2$ and it is divergent to $+\infty$ for $c < k$ and $c = k$ in the case $k \leq 1$.
(c) It is divergent to $-\infty$ for $b > 1$, it is convergent for $0 < b \leq 1$; it is divergent to $+\infty$ for $b < 0$.
(d) It is convergent for $b > 2$; it is divergent to $+\infty$ for $b \leq 2$.
(e) It is convergent for any $b \in [0, \pi/2)$ and for any $c \in \mathbb{R}$.

12.21 It is divergent to $+\infty$ for $\alpha = \pi/2 + 2k\pi$ and it is convergent otherwise.

12.22 (a) It is divergent for $\alpha \neq 1$, it is convergent otherwise. (b) s_n with $n \geq 23$.

12.23 It is convergent and $s_3 = 1/e + 2/e^4 + 3/e^9$ is such that $0 < S - s_3 < 10^{-3}$.

12.24 It is convergent and $s_3 = e/(e^2 + 1) + e^2/(e^4 + 1) + e^3/(e^6 + 1)$ is such that $0 < S - s_3 \leq 10^{-1}$.

Problems of Chapter 13

13.11 $y(x) = -\log(1 - \frac{x^2}{2} - 2x)$, defined for $x \in (-2 - \sqrt{6}, -2 + \sqrt{6})$.

13.12 $y(x) = -\log(1 - \frac{x^2}{2})$, defined for $x \in (-\sqrt{2}, \sqrt{2})$; it is not invertible.

13.13 (a) $y(x) = \log(\frac{1+e^{x^2}}{2})$, defined for all $x \in \mathbb{R}$. (b) 1.

13.14 $y(x) = \frac{1-\cos x^2}{3-\cos x^2}$, defined for all $x \in \mathbb{R}$.

13.15 (a) $y(x) = \tan(\frac{\pi}{4}e^{\frac{x^2}{2}})$, defined for $-\sqrt{\log 4} < x < \sqrt{\log 4}$. (b) $y(x) = 0$, $x \in \mathbb{R}$.

13.16 (a) There is a unique local solution for $x_0 > -1$, $y_0 \in \mathbb{R}$.
(b) The order is 1.
(c) $y(x) = \tan(\frac{2}{3}(x+1)^{\frac{3}{2}} - \frac{2}{3})$, defined for $-1 \le x < (1 + \frac{3}{4}\pi)^{\frac{2}{3}} - 1$.

13.17 (a) There is a unique local solution for $y_0 \in \mathbb{R}$.
(b) $y(x) = 0$, defined for $(0, +\infty)$.
(c) $y(x) = \frac{x}{2-x}$, defined for $(0, 2)$.

13.18 $y(x) = 1 + \sqrt{x^2 + 4x + 1}$, defined for $x \in (-2 + \sqrt{3}, +\infty)$.

13.19 (a) There is a unique solution.
(b) $y(x) = \arccos(\sqrt{2}e^{\frac{1-x}{x}}/2)$, defined for $(1/(1 + \log\sqrt{2}), +\infty)$.

13.20 (a) $\left(2/\sqrt{16 - \pi^2}\right)^3$ for $k = 3$; 0 for $k > 3$; $+\infty$ for $k < 3$. (b) No.

13.21 (a) No. (b) All $x_0 \in \mathbb{R}$, $y_0 > 0$, $y_0 \ne 1/e$.
(c) $y(x) = e^{-1-\sqrt{9-2x-x^2}}$, defined for $x \in (-1 - \sqrt{10}, -1 + \sqrt{10})$.

13.22 (a) There is a unique solution.
(b) A local graph is in Fig. B.50.
(c) $p(x) = \frac{\pi}{4} + x - 1$
(d) $a = \sqrt{2}/2$.

13.23 (a) Yes. (b) $y(x) = -\sqrt{e^{\sqrt{4x+\log^2 2}} - 1}$, defined for $x \in (-\frac{1}{4}\log^2 2, +\infty)$.

13.24 (a) $1/e < y_0 < e$. (b) Yes.
(c) $y(x) = e^{\sin(e^{\frac{x^2}{2}} + \frac{\pi}{6} - 1)}$, and the domain is $\left(-\sqrt{2\log(\frac{\pi}{3} + 1)}, \sqrt{2\log(\frac{\pi}{3} + 1)}\right)$

13.25 (a) Yes. (b) Yes; $0 < y(x) < 2$.
(c) The solution has no local extreme point and is monotonic.
(d) Yes; $P_2(x) = 1 - x - \frac{x^2}{2}$.
(e) A local graph is in Fig. B.51.

13.26 $y_1(x) = 1$
$y_2(x) = e^{\frac{x^4}{16}}$
$y_3(x) = \begin{cases} e^{\frac{x^4}{16}} & x > 0 \\ 1 & x \le 0 \end{cases}$
$y_4(x) = \begin{cases} 1 & x \ge 0 \\ e^{\frac{x^4}{16}} & x < 0. \end{cases}$

13.27 (a) For all $k \in \mathbb{R}$. (b) $y(x) = -1$.

13.28 (a) Yes. (b) $P_2(x) = 1 + 2x + 2x^2$ (c) It is strictly increasing.
(d) $y(x) = \sqrt{\tan(2\arcsin x + \frac{\pi}{4})}$, defined for $x \in (-\sin(\pi/8), \sin(\pi/8))$.

13.29 $y(x) = \sqrt{4\sqrt{x} - 3}$, $I = (9/16, +\infty)$.

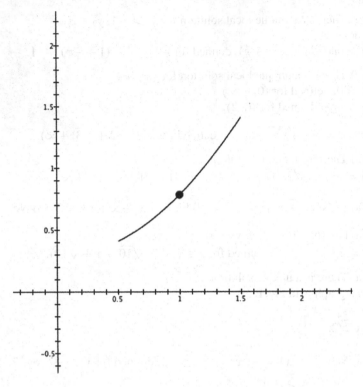

Fig. B.50 Graph of of the solution in a neighborhood of $x_0 = 1$, Problem 13.22

Problems of Chapter 14

14.10 $y(x) = -4e^{x+1} + 4x^2 + 8x + 8$ for $x \in \mathbb{R}$.

14.11 $y(x) = (x - 1)\log(1 - x)$ for $x \in (-\infty, 1)$.

14.12 (a) For all $y_0 \in \mathbb{R}$ and for any $x_0 \geq -1$ there is a unique solution in $[-1, +\infty)$.
(b) $y(x) = e^x \int_0^x \sqrt{t + 1} e^{-t}\, dt$.

14.13 (a) Yes. (b) $A = \frac{1}{2}$, $B = -\frac{1}{2}$, $C = 0$.

14.14 $k > 0$.

14.15 The solution is differentiable only once at 0.

14.16 The problem can be reduced to the following

$$\begin{cases} z'(x) = xz(x) \\ z(0) = 1. \end{cases}$$

Fig. B.51 Graph of of the solution in a neighborhood of $x_0 = 0$, Problem 13.25

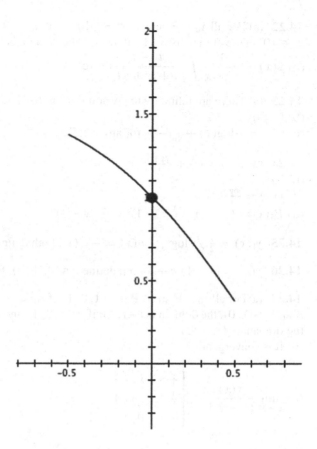

Hence the solution is: $y(x) = 1 + \int_0^x e^{\frac{t^2}{2}}\, dt$; $\displaystyle\lim_{x \to 0^+} \frac{y(x) - x - 1}{(\sin x)^k} = \begin{cases} +\infty & k > 3 \\ 1/6 & k = 3 \\ 0 & k < 3. \end{cases}$

14.17 (a) The solution is: $y(x) = \int_0^x e^{-\frac{t^2}{2}}\, dt$. (b) $\displaystyle\lim_{x \to +\infty} y(x) = \int_0^{+\infty} e^{-\frac{t^2}{2}}\, dt \in \mathbb{R}$.

14.18 (a) $y(x) = e^{\sin x} \int_0^x e^{-\sin t}\, dt$. (b) 0.

14.19 (a) $y(x) = x^2(c + e^x)$.

(b) $y(x) = \begin{cases} x^2(a + e^x) & x \geq 0 \\ x^2(b + e^x) & x < 0 \end{cases}$ as a and b range in $\mathbb{R}$.

14.20 (a) $\displaystyle\lim_{x \to 0^+} \frac{(y(x) - x - 1)^2}{1 - \cos x} = 2(k-1)^2$. (b) $\displaystyle\lim_{x \to +\infty} y(x) = +\infty$, $\displaystyle\lim_{x \to -\infty} y(x) = +\infty$.

14.21 (a) For every $x_0, y_0 \in \mathbb{R}$. (b) $\sqrt{2}/3$. (c) $P_2(x) = x$.

14.22 (a) For all $y_0 \in \mathbb{R}$ and $x_0 \in \mathbb{R} \setminus \{0, 1\}$; if $x_0 < 0$ the domain is $(-\infty, 0)$; if $x_0 \in (0, 1)$ the domain is $(0, 1)$; if $x_0 > 1$ the domain is $(1, +\infty)$.

(b) $y(x) = \dfrac{1}{x^2 - x} \displaystyle\int_{-1}^{x} \dfrac{t^2 - t}{t + \arctan t}\, dt$. (c) $1/2$.

14.23 (a) There are infinitely many solutions in $(0, +\infty)$.
(b) $k = 0$.
(c) $y(x) = c \log(\frac{x}{2}) + \frac{x^4}{16} - 1$, for any $c \in \mathbb{R}$.

14.24 (a) $y(x) = c\frac{1}{x} + \frac{\log|x|}{x}$.
(b) No.
(c) $y(x) = \frac{\log x}{x}$.
(d) $P_3(x) = (x - 1) - \frac{3}{2}(x - 1)^2 + \frac{11}{6}(x - 1)^3$.

14.25 $y_1(x) = \frac{1}{2}\sqrt{x} \log^2 x$, $y_2(x) = -y_1(x)$, both defined in $(0, +\infty)$.

14.26 $y(x) = (x - 1)[c + \log x]$, defined in $(0, +\infty)$, for any $c \in \mathbb{R}$.

14.27 (a) For all $y_0 \in \mathbb{R}$, $x_0 \in \mathbb{R} - \{-1, 0, 1\}$; if $x_0 < -1$, the domain is $(-\infty, -1)$; if $x_0 \in (-1, 0)$, the domain is $(-1, 0)$; if $x_0 \in (0, 1)$, the domain is $(0, 1)$; if $x_0 > 1$, the domain is $(1, +\infty)$.
(b) It is convergent.
(c) -1.

(d) $\displaystyle\lim_{x \to 2^+} \frac{y(x)}{(x - 2)^k} = \begin{cases} +\infty & k > 1 \\ \frac{4}{\pi} & k = 1 \\ 0 & 0 < k < 1. \end{cases}$

14.28 (a) No. (b) $y(x) = \dfrac{e^{x^2}}{(2x^2 + 1)^2}\left(c + \displaystyle\int_0^x (4t^4 - 8t^3 - 1)(2t^2 + 1)e^{-t^2}\, dt\right)$. (c) No.

14.29 (a) No. (b) $y(x) = x\left(c + \displaystyle\int_1^x \frac{\log(1 + t^2)}{t^2}\, dt\right)$ with $c < -\displaystyle\int_1^{+\infty} \frac{\log(1 + t^2)}{t^2}\, dt$.

Problems of Chapter 15

15.13 $y(x) = c_1 e^x + c_2 e^{\frac{1-\sqrt{5}}{2}x} + c_3 e^{\frac{1+\sqrt{5}}{2}x}$.

15.14 (a) $y(x) = c_1 e^{\frac{x}{3}} + c_2 e^{\frac{2}{3}x} + e^{\frac{x}{3}}(-3x^2 - 18x)$.
(b) $y(x) = c_1 e^x + c_2 \cos\sqrt{2}x + c_3 \sin\sqrt{2}x + \frac{1}{6}e^{2x}$.
(c) $y(x) = c_1 e^x + c_2 e^{(-2-\sqrt{2})x} + c_3 e^{(-2+\sqrt{2})x} + e^x(\frac{x^2}{7} - \frac{12}{49}x) - \frac{1}{2}$.

15.15 (a) If $k > -\frac{1}{8}$ and $k \neq 0$, then $y(x) = c_1 + c_2 e^{(\frac{-1-\sqrt{1+8k}}{2k})x} + c_3 e^{(\frac{-1+\sqrt{1+8k}}{2k})x} - \frac{x}{2}$. If $k = -\frac{1}{8}$ then $y(x) = c_1 + c_2 e^{4x} + c_3 x e^{4x} - \frac{x}{2}$.

If $k < -\frac{1}{8}$ then $y(x) = c_1 + c_2 e^{-\frac{1}{2k}x} \cos \frac{\sqrt{-1-8k}}{2k}x + c_3 e^{-\frac{1}{2k}x} \sin \frac{\sqrt{-1-8k}}{2k}x$.

If $k = 0$ then $y(x) = c_1 + c_2 e^{2x} - \frac{x}{2}$.

(b) $y(x) = c_1 + \frac{1}{2(e^2-1)}e^{2x} - \frac{x}{2}$.

15.16 (a) For $k = -81$.

(b) $y(x) = e^{\frac{\sqrt{2}}{2}x}(c_1 \sin \frac{\sqrt{2}}{2}x + c_2 \cos \frac{\sqrt{2}}{2}x) + e^{-\frac{\sqrt{2}}{2}x}(c_3 \sin \frac{\sqrt{2}}{2}x + c_4 \cos \frac{\sqrt{2}}{2}x)$.

15.17 (a) The general solution is:

$$y(x) = c_1(e^{\frac{\sqrt{2}}{2}x} \cos \frac{\sqrt{2}}{2}x - 1) + c_2 e^{\frac{\sqrt{2}}{2}x} \sin \frac{\sqrt{2}}{2}x+$$
$$+ c_3(e^{-\frac{\sqrt{2}}{2}x} \cos \frac{\sqrt{2}}{2}x - 1) + c_4 e^{-\frac{\sqrt{2}}{2}x} \sin \frac{\sqrt{2}}{2}x+$$
$$+ \frac{8}{5\sqrt{2}}(1 - \cos \frac{\sqrt{2}}{2}x) - x.$$

(b) No.

15.18 $y(x) = c_1 + c_2 x + c_3 \sin \sqrt{3}x + c_4 \cos \sqrt{3}x + \frac{1}{18}x^3$.

15.19 $y(x) = c_1(e^x - 1 - xe^x) + x^3 + \frac{13}{2}x^2 + 20x - 19xe^x$.

15.20 (a) No. (b) $y^{(4)}(x) - 2y^{(3)}(x) + y^{(2)}(x) = 0$.

15.21 (a) $y(x) = c_1 \sin x + c_2 \cos x + (\alpha^2 - 1)e^x + (\alpha - 1)e^{2x}$. (b) $\alpha = 1$.

15.22 No, in fact a such an equation would have as solution also $y_3(x) = \cos x$ and so it would be at least of order 3.

15.23 (a) $P_3(x) = \frac{x^2}{2} - \frac{kx^3}{6}$. (b) 1.

15.24 (a) No. (b) $y(x) = \begin{cases} c(e^x - e^{-x}) + x & x < 0 \\ (c+1)(e^x - e^{-x}) - x & x \geq 0 \end{cases}, c \in \mathbb{R}$.

15.25 $y'''(x) = 0$.

15.26 (a) $y^{(6)} - 4y^{(5)} + 7y^{(4)} - 6y^{(3)} + 2y^{(2)} = 0$. (b) $y(x) = c_1 + c_2 x + c_3 e^x + c_4 x e^x + c_5 e^x \sin x + c_6 e^x \cos x$.

15.27 (a) $y^{(5)} - 4y^{(4)} + 5y^{(3)} = 0$. (b) $y(x) = c_1 + c_2 x + c_3 x^2 + c_4 e^{2x} \sin x + c_5 e^{2x} \cos x$.

15.28 $y'''(x) + 4y'(x) = 0$.

15.29 $y'''(x) - 3y''(x) + 3y'(x) - y(x) = 0$.

15.30 $y(x) = c_1 + c_2 e^{-x} + x(\log x - 1) - e^{-x} \int_1^x e^t \log t \, dt$.

Further Reading

The literature in Calculus, Advanced Calculus or, more generally, Mathematical Analysis, is vaste and a growing number of excellent online documents is also available. Below is a list of books that can profitably be consulted by the reader. Some of them are classics, some are the books with which the authors grew up mathematically, some are commonly used nowadays, especially in Italian universities.

References

1. T.M. Apostol, *Calculus. Vol. I: One-variable calculus, with an introduction to linear algebra*, 2nd edn. (Blaisdell Publishing Co. Ginn and Co., Waltham, Mass.-Toronto, Ont.-London, 1967).
2. A. Bacciotti, F. Ricci, *Lezioni di Analisi Matematica 1*, (Levrotto & Bella, Torino, 1991).
3. J. Cecconi, G. Stampacchia, *Analisi Matematica, volume 1*, (Liguori Editore, Napoli, 1974).
4. C. Canuto, A. Tabacco, *Mathematical Analysis, I* (Springer-Verlag Italia, Milano, 2003).
5. R. Courant, F. John *Introduction to calculus and analysis. Vol. I.*, Reprint of the 1965 edn. (Springer-Verlag, New York, 1989).
6. G. De Marco, *Analisi uno*, (Decibel Editrice, Padova, 1992).
7. P.M. Fitzpatrick, *Advanced Calculus*, (American Mathematical Society, 2006)
8. E. Giusti, *Analisi Matematica 1*, 2nd edn., (Bollati Boringhieri, Torino, 1988).
9. C.D. Pagani, S. Salsa, *Analisi Matematica 1*, (Zanichelli, 2015).
10. T. Tao, *Analysis I*, (Hindustani Book Agency, New Delhi, 2006).

© Springer International Publishing Switzerland 2016
M. Baronti et al., *Calculus Problems*, UNITEXT - La Matematica per il 3+2 101,
DOI 10.1007/978-3-319-15428-2

Index

© Springer International Publishing Switzerland 2016
M. Baronti et al., *Calculus Problems*, UNITEXT - La Matematica per il 3+2 101,
DOI 10.1007/978-3-319-15428-2

Printed in the United States
By Bookmasters